MORE PRAISE FOR *UNDER MY SKIN*:

"Intriguing. . . . Set down in quick, fluent prose, *Under My Skin* offers the reader a beautifully observed portrait of the African landscape that's often as sensually resonant as the one Isak Dinesen created in *Out of Africa*."

—Michiko Kakutani, *New York Times*

"Remarkable. . . . Endlessly fascinating. . . . Her own emotional and intellectual history serves as a brilliant, febrile recapitulation of the history of this bloody century and its competing ideologies."

—*Elle* magazine

"An act of intuition . . . juggling history and memory, allowing us to see both past and present afresh. . . . Even readers unfamiliar with Lessing's fiction will find *Under My Skin* inviting."

—Michael Upchurch, *Atlanta Journal-Constitution*

"The rhythm and exactitude of her reporting . . . are unmatched in modern letters."

—Bob Targett, *Cleveland Plain Dealer*

"An unwillingness to adopt any phrase, ism or attitude because it is current informs this story of Mrs. Lessing's extraordinary life."

—Janet Burroway, *New York Times Book Review*

"*Under My Skin* is fascinating. It records and recovers worlds now gone, each as distant and significant as an exploding sun whose impact rocks us still."

—Roz Spafford, *San Francisco Chronicle*

"A brutally frank examination of how Lessing became what she is—a distinguished writer, a woman who has lived life to the full, and a constant critic of cant."

—*Kirkus Reviews*, starred review

"I would send the complete works of Doris Lessing . . . up in a rocket to speak on our behalf to whatever alien civilizations might be there in outer space."

—John Leonard, *The Nation*

Also by Doris Lessing

Novels

The Grass Is Singing
The Golden Notebook
Briefing for a Descent into Hell
The Summer Before the Dark
The Memoirs of a Survivor
The Diaries of Jane Somers:
The Diary of a Good Neighbor
If the Old Could...
The Good Terrorist
The Fifth Child

"Canopus in Argos: Archives" series

Re: Colonized Planet 5-Shikasta
The Marriages Between Zones Three, Four, and Five
The Sirian Experiments
The Making of the Representative for Planet Eight
Documents Relating to the Sentimental Agents in the Volyen Empire

"Children of Violence" series

Martha Quest
A Proper Marriage
A Ripple from the Storm
Landlocked
The Four-Gated City

Short Stories

This Was the Old Chief's Country
The Habit of Loving
A Man and Two Women
The Temptation of Jack Orkney and Other Stories
Stories
African Stories
The Real Thing: Stories and Sketches

Opera

The Making of the Representative for Planet Eight (Music by Philip Glass)

Poetry

Fourteen Poems

Nonfiction

In Pursuit of the English
Particularly Cats
Going Home
A Small Personal Voice
Prisons We Choose to Live Inside
The Wind Blows Away Our Words
Particularly Cats...And Rufus
African Laughter

The Doris Lessing Reader

Doris Lessing aged fourteen.

DORIS LESSING

UNDER MY SKIN

Volume One of
My Autobiography, to 1949

HarperPerennial
A Division of HarperCollins*Publishers*

A hardcover edition of this book was published in 1994 by HarperCollins Publishers.

HarperCollins books may be purchased for educational, business, or sales promotional use. For information, please write: Special Markets Department, HarperCollins Publishers, Inc., 10 East 53rd Street, New York, NY 10022.

First HarperPerennial edition published 1995.

The Library of Congress has catalogued the hardcover edition as follows:

Lessing, Doris May, 1919–
Under my skin/Doris Lessing.—1st ed.
p. cm.
"Volume one of my autobiography, to 1949."
ISBN 0-06-017150-2
1. Lessing, Doris May, 1919– —Biography. 2. Women authors, English—20th century—Biography I. Title
PR6023.E833Z477 1994
823'.914—dc20 94-20051

ISBN 0-06-092664-3 (pbk.)
95 96 97 98 99 RRD 10 9 8 7 6 5 4 3 2 1

Acknowledgements

The author and publisher are grateful to the proprietors listed below for permission to quote extracts from the following material:

(p.ix) *Coercive Agencies*. Reprinted by permission from *Caravan of Dreams* by Idries Shah (The Octagon Press Ltd.); (p.ix) *The Dance of Life* by Edward T. Hall. Copyright © 1983 by Edward T. Hall. Used by permission of Doubleday, a division of Bantam Doubleday Dell Publishing Group, Inc.; (p.ix) *Cold Warrior* by Tom Mangold. Copyright © Tom Mangold 1991. Reprinted with permission of Simon & Schuster Ltd.; (p.360-1) "The Age" from Osip Mandelstam: *Selected Poems* translated by Clarence Brown and W.S. Merwin (OUP 1973), pp.44-45. Reprinted by permission of Oxford University Press; (p.vii & p.204) "I've Got You Under My Skin" (Porter), (p.211) "Swinging on a Star" (Van Heusen/Burke), (p.272) "Dancing With Tears In My Eyes" (Burke & Dubin), (p.273) "It's a Sin to Tell A Lie" (Mayhew), (p.283) "Brother Can You Spare A Dime?" (Gorney/Harburg), (p.303) "Night and Day" (Porter), (p.303 & 359) "Man I Love" (Gershwin), (p.359) "Somebody Loves Me" (DeSylva/MacDonald), and (p.376) "There's A Small Hotel" (Rogers/Hart) © Warner Chappell Music Ltd., London W1Y 3FA. Reproduced by permission of International Music Publications Ltd.; (p.122) "Blue Skies" (Berlin) Copyright © 1927, and (p.205) "Cheek to Cheek" (Berlin) Copyright © 1935 by Irving Berlin. Copyright renewed. International Copyright Secured. Used by Permission of Irving Berlin Music Company. All Rights Reserved; (p.122) "Red Sails in the Sunset" (music by Hugh Williams and words by Jimmy Kennedy) © 1935, reproduced by permission of Peter Maurice Music Co. Ltd., London WC2H 0EA; (p.273) "Goodbye-ee" (words and music by R.P. Weston and Bert Lee) © 1917, reproduced by permission of Francis Day and Hunter Ltd., London WC2H 0EA; (p.273) "We'll Meet Again" (Ross Parker/Hughie Charles) copyright © 1939 by Dash Music Co. Ltd., 8-9 Frith Street, London W1V 5TZ. Used by permission. All rights reserved; (p.359) "Smoke Gets in Your Eyes" (Otto Harback/Jerome Kern) © Reproduced by kind permission of Polygram Music Publishing Ltd., 347-353 Chiswick High Road, London W4 4HS.

And with thanks to my researcher, Elizabeth Murray, two of whose nutshell biographies I have quoted.

A Note on Population

It is believed that when the whites arrived in the area that later became Southern Rhodesia, there were a quarter of a million black people. By about 1924 there were half a million. When I left the country in 1949 there were one and a half million. In 1982 the estimate was nine or ten million. In 1993 they think there are twelve to thirteen million. Some demographers believe there will be thirty million by 2010. Now, in 1993, ninety per cent of the population are under the age of fifteen.

It is currently thought by most experts that the continual increase of population since the whites arrived is because the Portuguese introduced maize which is easily grown, abundant, easily stored and nourishing.

Glossary

assegai	a spear
biltong	dried meat
gymkhana	a day of sporting events, with all kinds of games and competitions, but with the emphasis on horse racing and horse jumping
kaross	a coverlet or blanket made of animal hide
kopje	a hill
kraal	an enclosure of cattle or other animals; or a village, as in 'Are you going home to your kraal to visit your family?'
the lands	where did this manorial phrase come from?
mombies	Shona word for cattle
piccanin	a small black child
rimpi	a strip of cured hide
sjambok	a rhinoceros hide whip
skellum	someone or something mischievous or wicked
rondaavel	a round brick or mud-and-pole hut, usually thatched
veldschoen	shoes made of cow hide
vlei	a valley, usually with a watercourse running through it

I've got you under my skin
I've got you deep in the heart of me
So deep in my heart you're really a part of me,
I've got you under my skin.
I've tried so not to give in . . .

COLE PORTER

The individual, and groupings of people, have to learn that they cannot reform society in reality, nor deal with others as reasonable people, unless the individual has learned to locate and allow for the various patterns of coercive institutions, formal and also informal, which rule him. No matter what his reason says, he will always relapse into obedience to the coercive agency while its pattern is within him.

IDRIES SHAH, *Caravan of Dreams*

No matter where one looks on the face of the earth, wherever there are people, they can be observed syncing when music is played. There is popular misconception about music. Because there is a beat to music, the generally accepted belief is that the rhythm originates in the music, not that music is a highly specialized release of rhythms already in the individual. Otherwise how can one explain the close fit between ethnicity and music?

Rhythm patterns may turn out to be one of the most basic personality traits that differentiates one individual from another.

. . . when people converse . . . their brain waves even lock into a single unified sequence. When we talk to each other our central nervous systems mesh like two gears in a transmission.

The power of rhythmic message within the group is as strong as anything I know. It is . . . a hidden force, like gravity, that holds groups together.

I can remember being quite overwhelmed when I first made cinematographic recordings of groups of people in public. Not only were small groups in sync, but there were times when it seemed that all were part of a larger rhythm.

EDWARD T. HALL, *The Dance of Life*

1

'SHE WAS VERY PRETTY but all she cared about was horses and dancing.'

This refrain tinkled through my mother's tales of her childhood, and it was years before it occurred to me, 'Wait a minute, that's her mother she's talking about.' She never used any other words than those, and they could not have been her words, since she did not remember her mother. No, this was what she had heard from the servants, for she unconsciously put on a kitchen face, with a condemning look about her mouth, and she always gave a disapproving sniff. That little sniff evoked for me a downstairs world as exotic as the people in it would have found tales of cannibals and the heathen. Servants and nursemaids brought the little children up, after the frivolous Emily McVeagh died, in childbed, of peritonitis, with her third, when her first, my mother, was still only three. There is not even a photograph of Emily. She is Nobody. She is nothing at all. John William McVeagh would not talk about his first wife. What can she have done? – I asked myself. After all, to be light-minded is not a crime. At last it came to me. Emily Flower was common, that must have been it.

Then a researcher was invited to throw light into those distant places and she came up with a mass of material that would do very well as a basis for one of those Victorian novels, by Trollope perhaps, where the chapter about Emily Flower, called 'What Can Have Been Her Fault?', could only be a short one, if the saddest.

'The information on the Flower family was got through birth, marriage and death certificates, parish records, census records, apprentice records, barge owners' records, lightermen and watermen records, local history and wills,' says the researcher, evoking Dickens' England in a sentence.

There was a Henry Flower who in 1827 was described as Mariner, and in the 1851 census as a Victualler. He was born in

Somerset and his wife Eleanor was born in Limehouse. Their son, George James Flower, delinquent Emily's father, was apprenticed to a John Flower, presumably a relative. The Flower family were barge owners, and on Emily's birth certificate her father was described as lighterman.

The Flower clan lived in and around Flower Terrace, now demolished, and George James and his wife Eliza Miller lived at Number 3 Flower Terrace. This was in Poplar, near what is now Canary Wharf. There were four children. Eliza was widowed, aged thirty-five, and the closeness and mutual helpfulness of the clan is shown by how, although women did not do this then, the lightermen and watermen allowed her to be a barge owner and take apprentices. She made her son Edward an apprentice and he later became lighterman and barge owner in her place. Her children did well, and she ended in a pleasant house, with an annuity. Emily was the youngest child and she married John William McVeagh in 1883.

My mother described the house she was brought up in as tall, narrow, cold, dark, depressing, and her father as a disciplinarian, strict, frightening, always ready with moral exhortations.

The well-off working class had a good life in late Victorian times, with jaunts to the races, all kinds of parties and celebrations. They most heartily ate and drank. Nothing dreary or cold about Flower Terrace and its companion streets, full of relatives and friends. Emily came from this warm clan life into the doubtless ardent arms of John William McVeagh – he must have been very much in love to marry her – but she was expected to match herself to his ambitions, to the frightful snobberies of a man fighting to leave the working class behind. I imagine her running back home when she could to her common family, for dances, good times and going to the races. She must have lived in her husband's house under a cold drizzle of disapproval, from which, or so I see it, she died, aged thirty-two.

My mother never mentioned her grandfather, John William's father, and that meant John William did not talk about him any more than he did about Emily.

'The information for this family,' says the researcher, 'comes from births, deaths and marriages, the clerical directory, the Public Record Office, army records and books on the Charge of the Light Brigade, census reports, wills and local directories. John McVeagh's date of birth and place of birth conflict in the records. Army records of birth and occupation are frequently incorrect as men enlisting, for reasons of their own, gave wrong information, and it would have been difficult to check up in the pre-1837 registration time. In any case, recruiting stations were not particular in the army of the nineteenth century.'

John McVeagh was born in Portugal, and his father was a soldier. He was in the 4th Light Dragoons, and was a Hospital Sergeant Major when he left the army in 1861. He was in the Crimea and East Turkey and in the Charge of the Light Brigade – he really was, for soldiers made that claim who had no right to it. But why did they want to have been part of such carnage? John McVeagh's conduct as a soldier was exemplary. When his horse was shot under him in the Charge he continued to tend the wounded though wounded himself. He received various medals. Here is an entry for March 1st 1862, the *United Service Gazette*:

> 4th (Queen's) Hussars – Cahir. On Friday the 21st ult. Serjt-Major J. McVeagh late of this regiment, now Yeoman Warder of the Tower, was presented by the officers of his late corps with a purse containing 20 guineas, a silver snuff box beautifully engraved, showing his former services. Few men have been more honoured for their good conduct than Serjt-Major McVeagh on leaving his regiment, then at the Curragh, a few months back, to take his new appointment after 24 years service. The non-commissioned officers and privates presented him with a splendid tea service with the following inscription: 'To Hospital Serjt-Major John McVeagh, as a token of respect for his general kindness.' During the Crimean War he was at all times with his regiment in the field, attending both sick and wounded, and for such distinguished conduct received a medal, with an annuity of £20, besides a Turkish and a Crimean one with 4 clasps.

His wife was Martha Snewin, and her father was a bootmaker. She was born in Kent. She travelled all around the country with her husband when he was an army recruiter. That is all we know about her. He saw to it the children had a good education. Their daughter Martha, who looked after him when his wife died, was left well provided-for, but she is one of the invisible women of history.

My grandfather John William was the youngest son. First he was a clerk in the Meteorological Office, and by 1881 he was a bank clerk. Then he became a bank manager, in the Barking Road, but he died in Blackheath. He bettered himself, house by house, as he moved, and this son of a common soldier married his second wife, Emily's successor, in St George's, Hanover Square. This stepmother was not, as I imagined – because of her elegant beaked face – Jewish, but was the daughter of a dissenting cleric, who later became a priest in the Anglican church. She came from a middle-class family. Her name was Maria Martyn. My mother described her, with dislike, as a typical stepmother, cold, dutiful and correct, unable to be loving or even affectionate with the three children. They preferred life downstairs with the servants for as long as it was allowed, but my mother and her brother John became snobbishly, not to say obsessively middle-class, while the third child, Muriel, married back into the working class. Although my mother kept tenuous contact with her, the father would have nothing to do with her. It was her mother coming out in her, the servants said.

So he was disappointed in both his daughters. When my mother decided to be a nurse, instead of going to university – John William was ambitious for her – she was similarly cut off from his approval. Until, that is, she did well, but it was too late, the bonds had snapped. Never, ever, did my mother speak of her father with affection. Respect, yes, and gratitude that he did well by her, for he made sure they were given everything proper for middle-class children. She went to a good school, and was taught music, where she did so well the examiners told her she could have a career as a concert pianist.

The chapter heading for my mother in this saga would be a sad one, and the older I get, the more sorrowful her life seems. She did not love her parents. My father did not love his. It took me years to take in that fact, perhaps because it was always a joke when he said he left home the moment he could and went off as far as possible from them, as a bank clerk in Luton.

My paternal great-grandfather, a James Tayler, appears in the 1851 census as a farmer with 130 acres, employing five men, at East Bergholt. He went in for melancholy and philosophical verse, which is perhaps why he was not successful. He married a Matilda Cornish. The Tayler family worked in various capacities in banks, were civil servants, minor literary figures, often farmers, all over Suffolk and Norfolk. During the migrations of the nineteenth century they went off to Australia and to Canada, where many live still. But my grandfather Alfred decided not to be a farmer. He was a bank clerk in Colchester. His wife was Caroline May Batley.

This was the woman my father disliked so much – his mother.

The picture he presented of his father, Alfred Tayler, was of a dreaming unambitious man who spent his spare time playing the organ in the village church, driving his ambitious wife mad with frustration. But by the time I heard this my father was also a dreaming unambitious man who drove his poor wife mad with frustration. And the fact was, my grandfather Alfred ended up as manager of the London County Westminster Bank, Huntingdon, but whether he went on playing the organ in the local church I do not know. When Caroline May died he at once married again, in the very same year, a woman much younger than he was, Marian Wolfe, thirty-seven to his seventy-four. She, too, was the daughter of a minister of religion.

Ministers of religion and bank managers, there they are, in the records on both sides of the family.

Caroline May Batley, my father's mother, is almost as much of a shadow as poor Emily. The only pleasant thing my father remembered about her was that she cooked the delicious, if solid, food described by Mrs Beeton. The tale he told, and retold, and with relish shared by my mother, was how his mother came to the Royal Free Hospital to confront the newly engaged pair, both of them rather ill, to tell him that if he married that battleaxe Sister McVeagh he would always regret it. But I daresay Caroline May would have something to say for herself, if asked. It is probable she was related to Constable the painter. I like to think so.

My mother's childhood and girlhood were spent doing well in everything, because she had to please her stern father. She excelled in school, she played hockey and tennis and lacrosse well, she bicycled, she went to the theatre and music hall and musical evenings. Her energy was phenomenal. And she read all kinds of advanced books, and was determined her children would not have the cold and arid upbringing she did. She studied Montessori and Ruskin, and H. G. Wells, particularly *Joan and Peter*, with its ridicule of how children were deformed by upbringing. She told me all her contemporaries read *Joan and Peter* and were determined to do better. Strange how once influential books disappear. Kipling's 'Baa, Baa, Black Sheep' made her cry because of her own childhood.

Then she became a nurse, and had to live on the pay, which was so little she was often hungry and could not buy herself gloves and handkerchiefs or a nice blouse. The World War started, the first one, and my badly wounded father arrived in the ward where she was Sister McVeagh. He was there for over a year, and during that time her heart was well and truly broken, for the young doctor she loved and who loved her was drowned in a ship sunk by a torpedo.

While my mother was being an exemplary Victorian and then Edwardian girl, the pattern of a modern young woman, my father was enjoying a country childhood, for he spent every minute out of school (which he hated, unlike my mother, for she loved school where she did so well) with the farmers' children around Colchester. His parents beat him – spare the rod, spoil the child – and until he died he would talk with horror about the Sundays, when there were two church services and Sunday school. He dreaded Sundays all week, and would not go near a church for years. Butler's *The Way of All Flesh* – that was what his childhood was like, he said, but luckily he could always escape into the fields. He wanted to be a farmer, always, but the moment he left school put distance between himself and his parents, went into the bank, which he hated, but worked hard there, for people did work harder then than now, and above all, played hard. He loved every kind of sport, played cricket and billiards for his county, rode, and danced, walked miles to and from a dance in another village or town. If when my mother talked about her youth it sounded like *Ann Veronica* or the New Women of Shaw, my father's reminiscences were like D. H. Lawrence in *Sons and Lovers*, or *The White Peacock*, young people in emotional and self-conscious literary friendships, improving themselves by talk and shared books. He used to say that from the moment he got away from his parents and was independent he had a wonderful time, he enjoyed every minute of it, no one could have had a better life than he had for ten years. He was twenty-eight when the war began. He was lucky twice, he said, once when he was sent out of the Trenches because of a bad appendix, thus missing the Battle of the Somme when all his company was killed, and then, having a shell land on his leg a couple of weeks before Passchendaele, when, again, no one was left of his company.

He was very ill, not only because of his amputated leg, but because he was suffering from what was then called shell shock. He was in fact depressed, the real depression which was like – so he said – being inside a cold, dark room with no way out, and where no one could come in to help him. The 'nice doctor man' he was sent to said he had to stick it out, there was nothing medicine could do for him, but the anguish would pass. The 'horrible things' that my father's mind was assailed by were not as uncommon as he seemed to think: horrible things were in everybody's mind, but the war had made them worse, that was all. But my father remembered and spoke often about the soldiers who, 'shell-shocked' or unable to get themselves out of their mud holes to face the enemy, might be shot for cowardice. 'It could have been me,' he might say, all his life. 'It was just luck it wasn't.'

So there he was, in my mother's ward in the old Royal Free Hospital in East London. He saw her unhappiness when her great love was drowned, he knew she had been offered the matronship of St George's, a famous teaching hospital, an honour, for usually this job was offered to older women. But they decided to get married, and there was no conflict in it for him, though there was for her, because later she said so. He said, often, that he owed her his sanity, owed her everything, for without her devoted nursing he would not have come through that year of illness. Marriages for affection were best, he might add. As for her, she enjoyed her efficiency and her success, and knew she would make a wonderful matron of a great teaching hospital. But she wanted children, to make up to them what she had suffered as a child. So she put it.

My father was not the only soldier never, ever, to forgive his country for what he saw as promises made but betrayed: for these soldiers were many, in Britain, in France and in Germany, Old Soldiers who kept that bitterness till they died. They were an idealistic and innocent lot, those men: they actually believed it was a war to end war. And my father had been given a white feather in London by women he described as dreadful harridans – and that was when he already had his wooden leg under his trouser leg, and his 'shell shock' making him wonder if it was worth staying alive. He never forgot that white feather, speaking of it as yet another symptom of the world's ineradicable and inevitable and hopeless insanity.

He had to leave England, for he could not bear England now, and he got his bank to send him out to the Imperial Bank of Persia, to Kermanshah. Now I use the name Imperial Bank of . . . to watch the reaction, which is incredulity, and then a laugh, for so much of that time now seems as delightfully absurd as – well, as something or other we now take for granted will seem to our children.

My mother was having a breakdown, I think because of the difficulties of that choice, marriage or the career where she was doing so well. And because of her lost love, whom she never forgot. And because she had worked so very hard during the war, and because of the many men she had watched die and because . . . it was 1919, the year when 29 million people died of the flu epidemic which for some reason gets left out of the histories of that time. Ten million were killed in the Great War, mostly in the Trenches, a statistic we remember now on the 11th November of every year, but 29 million people died of the flu, sometimes called the Spanish Lady.

My father was still in breakdown, though the worst of the depression he had suffered from was over. They had been advised by the doctors not to have a child yet. They joked my mother must have got pregnant on the first night. In those days people actually often did wait until the first night of marriage. But there is another thing. In 1919 my mother was thirty-five and in those days it was considered late to have a first baby. And as a nurse she must have been aware of the dangers of waiting. Perhaps a part of my mother's mind she did not know about was making sure she got pregnant then.

And so they arrived, the two of them, both ill, in the great stone house on a plateau surrounded by snowtopped mountains, in that ancient trading town, Kermanshah – which was much damaged, parts of it bombed into dust, during the war between Iraq and Iran in the 1980s.

And there I was born on the 22nd October 1919. My mother had a bad time. It was a forceps birth. My face was scarred purple for days. Do I believe this difficult birth scarred me – that is to say, my nature? Who knows. I do know that to be born in the year 1919 when half of Europe was a graveyard, and people were dying in millions all over the world – that was important. How could it not be? Unless you believe that every little human being's mind is quite separate from every other, separate from the common human mind. An unlikely thing, surely.

That war does not become less important to me as time passes, on the contrary. In 1990, the year I began to write this book, I was in the south of France, in that hilly country behind the Riviera, visiting the delicious little towns and villages which began centuries ago as hill forts, and in every town or village is a war memorial. On one face is a list of the twelve or twenty young men killed in World War One, and this in tiny villages that even now have only half a hundred inhabitants. Usually every one of the young men of a village was killed. All over Europe, in every city, town, and village is a war memorial, with the names of the dead of World War One. On another face of the shaft or obelisk are the two or three names of the dead of World War Two. By 1918, all the healthy young men of Europe, dead. In 1990 I was in Edinburgh where in a cold, grey castle are kept the lines of books recording the names of the young men from Scotland killed between 1914 and 1918. Hundreds of thousands of names. And then in Glasgow – the same. Then, Liverpool. Records of that slaughter, the First World War. Unlived lives. Unborn children. How thoroughly we have all forgotten the damage that war did Europe, but we are still living with it. Perhaps if 'The Flower of Europe' (as they used to be called) had not been killed, and those children and grandchildren had been born, we would not now in Europe be living with such second-rateness, such muddle and incompetence?

Not long ago, in a cinema in Kilburn, they showed *Oh What a Lovely War!*, that satire on the silliness of World War One. As we came out of the dark into the street, an old woman stood alert and alive at the exit, and she looked hard into every face, impressing herself on every one of us. That film ends with two females stumbling, wandering through acres, miles, of gravestones, war graves, women who never found men to marry and have children with. This old woman, there was no doubt, was one of them, and she wanted us to know it. That film expressed her: she was telling us so.

There were also the wounded from the war, of whom my father was one, and the people whose potential was never used because their lives were wrenched out of their proper course by the war – my mother was one.

During that trip through the villages of France, then in Scotland and towns in England, were revived in me the raging emotions of my childhood, a protest, an anguish: my parents'. I felt, too, incredulity, but that was a later emotion: *how could it have happened?* The American Civil War, less than half a century before, had shown what the newly invented weapons could do in the way of slaughter, but we had learned nothing from that war. That is the worst of the legacies from the First World War: the thought that if we are a race that cannot learn, what will become of us? With people as stupid as we are, what can we hope for? But the strongest emotion on that trip was the old darkness of dread and of anguish – my father's emotion, a very potent draught, no homeopathic dose, but the full dose of adult pain. I wonder now how many of the children brought up in families crippled by war had the same poison running in their veins from before they could even speak.

We are all of us made by war, twisted and warped by war, but we seem to forget it.

A war does not end with the Armistice. In 1919, all over a Europe filled with graves, hung miasmas and miseries, and over the whole world too, because of the flu and its nearly thirty million deaths.

I used to joke that it was the war that had given birth to me, as a defence when weary with the talk about the war that went on – and on – and on. But it was no joke. I used to feel there was something like a dark grey cloud, like poison gas, over my early childhood. Later I found people who had the same experience. Perhaps it was from that war that I first felt the struggling panicky need to escape, with a nervous aversion to where I have just stood, as if something there might blow up or drag me down by the heel.

2

YOU CANNOT SIT DOWN to write about yourself without rhetorical questions of the most tedious kind demanding attention. Our old friend, the Truth, is first. The truth . . . how much of it to tell, how little? It seems it is agreed this is the first problem of the self-chronicler, and obloquy lies in wait either way.

Telling the truth about yourself is one thing, if you can, but what about the other people? I may easily write about my life until the year I left Southern Rhodesia in 1949, because there are few people left who can be hurt by what I say; I have had to leave out, or change – mostly a name or two – very little. So Volume One is being written without snags and blocks of conscience. But Volume Two, that is, from the time I reached London, will be a different matter, even if I follow the example of Simone de Beauvoir who said that about some things she had no intention of telling the truth. (Then why bother? – the reader must be expected to ask.) I have known not a few of the famous, and even one or two of the great, but I do not believe it is the duty of friends, lovers, comrades, to tell all. The older I get the more secrets I have, never to be revealed and this, I know, is a common condition of people my age. And why all this emphasis on kissing and telling? Kisses are the least of it.

I read history with conditional respect. I have been involved in a small way with big events, and know how quickly accounts of them become like a cracked mirror. I read some biographies with admiration for people who have chosen to keep their mouths shut. It is, I have observed, a rule that people who have been on the periphery of events or a life are those who rush forward to claim first place: the people who do know often say nothing or little. Some of the most noisy, not to say noisome, scandals or affairs of our time, that have had a searchlight on them for years, are reflected wrongly in the public mind because the actual participants keep their counsel, and watch, ironically, from the shadows. And there is another thing, much harder to see. People who have been real movers and exciters get left out of histories, and it is because memory itself decides to reject them. These instigators are flamboyant, unscrupulous, hysterical, or even mad, certainly abrasive; but the real point is that they are apparently of a different substance from the smooth, reasonable and sane people who have been inspired by them, and who do not like to remember temporary submersions in lunacy. Often, reading histories, there are events which stick out, do not make enough sense, and one may deduce the existence of some lunatic, male or female, who was equipped with the fiery stuff of inspiration – but was quickly forgotten, since always and at all times the past gets tidied up and made safer. 'A rough beast' is usually the real begetter of events. There would have been no 'communist party' in Southern Rhodesia without such an inspirational character.

Women often get dropped from memory, and then history.

Telling the truth or not telling it, and how much, is a lesser problem than the one of shifting perspectives, for you see your life differently at different stages, like climbing a mountain while the landscape changes with every turn in the path. Had I written this when I was thirty, it would have been a pretty combative document. In my forties, a wail of despair and guilt: oh my God, how could I have done this or that? Now I look back at that child, that girl, that young woman, with a more and more detached curiosity. Old people may be observed peering into their pasts, *Why?* – they are asking themselves. *How did that happen?* I try to see my past selves as someone else might, and then put myself back inside one of them, and am at once submerged in a hot struggle of emotion, justified by thoughts and ideas I now judge wrong.

Besides, the landscape itself is a tricky thing. As you start to write at once the question begins to insist: Why do you remember this and not that? Why do you remember in every detail a whole week, month, more, of a long ago year, but then complete dark, a blank? *How do you know that what you remember is more important than what you don't?*

Suppose there is no landscape at all? This can happen. I sat next to a man at dinner who said he could never write an autobiography because he didn't remember anything. What, nothing? Only a little scene here and there. Like, so he said, those small washes and blobs of colour that stained-glass windows lay on the dark of a stone floor in a cathedral. It is hard for me to imagine such a darkening of the past. Once even to try would have plunged me into frightful insecurity, as if memory were Self, Identity – and I am sure that isn't so. Now I can imagine myself arriving in some country with the past wiped clean out of my mind: I would do all right. It is after all only what we did when we were born, without memories, or so it seems to the adult: then we have to create our lives, create memory.

'Besides' – said this dinner companion – who seemed perfectly whole and present, despite his insufficient hold on his past, 'the little blobs of colour move all the time, because the sun is moving outside.'

True. Move they do. You forget. You remember. As I brooded over the material for this book, faces and places emerged from the dark. 'Good Lord! So there you are! Haven't thought about you for years!' Not only the perspective but what you are looking at changes.

When you write about anything – in a novel, an article – you learn a lot you did not know before. I learned a good deal writing this. Again and again I have had to say, 'That was the reason was it? Why didn't I think of that before?' Or even, 'Wait . . . it wasn't like that.' Memory is a careless and lazy organ, not only a self-flattering one. And not always self-flattering. More than once I have said: 'No, I wasn't as bad as I've been thinking,' as well as discovering that I was worse.

And then – and perhaps this is the worst deceiver of all – we make up our pasts. You can actually watch your mind doing it, taking a little fragment of fact and then spinning a tale out of it. No, I do not think this is only the fault of story-tellers. A parent says, 'We took you to the seaside, and you built a sandcastle, *don't you remember?* – look, here is the photo.' And at once the child builds from the words and the photograph a memory, which becomes hers. But there are moments, incidents, real memory, I do trust. This is partly because I spent a good part of my childhood 'fixing' moments in my mind. Clearly I had to fight to establish a reality of my own, against an insistence from the adults that I should accept theirs. Pressure had been put on me to admit that what I knew was true was not so. I am deducing this. Why else my preoccupation that went on for years: *this* is the truth, *this* is what happened, hold on to it, don't let them talk you out of it.

Why an autobiography at all? Self-defence: biographies are being written. It is a jumpy business, as if you were walking along a flat and often tedious road in an agreeable half-dark but you know a searchlight may be switched on at any minute. Yes, indeed there are good biographers, nearly all of them in Britain now, for we are enjoying a golden age of biography. What is better than a really good biography? Not many novels.

In the year just finished, 1992, I heard of five American biographers writing about me. One I had never met or even heard of. Another, I was told by a friend in Zimbabwe, is 'collecting material' for a biography. From whom? Long dead people? A woman I met twice, once when she asked me carefully casual questions, has just informed me she has written a book about me which she is about to get published. Yet another can only be concocting a book out of supposedly autobiographical material in novels and from two short monographs about my parents. Probably interviews, too, and these are always full of misinformation. It is an astonishing fact that you may spend a couple of hours with an interviewer, who is recording every word you say, but the article or interview always has several major errors of fact. But less and less do facts matter, partly because writers are like pegs to hang people's fantasies on. If writers do care that what is written about them should somewhere connect with the truth, does that mean we are childish? Perhaps it does, and certainly I feel every year more of an anachronism. Returning to Paris after a year's interval, I was interviewed by a young woman who had done me before. I said her previous article had been a tissue of invention, and she replied, 'But if you have to get an article in to a deadline, and you didn't have enough material, wouldn't you make it up?' Clearly she would not have believed me if I had said no. And that brings me straight to the heart of the problem. Young people brought up in today's literary climate cannot believe how things were. You get sceptical looks if you say something like this: 'Once serious publishers tried to find serious biographers for their serious authors.' Now everyone takes it for granted that all they are concerned about is to publish as many biographies as possible, no matter how second-rate, because biographies sell well. Writers may protest as much as they like: but our lives do not belong to us.

If you try and claim your own life by writing an autobiography, at once you have to ask, But is this the truth? There are aspects of my life I am always trying to understand better. One – what else? – my relations with my mother, but what interests me now is not the narrowly personal aspect. I was in nervous flight from her ever since I can remember anything, and from the age of fourteen I set myself obdurately against her in a kind of inner emigration from everything she represented. Girls do have to grow up, but has this battle always been so implacable? Now I see her as a tragic figure, living out her disappointing years with courage and with dignity. I saw her then as tragic, certainly, but was not able to be kind. Every day you may watch, hear of, some young person, usually a girl, giving parents, often a mother, such a bad time that it could be called cruelty. Later they will say, 'I am afraid I was difficult when I was an adolescent.' A quite extraordinary degree of malice and vindictiveness goes into the combat. Judging from histories and novels from the past, things were not always like this. So what has happened, why now? Why has it become a right to be unpleasant?

I have a woman friend who in the Second World War went to New York with her young child, having no support in Britain, her home. She earned her living precariously as a model for artists, and sometimes modelling clothes. She lived in a small town outside New York. She was poor, isolated, and being twenty years old, yearned for some fun. Once, just once, exactly once, she left the little boy with a friend, spent the evening in New York, and did not get home until dawn. I used to listen to this boy, now adolescent, accuse her most bitterly, 'You left me alone night after night and went off to enjoy yourself.' A small boy, the son of parents who did not approve of smacking, had his fingers smacked once when he persisted in putting them through the paper covering jam pots. This became, 'And you used to hit me when I was small.' These petty recollections are to the point.

For years I lived in a state of accusation against my mother, at first hot, then cold and hard, and the pain, not to say anguish, was deep and genuine. But now I ask myself, against what expectations, what promises, was I matching what actually happened? And this is the second area of my preoccupation, which has to be linked with the first.

Why is it I have lived my whole life with people who are automatically against authority, 'agin the government', who take it for granted that all authority is bad, ascribe doubtful or venal motives to government, the Establishment, the ruling class, the local town council, the headmaster or mistress? So deep-rooted is this set of mind that it is only when you begin to climb out of it you see how much of your life has been determined by it. This week I was with a group of people of mixed ages, all on the left (or who had been once), and someone happened to mention that the government was doing something – quite a good thing, but that isn't the point – and at once every face put on a look of derision. Automatic. Push-button. This look is like a sneer or a jeer, a *Well, what can one expect?* It can only come out of some belief, one so deep it is well out of sight, that a promise of some kind has been made and then betrayed. Perhaps it was the French Revolution? Or the American Revolution, which made the pursuit of happiness a right with the implication that happiness is to be had as easily as taking cakes off a supermarket counter? Millions of people in our time behave as if they have been made a promise – by whom? when? – that life must get freer, more honest, more comfortable, always better. Has advertising only set our minds more firmly in this expectant mode? Yet nothing in history suggests that we may expect anything but wars, tyrants, sickness, bad times, calamities, while good times are always temporary. Above all, history tells us nothing stays the same for long. We expect gold at the foot of always renewable rainbows. I feel I have been part of some mass illusion or delusion. Certainly part of mass beliefs and convictions that now seem as lunatic as the fact that for centuries expeditions of God-lovers trekked across the Middle East to kill the infidel.

I have just read of a historian who claims that the distrust, even contempt, of government and authority is precisely because of the First World War, because of the stupidity and incompetence of its generals, because of the slaughter of Europe's young men.

When journalists or historians come to ask about something in the past the hardest moment is when I see on their faces the look that means, But how *could* you have believed this, or done that? Facts are easy. It is the atmospheres that made them possible that are elusive. 'You see, we believed . . .' (You must have been pretty stupid then!) 'No, you don't understand, it was such a fevered time . . .' (Fever you call it, do you!) 'I know it's hard to understand, without being immersed in the poisonous air of then.'

A subsidiary question, not without general relevance: how to account for the fact that all my life I've been the child who says the Emperor is naked, while my brother never, not once, doubted or criticized authority?

Mind you, a talent for seeing the Emperor's nakedness can mean his other qualities are not noticed.

I am trying to write this book honestly. But were I to write it aged eighty-five, how different would it be?

3

A TINY THING AMONG TRAMPLING, knocking careless giants who smell, who lean down towards you with great ugly hairy faces, showing big dirty teeth. A foot you keep an eye on, while trying to watch all the other dangers as well, is almost as big as you are. The hands they use to grip you can squeeze the breath half out of you. The rooms you run about in, the furniture you move among, windows, doors, are vast, nothing is your size, but one day you will grow tall enough to reach the handle of the door, or the knob on a cupboard. These are the real childhood memories and any that have you level with grown-ups are later inventions. An intense physicality, that is the truth of childhood.

My first memory is before I was two, and it is of an enormous dangerous horse towering up, up, and on it my father still higher, his head and shoulders somewhere in the sky. There he sits with his wooden leg always there under his trousers, a big hard slippery hidden thing. I am trying not to cry, while being lifted up in tight squeezing hands, and put in front of my father's body, told to grip the front of the saddle, a hard jutting edge I must stretch my fingers to hold. I am inside the heat of horse, the smell of horse, the smell of my father, all hot pungent smells. When the horse moves it is a jerking jolting motion and I lean back my head and shoulders into my father's stomach and feel there the hard straps of the wooden-leg harness. My stomach is reeling because of the swoop up from the ground now so far below me. Now, that is a real memory, violent, smelly – physical.

'Daddy used to put you in front of him on the horse when he rode to the Bank, and Marta waited at the gate to bring you back. You absolutely loved it.' And perhaps I did, perhaps it was only the first ride, which I did not love, that has stayed in my memory. The gate is in a photograph, a graceful arch, and I have added it to the real memory. Of being lifted down into the hands of

Marta, whom I disliked, there is nothing in my mind. Those rides had to be in Kermanshah, and I was two and a half when we left.

Sharp steep stone steps, like boulders on a mountainside; they are in a photograph, too, but the memory is of dangerous descent, threatened by sharp edges.

Another memory, a real one, not what was told me, or what is in the photograph album. A swimming bath, a large tank, full of great naked pallid people shouting and laughing and splashing me with hard slaps of cold water. The naked bodies were my mother, rowdy and noisy, enjoying herself, my father holding on to the edge of the tank, because that pitiful shrunken stump of a leg with its shrapnel scars, waving or jerking about in the water, made it hard for him to swim. And others, for the tank seems crowded with people. They are not naked, for they wear the serious swimming costumes of the time, but if adults are always dressed in the daytime, and then wear long-sleeved clothes in bed, when in bathing costumes they seem all pale flesh and unpleasant revelation. Loose bulging breasts. Whiskers of hair under arms, matting or streaming water like sweat. Sometimes snot on a face that is grinning and shouting with pleasure. Snot running into the water that already has dying or rotting leaves in it, as well as the broken reflections of clouds, down here, not up there in the sky. Small children are always trying to keep things in their proper places, their world is always coming apart, things in it move about, deceive, lie. 'We used to swim every afternoon in the summer. And we had swimming parties at the weekend. Oh they were such fun. You always loved it when we had parties.' Thus spoke my mother, mourning the best years of her life, in Persia. 'We used to lift you in with us, but you screamed and had to be put back on the side. The water was so cold! It was mountain water. It came running down from the mountains in stone channels. You simply had to shout as you jumped in! There were beds of asters all around the tank. The Persian gardeners were wonderful, they grew everything.' And so you imagine jumping in, all jolly and laughing, and being lifted out, you see the asters, in paintbox colours, and hear the scolding Persian gardeners, who would not let you pick the asters, mother said so. But the real memory, the authentic one, was of enormous pale bodies, like milk puddings, sloshing about in out-of-control water that smelled cold, the flailing large pale arms, the hard breath-stopping slap of water on your face. 'Go on, be a sport, brave girls don't cry about a silly little thing like that.'

Two memories, concocted ones, or induced, but probably true enough. In the 1960s, when we were experimenting with drugs, I tried one absolutely not to be recommended. You eat morning glory seeds, previously soaked in hot water to an acid jellyish state, but you have to eat a lot, in my case sixty or more. I felt sick, and as for the revelations I was doing as well using my novelist's mind. I had been thinking, why had so little remained in me of that big stone house, with its big high stone rooms? I was born there. I learned to walk there. And imagined that I lay in a cot with bars, like a prison cell for size, and heard large feet clanging on stone. I knew the floors were stone and that there were few rugs, that the windows were large and showed mountains, that the house was cold in winter. The cot was bound to be something of the sort, and a small child hears every sound with new ears, nothing shut off, as adults shut off sound.

The other invited memory was useful, and has been ever since. I took mescalin – just once. Two friends monitored the dosage and then sat with me. They were concerned that I would jump out of a window or something of the kind, because someone they knew had done that a short time before. What I learned then was how strong in me was the personality I call the Hostess, for I was presenting my experience to them, chatting away, increasingly scatty, but in control, but all that was a protection for what went on within. This Hostess personality, bright, helpful, attentive, receptive to what is expected, is very strong indeed. It is a protection, a shield, for the private self. How useful it has been, is now, when being interviewed, photographed, a public person for public use. But behind all that friendly helpfulness was something else, the observer, and it is here I retreat to, take refuge, when I think that my life will be public property and there is nothing I can do about it. *You will never get access here, you can't, this is the ultimate and inviolable privacy*. They call it loneliness, that here is this place unsharable with anyone at all, ever, but it is all we have to fall back on. Me, I, this feeling of me. The observer, never to be touched, tasted, felt, seen, by anyone else.

That day, chatting away, telling them this is happening, that is happening, I was protecting an experience I had induced. I was being born. In the 1960s this kind of 'religious' experience was common. I was giving myself 'a good birth' – in the jargon of the time. The actual birth was not only a bad one, but made worse by how it was reported to me, so the storyteller invented a birth as the sun rose with light and warmth coming fast into the enormous lamplit room. Why not? I was born early in the morning. Then I invented a chorus of pleasure that I was a girl, for my mother had been sure I was a boy and had a boy's name ready. In this 'game' my girl's name had been planned for months, instead of given me by the doctor. My father – well, where was he, in reality? He was ill because of his imaginative participation in the birth and had gone to sleep after being informed I was safely born.

Probably this 'good' birth was therapeutic, but it was the revelation of the different personalities at work in me I valued and value now. One had to be authentic and not invented, because it was unexpected. Before my eyes, through the whole experience that is, for hours, ran a picture show of beautiful and smart clothes, fashionable clothes, as if a fashion designer inside me was being given her head. They were not on me, but on fashion models: I have never worn this kind of garment. The other person, or personality, was a sobbing child. I wept, and wept, much to the concern of my companions, but I knew it was not important, my weeping. I do not cry enough; that has always been true, and to weep without constraint was a bonus and a bliss. I could easily have cradled that poor baby and comforted her, if I had not been so fascinated by the parallel picture gallery of wonderful clothes, and by the gracious protective chat of the hostess.

That weeping child . . . now she's a real enemy. She transmogrifies into a thousand self-pitying impostors, grabbing and sucking, and when I cut off a long clutching tentacle, at once another appears, just where I don't expect it.

An intensity of the senses accompanies drug-taking, a reminder of how small children experience tastes, textures, smells. While the drug was wearing off they took me out to a meal and I remembered how food tasted in childhood. The omelette exploded on my tongue into a hundred nuances of butter and egg and herb. Already, half-way through my life – I was in my forties – I had lost so much of my capacity for taste. We all fear old age because we are going to lose pleasure, be sans taste. But you lose it all slowly and unnoticed as you live. A small child does not taste anything like the same omelette an adult does. Heat suffocates and burns, pricking the skin, making small limbs wriggle and shrink. Cold attacks like freezing water. Smells expand the nose in delight, shrivel it in disgust. Noises, sounds, fill the inner ear, clamouring, insisting, threatening, *listen to me*. Children and grown-ups do not live in the same sensory world.

I do not actually remember, I was only told, that the climate in Kermanshah was all extremes. It was very hot. It was very cold. It was nearly always very dry. 'The air was so dry the servants threw out the household slops on to the ground behind the house and by lunchtime it was just dust.' 'In Kermanshah the washing was hung out in the early morning and it was bone dry by ten.'

There were three adults in that house, not counting the Persian servants. One was a friend, an American, working in oil. For years I wondered why the American male voice seduced and cajoled, soothed, promised more than any reasonable woman could believe in. At last I saw the obvious explanation and with what reluctance had to accept – again – that our lives are governed by voices, caresses, threats we cannot remember.

A fourth absolutely valid hallmarked memory is of the journey from Kermanshah to Tehran, by car. There were not many cars then, in Persia. We drove through mountains on roads made for caravans, horses, mules, donkeys. It was an open car. I looked over the side, gripping rough canvas, down, down over cliffs to valleys that were all rock, and in particular one a rocky abyss with a village like one of my toys perched beside it. I would recognize that valley now, because terror imprinted it on me for ever. The car ground along the edge of the track that wound around the mountain, wheels on the edge of a void. Then a rocky corner blocked the car. The grown-ups got out with difficulty because my mother was very pregnant, and my father had to manoeuvre his clumsy wooden leg. I was handed over the canvas hood at the back of the car, and I stood behind the screen of my father's legs, one of my arms around a real warm human leg, the other around the hard wood of the dead leg, and I peered down through the legs. Meanwhile the driver (who?) ground the car forward, one wheel on the collapsing outer edge of the track. He was driving, it seemed, into blue air . . . the terror of it, watching the car, would it go over, roll down that mountain? Just above us balanced an eagle large enough to snatch a child, looking down at me. 'Daddy, Daddy, look at the big bird,' but the bird did not swoop off with me, and the car did not go over the edge, for the next thing was, we were in the Edwardian nursery in Tehran, where my brother was soon born.

My mother planned to use the loving coercions of Montessori for our upbringing, but meantime it was the harsh disciplines of one Doctor Truby King that ruled the nurseries both in Kermanshah and in Tehran. He was a New Zealander, whose book was law for innumerable parents, and whose influence can still be heard in the voices of older nurses and nannies. 'You must have discipline – that's the important thing.' Truby King was the continuation of the cold and harsh discipline of my mother's childhood and my father's childhood. I am sure my mother never saw this: she was only doing what all good parents did. Even to read that guide to excellence in family relations is painful.

Take feeding. The infant was supposed to be fed every two hours, and then every three hours, day and night, and the consummation and crown of this clockwork provisioning was to achieve a four-hourly, or three-hourly, pattern of four or six feeds a day, while between them the baby must be left to howl and scream, otherwise the baby will call the tune, the baby will rule the roast, the baby's character will be ruined for life, the baby will become spoiled, soft, self-indulgent, and above all, the baby will 'get on top' of the mother. The baby must never be picked up between feeds. The baby must learn what's what and who is boss right from the start, and this essential instruction must be imparted while the infant is lying alone in a cot, in its own room, never in the parents' bedroom. He, she, must learn its place, understand its position in the universe – alone.

In my case, as my mother cheerfully told me, again and again, I was starved for the first ten months of my life since, because she could not feed me, being too run-down after the war, she fed me cows' milk, diluted to English standards, and cows' milk in Persia had only half the goodness of cows' milk in England. 'You just screamed and screamed all day and all night.'

Well, perhaps, but in the photographs I do not seem to be a mere rack of bones. I look quite plump and cheerful. Why did my mother need to tell her little daughter, so often, and with such enjoyment, that she had been starved by her mother all through her infancy? I think her sense of the dramatic might have contributed here. It used to drive me wild with irritation – and my father too – that everything, always, was presented to the world as a drama. I did not mind that she acted out everything, but that she seemed unaware she was doing it. But have it her way: if I was a permanently hungry baby, it did not seem to do me much harm.

Now, toilet training, that key to character building. Believe it or not, it was recommended the infant must be held over the pot from birth, at regular times every day. 'You were clean by the time you were a month old!' Do I believe this? I do not, but the triumph in her voice spoke of victories over much more than an infant's bowels. Cleanliness is next to Godliness. (The Koran has something on these lines too.) A small baby has no control over its functions. But if you 'hold out' an infant, with encouraging words, using the cold edge of the pot firmly, as a reminder, pouring water from a jug held high enough to make a tinkling sound into a basin, all the while gently rubbing the stomach, then the infant is likely to oblige. Just imagine it, from one end of the British Empire to the other, wherever the map of the world was coloured pink, British matrons or their nurses were 'holding out' tiny infants.

You would think all this must have left me with obsessive cleanliness, tidiness, need for order. No. I am untidy, tolerate disorder, but am obsessive in small useful ways, like keeping a diary.

The vividest early memory was – not the actual birth of my brother – but my introduction to the baby. I was two and a half years old. The enormous room, lamplit, the ceiling shadowed and far above; the enormous bed, level with my head, on which my father lay, for he was ill again: these days they would be making jokes about couvade. Women were supposed to stay in bed for at least a month after childbirth, preferably six weeks, all the time bound tightly from waist to knee with rigid linen – hard to believe that my energetic mother would submit to this, and she was standing by an enormous cot that was all ebullient white flounces of dotted white muslin. The cot was well above my head, and she was bending past it and saying persuasively, 'It is your baby, Doris, and you must love it.' From the depths of the white flounces she lifted a bundle of baby and this was held close to me so that, if I were stupid, I could believe I held it. The baby I do not remember. I was in a flame of rage and resentment. It was not my baby. It was their baby. But I can hear now that persuasive lying voice, on and on and on, and it would go on until I gave in. The power of that rebellious flame, strong even now, tells me it was by no means the first time I was told, lyingly, what I must feel. For it was not my baby. Obviously it was not. Probably Truby King or even Montessori had prescribed that the older dispossessed child must be tricked into love, thus cleverly outwitting jealousy. I hated my mother for it. I hated her absolutely. But I was helpless. Love the baby I did. I loved that baby, and then the infant, and then the little boy with a most passionate protective love. This is not only an authentic memory, every detail present after all this time, but deduction too. By this event and others of the same kind my emotional life was for ever determined.

All you need is love. Love is all you need. A child should be governed by love, as my mother so often said, explaining her methods to us. She had not known love as a child, and was making sure we would not be similarly deprived. The trouble is, love is a word that has to be filled with an experience of love. What I remember is hard bundling hands, impatient arms and her voice telling me over and over again that she had not wanted a girl, she wanted a boy. I knew from the beginning she loved my little brother unconditionally, and she did not love me.

The fact was, my early childhood made me one of the walking wounded for years. A dramatic remark, and pretty distasteful, really, but used with an exact intention although it makes me easy victim to the current obsessionalists who see evidences of 'abuse' everywhere. They mean, usually, sexual abuse. If you say, I wasn't abused, they at once put on that knowing-better smile used by certain kinds of analyst. But these hysterical mass movements surge past, die, change into something else, perhaps even into an examination not of sexual handling or using of children (which I think are not as common as some people want to believe), rather into the emotional hurts which are common, are the human condition, part of everyone's infancy. I think that some psychological pressures, and even well-meant ones, are as damaging as physical hurt. However that may be, all my life I have understood, felt at home with, sometimes lived with, people who had bad childhoods (I nearly wrote, conventionally bad childhoods). They were adopted and then neglected, spent time in care or in orphanages, were bargaining counters in savage power games between parents, were sent too young to cruel or cold schools – now we might be getting somewhere, but that was a late hurt, not an original one. All these people had put themselves together after panic flight from home, or a collapse. For years my friends were nearly all people who had created their own families. Then, it was not all that common, but now it is. The world is so full of war, civil war, famines, epidemics, that waifs and strays are bred, it seems, by the million. They create for themselves a family. In every one of them is a place, large or small, that is an emotional wasteland.

Yet my mother was conscientious, hardworking, always doing the best as she saw it. She was a good sort, a good sport. She never hit or even slapped a child. She talked about love often. The tenderness she had never been taught came out in worrying and fussing and – in the case of my brother – making him 'delicate' so she could nurse him; in my case, actually making me sick for a time.

My father was affectionate but he was not tender. Neither parent liked displays of emotion. If my mother's daughter had been like her, of the same substance, everything would have gone well. But it was her misfortune to have an over-sensitive, always observant and judging, battling, impressionable, hungry-for-love child. With not one, but several, skins too few.

The Tehran nursery was English, Edwardian, and could have been in London. An enormous room, square, high, filled like a lumber room with heavy furniture. In the wall burns a fierce and exuberant fire, held safe from the room and from curious children by a brass fireguard like a gate. On the brass rails are folded ironed clothes and nappies, airing. A wooden folding stand holds wads and pads and swaddles of clothes, more and more bibs, nappies, vests, binders, woollies, robes, dresses, socks, caps, jackets, shawls. All that side of the room is screened by a wall of these clothes, and behind them in the wall itself are cupboards packed with piles of jackets and dresses and petticoats in wool and in lawn, in nun's veiling and in silk, in cotton and in flannel. Hundreds of them, dozens of everything. This wardrobe is needed for two tiny children, who are sitting on chamber pots low down among the vast chairs and a high chair like scaffolding. The air in that room is all smells. The scorch of newly ironed cloth, vaseline, Elliman's Embrocation, cod-liver oil, almond oil, camphorated oil, Pears soap, the nostril-expanding tang from the copper jug and basin on the washstand, the airless smell of flames, paraffin from the little stove that heats bottles and milk, the smell of the contents of the two pots that are only partially kept confined by the small bottoms. Heavy curtains hold dust, behind them muslin curtains with their smell of soap, and the wood smells of furniture polish. The curtains have blue and pink Bo-peeps and lambs, but otherwise everything, but everything, is white. A suffocation of smelly whiteness.

First the tiny girl and then the baby, who always did what she did, lift a bottom off the pot and the women in the room exclaim and coo, Harry is a *good* little baba, Doris is a *good* little baba.

So rewarding was this continuous daily and nightly approval, that Doris actually arrived at a formal Legation dinner party holding out a pot and announcing, 'Doddis is a *good* little baba.' I would not have paid this memory much respect if, decades later, this same Doris, having finished a novel which was to arrive at the publisher's next day, had not dreamed she walked into the publisher's office – Jonathan Cape, as it happened – holding out a pot that contained a manuscript. Doris had been a good little girl. She was full of the glow of achievement, of having proved herself worthy of loving affection.

I offer this as my contribution to understanding the far from simple relations between publishers and authors. (I think it is necessary for the sake of the uninstructed to insist that this dream, so say experts, is the best of auguries.)

There were two women in the nursery. My mother was enormous, solid, a vibrating column of efficiency and ruthless energy, and part of my attention was always on her, for I was afraid she would carelessly knock me over, tread on me. She was taller and larger than the other woman, whom an adult would judge as small. This was Marta, a Syrian, a cross old woman, the nurse. She spoke only French. This pleased my mother, bent on getting her children properly educated. Has this left me with a natural disposition for French, though I have never done more than read it, and use it on the restaurant, taxi, it-is-a-fine-day, where do you live, level? It could be said, yes, for any other language I attempt to learn, no matter how much effort I put in, is screened from me by French. The first word that comes is French, and has to be batted out of my brain. Often baby words, nursery talk.

Just as I now wonder about Emily Flower, who did not deserve even a photograph, and about Caroline May Batley, whose son disliked her and whose husband married again the year she died, I would like to know more about Marta, forced to be a nursemaid in the English family. 'Old Marta.' But she doesn't look so old in the photographs. What war, calamity, famine, personal misfortune forced her to work in the strict English nursery where her sufferings and loneliness goaded her tongue and made her hands hard and unkind? At least, with me. 'Bébé is my child, madame. Doris is not my child. Doris is your child. But Bébé is mine.' So she said. Often. And very often was I reminded of it, all through my childhood, with the relish that always accompanied such information. Now I see this pleasure in authenticating my inadequacies not only as insensitivity, which it was, but also as another expression of my mother's natural theatricality. She might have been an actress, but I am sure that did not occur to her. If it was shameful for a nice girl to be a nurse, how much worse to go on the stage? John William would have died from the disgrace of it. Yet it was born in her. Years after the Tehran nursery, she would bring to life Marta, an irritable scolding old woman. 'I had to stop her slapping and pinching you. She never slapped Baby. She loved him too much for that. "Méchante, tu es méchante!"' she snapped at me, in Marta's voice. And I knew how she experienced her father, for she became the cold angry man, his mouth full of self-righteous platitudes, and the frightened little girl standing stiffly in front of him, looking bravely up into the face of Authority.

She did not weep when her father was harsh: she stood up to him by being everything he demanded of her, and more. I on the other hand fought Marta for my rights in that nursery, and unloved children are not 'nice', not 'gentille'. Who did love the child? Her father. The smell of maleness, tobacco, sweat, the smell of father, enveloped her in safety.

When I wrote *Memoirs of a Survivor* I called it, 'An Attempt at an Autobiography', but no one was interested. Foreign publishers simply left it off the title page, and soon no one remembered to put it on reprints in English. People seemed embarrassed. They did not understand it, they said. For thousands upon thousands of years, we – humankind – have told ourselves tales and stories, and these were always analogies and metaphors, parables and allegories; they were elusive and equivocal; they hinted and alluded, they shadowed forth in a glass darkly. But after three centuries of the Realistic Novel, in many people this part of the brain has atrophied.

To me nothing seems more simple than the plan of this novel. A middle-aged person – the sex does not matter – observes a young self grow up. A general worsening of conditions goes on, as has happened in my lifetime. Waves of violence sweep past –

represented by gangs of young and anarchic people – go by, and vanish. These are the wars and movements like Hitler, Mussolini, Communism, white supremacy, systems of brutal ideas that seem for a time unassailable, then collapse. Meanwhile behind a wall, other things go on. The dissolving wall is an ancient symbol, perhaps the oldest. When you make up a story, and you need a symbol or analogy, it is always best to choose the oldest and most familiar. This is because it is already there, in the human mind, is an archetype, leads easily in from the daytime world to the other one. Behind my wall two different kinds of memory were being played, like serial dreams. There are the general, if you like, communal, dreams, shared by many, like the house you know well, but then find in it empty rooms, or whole floors, or even other houses you did not know were there, or the dream of gardens beneath gardens, or the visits to landscapes never known in life. The other kind was of personal memories, personal dreams. For years I had wondered if I could write a book, a personal history, but told through dreams, for I remember dreams well, and sometimes have kept notes of them. Graham Greene has tried something of the kind. This idea of a dream autobiography became the world behind the wall in *Memoirs of a Survivor.* I used the nursery in Tehran, and the characters of my parents, both exaggerated and enlarged, because this is appropriate for the world of dreams. I used that aspect of my mother which she herself described as 'I have sacrificed myself for my children.' Women in those days felt no inhibitions about saying this: most are too psychologically sophisticated now. She was the frustrated complaining woman I first met as my mother, but who has often appeared in my life, sometimes as a friend. She talks all the time about what a burden her children are to her, how they take it out of her, how much she is unfulfilled and unappreciated, how no one but a mother knows how much she has to give of herself to ungrateful children who soak up her precious talents and juices like so many avid sponges.

The point is, this kind of talk goes on in front of the children, as if they were not present, and cannot hear how she tells the world what a burden her children are, what a disappointment, how they drain her life from her. There is no need to look for memories of 'abuse', cruelty and the rest. I remember very well – though how old I was I do not know – leaning against my father's knee, the real one, not the metal-and-wood knee, while my mother chatted on and on in her social voice to some visitor about her children, how they brought her low and sapped her, how all her own talents were withering unused, how the little girl in particular (she was so difficult, so naughty!) made her life a total misery. And I was a cold flame of hatred for her, I could have killed her there and then. Then this was succeeded by a weariness, a bitterness. How could she talk about me as if I were not there? And about my little brother whom I so adored, as a burden? Hypocrisy – for she adored him, and said so. How could she diminish and demean and betray me like this? And to a mere visitor . . . I knew my father did not like her doing it: I could feel what he felt coming into me from him. He was suffering, because of this great lump of solid, heavy insensitivity, his wife, who did not seem to know what she was doing.

And yet, what was she doing? No more than other women did. Than women so often do. Everywhere, you can hear them at it on trains and on buses, on the streets, in shops, tugging their kids along by the hand or pushing them roughly in their pushchairs; they complain and they nag, while their children, assumed to be without ears, are told how they destroy her, how she does not want them and – for what else can she mean as she talks like this? – what a mistake she has made in having them at all.

I do not believe that even robust and insensitive children remain unaffected by this assault on their very existence.

But I was born with skins too few. Or they were scrubbed off me by those robust and efficient hands.

And my father, always suffering and shrinking because of the unawareness of his wife? Was a skin scrubbed off him by the efficient Caroline May? And what about all those other melancholy long-headed semi-poets of his family? Or is there such a thing as a gene for the condition, being born with a skin too few?

All I know is that I remember, sharp and clear and immediate, nothing invented or made up about it, how my father sat and watched the events and people around him with a slow, relishing, sardonic smile. (This same smile being the equivalent of the novelist's contemplation of the world.) And when the cross old nurse Marta and the great bustling woman who was my mother made me want to crawl off somewhere to hide, or made me hate them so much I would have killed them if I could, then it was with my father I took refuge.

And yet.

In that house in Tehran – not in the overcrammed nursery, but down in the drawing room, equally crammed and crowded with furniture but at least not white, white, deadly white – every night took place a ritual. We, the small children, were led down by the nurse for the bedtime game. We had pillowfights, were chased, caught, thrown up in the air – and tickled. This goes on now in many middle-class families, considered salutary, character building. I see now the inflamed, excited face of my mother, as her pillow flailed against mine, or my little brother's. I hear the excited cries from myself and my brother and my mother as the air filled with feathers and my head began to ache. And then the moment when Daddy captures his little daughter and her face is forced down into his lap or crotch, into the unwashed smell – he never did go in for washing much, and – don't forget – this was before easy dry-cleaning, and people's clothes smelled, they smelled horrible. By now my head is aching badly, the knocking headache of over-excitement. His great hands go to work on my ribs. My screams, helpless, hysterical, desperate. Then tears. But we were being taught how to be good sports. For being a good sport was necessary for the middle-class life. To put up with 'ragging' and with being hurt, with being defeated in games, being 'tickled' until you wept, was a necessary preparation.

It does not have to be like this, for you may watch a very little child being gently chased and tickled in a real game, not an exercise in disguised bullying. But I did not stop having nightmares about those great hands torturing my ribs until I was seven or eight. These nightmares are as clear in my mind now as they were then, though the emotion has long gone away. I became an expert on nightmares and how to outwit them when I was a small child, and that nightmare of being helpless and 'tickled' was the worst.

Yet my father was my ally, my support, my comforter. I wonder how many women who submit to physical suffering at the hands of their men were taught by 'games', by 'tickling'. No, I am not one of them. In all my life I have never been hit, slapped, or in any way at all physically maltreated by a man, and I am saying this because at this particular time it is hard even to pick up a popular paper without reading about women being physically bullied by men. There are worse kinds of bullying.

And now here is a deduced memory. In the big room where the bedtime rituals took place were heavy red velvet curtains. That they were heavy I know because of the memory of velvet dragging on my skin, my limbs, and I clung to folds that filled my small arms. That they were red I believe because when I was doing apprentice pieces in my twenties, several Poe-like stories appeared where red velvet curtains concealed threat. In one over-worked piece there was a man in a wheelchair who drove a child back and back across a room to a wall that was all red velvet, and when she took one step too far back through them, on the other side was no wall, only empty space. There are any number of childhood 'games' that could account for this one. The story was called 'Fear and Red Velvet'.

I have been writing of the tactile and sensuous subjective experience of a child, smelly, noisy, the rumble of a mother's stomach as she reads to you, the bubbling dottle in Daddy's pipe, the pounding of blood in your ears – all the din and stink and smother of life which a child soon learns to shut out, if she is not to be overwhelmed by it. But all that – and the battle for survival – went on side by side with what was being provided intelligently and competently by my mother, the daughter of John William, who had taught her what a good parent must provide for a child. For if my mother was an over-disciplined little girl frightened ever to defy her father – until she did, when she went to be a nurse – then she was also taken as a matter of course to Mafeking Night, and the celebrations at the end of the Boer War, and to all the Exhibitions, and to line the route when foreign kings and queens came on State Visits, and for trips on the new railways. She was taught to admire Darwin and Brunel, and to be proud of Britain's role as the great exemplar of progress. She was taught to take herself off to museums and to use libraries.

And in Tehran, she made sure her children experienced what they should. I was held high through the same velvet curtains to see the night sky. 'Moon, moon' – lisped attractively, for my mother as she reported this became a winsome little girl. 'Starth, starth' – she said I said. When my father, with no histrionic talent at all, tried to say a child's 'moon', but with a French 'u', for was it not also a *lune*? – then he failed. When it snowed – for it certainly snows heavily there, in Tehran, and I can see any time I want to the sheets of sparkling white over shrubs and walls – my mother built snowmen, with eyes of coal and noses of carrots, and cats of snow with green stone eyes. She was good at it, and made them well, and taught us how to say nose, and eyes, and paws and whiskers in French. She took us to mild slopes of snow, which I saw like the foothills of Everest, and pushed us off into snowdrifts while we clutched at teatrays, explaining that snow is water, which can also be ice and rain and hail. At holidays we were taken to the mountains, to Gulahek, whose name means a place of roses, and there in my mind now are the roses, red and white, pink and yellow, smelling of pleasure. And we were taken on picnics and to the Legation children's fancy dress parties. All these events were presented to us as our heritage, and our due, and, too, our responsibility. *This* was snow, *those* were stars, and here on this rocky face near the road was where Khosrhu on horseback had been carved thousands of years ago – and the thousands of years, as she said it, became yesterday, appropriated as our heritage. When we went to parties at the Legation her voice told us this was where we belonged, these were *nice people*, and we were nice people too. But my father did not like Mrs Nelligan, the senior lady of the British community. If my mother's voice had an orchestra of tones telling us what we must admire, then so did my father's, contradicting hers, for he never liked people because of their degrees of 'niceness', and if I did not then understand this, I knew very well he criticized her for liking others because of their position in society, not because they were likeable. To write about all this now, the terrible snobbery of the time, is to invite, 'Well what of it? That was then, it was that time . . .' But if the vocabulary of snobbery has changed, its structure has not, and the same mechanisms operate now, while people laugh (mindlessly, I think) about the old days.

The truth is, she did very well for us, my brother and me, in that country where she enjoyed the best years of her life, for she might have been frustrated in all that side of her nature which would have made her a brilliantly efficient matron of a big hospital, but there was never a woman who enjoyed parties and good times more than she did, enjoyed being popular and a hostess and a good sort, the mother of two pretty, well-behaved, well-brought-up, clean children.

She told us over and over again, for it was so important to her, long after, in Africa, how she had dressed up for a fancy dress ball at the Legation as a cockney flower girl (and did she know that she was for that evening her own poor mother Emily?) and while she was dancing with some young man on the Legation staff, he stopped in the middle of the dance floor and said, scarlet with shame, 'Good Lord, you aren't Maude Tayler, are you? You are so pretty I didn't recognize you.' And of course slunk away, because of his gaffe. For my mother was supposed to be plain, a plain Jane, all her life. I think it was the need to make sure she didn't become vain and flighty, like Emily. As a child, listening to the reminiscence (again and again), my heart hurt for her, and it went on hurting, as the story went on being told, for years – all her life – while her eyes glistened with real tears as she remembered the young man who thought she was so pretty.

There are memories that have about them something of the wonderful, the marvellous. A man, a gardener – Persian – stands over stone water channels, that come under the brick wall into the garden, bringing water from the snow-mountains, and he is pretending to be angry because I am jumping in and out of the delicious water, which splashes him too. I am sent by my parents into the kitchen to tell the servants that dinner may be served, and that is Tehran because I have my brother by the hand, and I look up, up, up at these tall dignified men and see that their faces are grave under their turbans, but their eyes smile.

And the most important, the one that has about it charm, magic, is also the most nebulous, and perhaps I dreamed it. I have lost my toy sheep, a bit of wood on wheels that has real sheepskin wrapped around. I am crying, and wander off and see a flock of sheep and the shepherd, a tall brown man in his brown robes, looking down at me. The dust is swirling around him and the sheep, and a sunset reddens the dust. That is all. In my *Tales from the Bible for Children* was a drawing of the Good Shepherd, but that could not have in it the dust, nor the smell of sheep and dust. The memory is charged with meaning, comes back and back, and I never know why.

Soon the tastes, textures, smells, of Persia faded because of the immediacy of the colours and smells and sounds of Africa, and it was only in the late 1980s that I went to Pakistan and there met a self still immersed in that early world. The voice of the man who chanted, or sang – what is the word for the most haunting of sounds, the Call to Prayer? . . . the slant of hot sun on a white-washed wall where reddish dust lived in the grain of the white . . . and the smells, the smells, a compound of sunheated dust, urine, spices, petrol, animal dung . . . and the sounds and voices of the bazaar and its colour, explosions of colour . . . and the sad bray of donkeys who, according to the ideas of Islam, are shameful because they cry only for food and sex, but I think they cry from loneliness, and prefer Chesterton's celebration of donkeys.

A cock crowing, a donkey braying, dust on a whitewashed wall – and there is Persia, and now, where I live in London, just down the hill a cock sometimes crows and at once I hardly know where I am.

Far away from England, in Persia, my parents were not as cut off from their family as they soon would be in Africa, for at least two relatives came to visit: one was Harry Lott, a cousin of my father's. It is strange that of this man he talked of so often, for so many years, I can say nothing, for I don't remember him. Uncle Harry Lott was the family's good friend: he sent presents and wrote letters, and that went on when we were in Africa, too, until he died. 'Oh he did love you kids, he couldn't get enough of you,' says Daddy, adding characteristically, 'God knows why.' And now I watch some little child in the arms of a loving friend, and know this will affect the child for always, like a little secret store of goodness, or one of those pills with a delayed reaction, releasing elixirs into the bloodstream all day – or for all of a life. But the child may remember nothing about it, not a thing. I find it a pretty uncomfortable experience, watching small children and what moulds and influences them, and they become adolescents, and you know exactly why they do this or that, while they often do not. And then they are young adults, still set in patterns of behaviour whose origins you know. Or, after a separation you meet this child grown or half-grown, and you find yourself searching in eyes that are unconscious of what you are looking for, or examine the way arms go around a friend, stiffly or warmly, or how a hand rests tenderly on the head of a dog.

The other visitor was Aunt Betty Cleverly, whose great love had been killed in the war – like all the women of her age in Europe then. She was a cousin of my father's, a big untidy woman with a buck-toothed smile. She, too, loved us, and for years and years my brother and I were told of it, but what I remember is being in her bed in the early morning, and on the bedside table the early morning tea tray, she in a long-sleeved, very pink woollen nightgown, her long hair filling the bed and tangling me in soap-smelling brown silk, while she is soaking Marie biscuits in strong tea, giving me fragments to taste, and laughing while I shudder at the bitter taste, and she gives me a new clean biscuit and cries, 'Don't tell Mummy, I'm spoiling your appetite for breakfast.' Then she sings 'Lead Kindly Light' and 'Rock of Ages' in a strong throaty voice, conducting herself with a teaspoon. Off she goes to China, for she is a missionary, and her letters to my parents report on the ways of the heathen who were being brought under control by Christianity, and on the London Missionary Society, and on parish matters back home in England.

When my father was due his leave at Home, after nearly five years of the Imperial Bank of Persia, first as branch manager in Kermanshah, and then as Assistant Manager in Tehran, he was expecting to return to Persia, and my parents' minds were full of anxieties about how to educate their children. To leave the older child, me, behind in England, aged five, would have been usual for the time, but my mother knew from Kipling's 'Baa Baa Black Sheep' what horrors of bullying and neglect small children could suffer because of ill-chosen parental substitutes. My father did not want to return to Persia. The social life bored him. He never had enjoyed working in a bank. The Persians were corrupt and when he said so no one seemed to think it mattered.

Meanwhile absence from England had not made his heart grow fonder. Nor did it, ever. Until he died he would see England – England, not Britain, or at least it was not Britain he apostrophized – as a country that had betrayed its promises to its people, as cynical, as corrupt. It was full of complacent crooks who had got rich out of the war and of stupid women who gave white feathers to men in civvies, half-dead from the Trenches, and then spat at them. And the people had no idea of what the Trenches had been like. And he would sing, all his life, his voice stiff with anger,

And when they ask us . . .
And they're certainly going to ask us . . .
We're going to tell them . . .

But they didn't ask, they never did, for the war had become the Great Unmentionable. Yet now he had to face six months' leave in the place. He would have to spend time with his brother Harry, whom he had always disliked, and who patronized him, for he was the successful one, a manager of the branch of the Westminster Bank, with a yacht and a smart car and a house my father hated, for it was the essence of smart suburbia. What matched his idea of himself, and where he had felt perfectly at home, was the great stone house in Kermanshah, with the snow-covered mountains all around. But he had lost that for ever. He did not like his brother's wife, Dolly, found her silly and suburban. He disliked his wife's sister-in-law, Margaret, and thought my mother's brother a bore. Six months of relatives, hell on earth, in snobby, self-important, provincial, parish pump, ignorant, little England. And then back to Tehran again, and its busy snobbish social life, the picnics and the Legation parties and the musical evenings where his wife played, while some young man sang 'The Road to Mandalay' and 'Pale Hands I Loved Beside the Shalimar'. 'Why can't people sit at home and be quiet?' he demanded, like the philosophers. But my mother merely smiled, for she knew she was in the right. The trouble was, his eccentricity was infecting her daughter.

'No I don't want to, I won't,' I weep, being forced into a Bo-peep costume. 'I don't want to be Bo-peep. Why can't I be a rabbit like Harry?' My mother laughs at me because of the ridiculousness, and the trouble is, I can feel my face wanting to laugh too. I change ground. 'I don't want to go to the party. I don't like parties.' 'Nonsense. Of course you like parties. Of course you want to be Bo-peep.' 'No, I don't, I *don't*.' 'Don't be silly. Tell her she's being silly, Michael.' 'Why should she go if she doesn't want to?' says Daddy, testy, irritable – difficult. 'I don't want to go either. Parties! Who thought of them first? Whoever it was should be hanged, drawn and quartered. The devil, I shouldn't be surprised.' 'Oh *Michael* . . .' 'No, I tell you, I've only got to think of a party and I want to upchuck. And that's what these kids are going to do. Well, don't they? They get overexcited, they eat too much, sick all over the place.' 'Oh rubbish, Michael, you like parties *really*.'

No hatred on earth is as violent as the helpless rage of a little child. And there was Gerald Nelligan, confronting his mother and shouting, 'No I don't want to, I won't dress up, why should I?' He was two years older than me, a big boy, but he flung himself down in the flailing white-faced yelling rage you see trapped children use every day. But they will be saying later, 'I had a wonderfully happy childhood.' Nature knows what it is doing, prescribing amnesia for early childhood.

And now, the cat: I wrote about this cat in *Particularly Cats*, but I know it needs more emphasis. 'You found that dirty cat in the gutter and brought it into the drawing room, and it was bigger than you were,' said my mother, being the child and the cat together. 'And you insisted on having it in your bed. We washed it in permanganate . . .' An essential prop of the British Empire, permanganate of potash. 'And old Marta came storming in and said, "Why is that dirty cat allowed here?"' But I was allowed the cat, and how much I loved it does not need much in the way of deduction. For years the death of a cat plunged me into grief so terrible I had to regard myself as rather mad. Did I feel anything as bad when my mother died, my father died? I did not. That old cat, rescued from slow death on the streets of Tehran, was my friend, and when we left Persia, what happened to it? They told me soothing lies, but I did not believe them, for I wept inconsolably. 'You were inconsolable,' says my mother.

I was getting on for being an old woman when I experienced grief which, on a scale of one to ten – ten being the real, frightful sodden depression that immobilizes, and which I have not myself experienced – was at nine. On this scale, grief for a dying cat is at four or five, while grief for parents and brother is at two. Clearly, the pulverizing pain over the cat is 'referred pain' as the doctors call it, when you have pain in one organ, but really another is the cause. Surely one has to ask, but why? And, at force nine, I was pulverized with a grief I did not know the origin of, and still don't.

But the question surely must be, why, of so many memories from that early time, there are so few that are jolly, pleasant, happy, even comfortable? That hungry, angry little heart simply refused to be appeased? Is there a clue in the business with the photographer? I was three and a half. There survives a photograph of a thoughtful little girl, a credit to everyone concerned, but as it happens I remember what I was feeling. There had been a long nag and fuss, and worry and trouble about the dress, of brown velvet, and it was hot and itchy. My stockings had been hard to get on, were twisted and wrinkled, and had to be hitched up with elastic. My new shoes were uncomfortable. My hair had been brushed, and done again and again. There was a padded stool I was supposed to sit on but it was hard to climb on to and then stay on, for it was slippery. I had also been put on a very large solid carved wooden chair, but then they said it was not right for me. They? – my mother and the photographer, a professional, whose studio was full of Japanese screens showing sunsets and lake scenes and flying storks, of chairs and tables and cushions and stuffed animals to set the scene for children. But I insisted on my own teddy, scruffy, but my friend. I felt low and nervous and guilty, because I was causing so much trouble: as usual it was as if my mother had tied, but too fast and awkwardly, a large clumsy parcel – me – and I did not fit in anywhere, and might suddenly come untied and fall apart and let her down. I felt weary. This small sad weariness is the base or background for all my memories. *Everything was too much*, that was the point, too high, or too heavy, or too difficult, or too loud or bright, and I could never manage it all, though they expected me to.

4

WHEN MY MOTHER DECIDED to travel to England via Moscow, across Russia, because she did not want to expose her little children to the heat of the Red Sea, she did not know what she was doing – as she often said herself. '*If I'd only known!*' She did know we would be the first foreign family to travel in an ordinary way since the Revolution. It was 1924. That it would be difficult, of course she knew, but difficulties are made to be overcome. The journey turned out to be horrendous, told and told again, the vividest chapter in the family chronicle. What I was told and what I remember are not the same, and the most dramatic moment of all is nowhere in my memory. At the Russian frontier, it turned out we did not have the right stamps in our passports, and my mother had to browbeat a bemused official into letting us in. Both my mother and my father loved this incident: she because she had achieved the impossible, he because of his relish for farce. 'Good Lord, no one would dare to put that on the stage,' he would say, recalling the calm, in-the-right, overriding British matron, and the ragged and hungry official who had probably never seen a foreign family with well-dressed and well-fed children.

The most dangerous part was at the beginning, when the family found itself on an oil tanker across the Caspian, which had been used as a troop carrier, and the cabin, 'not exactly everyone's idea of a cruise cabin', was full of lice. And, probably, of typhus, then raging everywhere.

The parents sat up all night to keep the sleeping children inside the circles of lamplight, but one arm, mine, fell into the shadow and was bitten by bugs, and swelled up, red and enormous. The cabin was usually shared by members of the crew, and was small. For me it was a vast, cavernous, shadowy place, full of menace because of my parents' fear, but above all, the smell, a cold stuffy metallic stink which is the smell of lice.

From the Caspian to Moscow took several days, and the tale went like this: 'There was no food on the train, and Mummy got off at the stations to buy from the peasant women, but they only had hard-boiled eggs and a little bread. The samovar in the corridor most of the time didn't have water. And we were afraid to drink unboiled water. There was typhoid and typhus, and filthy diseases everywhere. And every station was swarming with beggars and homeless children, oh it was horrible, and then Mummy was left behind at a station because the train just started without warning and we thought we would never see her again. But she caught us up two days later. She made the station master stop the next train, and she got on to it and caught us up. All this without a word of Russian, mind you.'

What I remember is something different, parallel, but like a jerky stop-and-start film.

The seats in the compartment, which was like a little room, were ragged, and they smelled of sickness and sweat and of mice, in spite of the Keating's Insect Powder my mother sprinkled everywhere. Mice scurried under the seats and ran between our feet looking for crumbs. The lamps on the wall were broken, but luckily my mother had thought of candles. At night I woke to see long pale dangerous flames swaying against the black panes where cracks let in air, warm in the south, cold in the north. I held my face in it, because of the smell. It was April. My father had flu, and lay on an upper bunk, away from the two noisy children and our demands. My mother was frightened: the great Flu Epidemic was over, but the threat of it would be heard in people's voices for years yet. There were little bloody dots and spatters on the seats, and that meant lice had been here. Years later I had to sit myself down and work out why the words flu and typhus made me afraid. Flu was easy, but typhus? It was from that journey. For years the word 'Russia' meant station platforms, for the train stopped all the time, at sidings as well as big towns, on the long journey from Baku to Moscow.

The train groaned and rattled and screamed and strained to a stop among crowds of people, and what frightening people, for they were nothing like the Persians. They were in rags, some seemed like bundles of rags, and with their feet tied in rags. Children with sharp hungry faces jumped up at the train windows and peered in, or held up their hands, begging. Then soldiers jumped down from the train and pushed back the people, holding their guns like sticks to hit them with, and the crowds fell back before the soldiers, but then swarmed forward again. Some people lay on the platforms, with their heads on bundles and watched the train, but not expecting anything from it. My parents talked about them, and their voices were low and anxious and there were words I did not know, so I kept saying, what does that mean, what does that mean? The Great War. The Revolution. The Civil War. Famine. The Bolsheviks. But why, Mummy, but why, Daddy? Because we had been told that the *besprizorniki* – the gangs of children without families – attacked trains when they stopped at stations, as soon as my mother got out to buy food, the compartment door was locked and the windows pushed up. The locks on the door were unsafe and suitcases were pushed against it. This meant my father had to come down from his high shelf. He wore his dark heavy dressing gown, bought for warmth in the Trenches, but under it he kept on all his gear and tackle for the wooden leg, so he could put it on quickly. Meanwhile the pale scarred stump sometimes poked out from the dressing gown, because, he joked, it had a life of its own, for it did not know it was only part of a leg, and in moments of need, as when he leaned forward to open the compartment door to let in my mother – triumphant, holding up her purchases, a couple of eggs, a bit of bread – it tried to behave like a leg, instinctively reaching out to take weight. The two little children fearfully watched our mother out there among those frightening crowds, as she held out money to the peasant women for the hard eggs, the half-loaves of the dark sour stuff that was called bread. The story said we were hungry because there was not enough food, but I don't remember feeling hungry. Only the fear and the anguish, looking at those swarms of people, so strange, so unlike us, and at the ragged children who had no parents and no one to look after them. When the train jerked forward, the soldiers jumped on to it, clutching what they had managed to buy from the women, and then turned to keep their guns pointed at the children who ran after the train.

The story says we were read to, we played with plasticine, we drew pictures with chalks, we counted telegraph wires and played 'I-Spy' out of the windows, but what is in my mind is the train rattling into yet another station – surely it was the same one? – the ragged people, the ragged children. And again my mother was out there, among them all. And then, when the train was pulling out, she did not appear in the corridor outside the compartment, holding up what she had bought to show us. She had been left behind. My sick father held himself upright in the corner and kept saying it was all right, she would come soon, nothing to worry about, don't cry. But he was worried and we knew it. That was when I first understood the helplessness of my father, his dependence on her. He could not jump down out of the train with his wooden leg and push through the crowds looking for food. 'You had to share an egg between you and there were some raisins we brought with us, but that was all.' She would have to reappear, she would have to, and she did, but two days later. Meanwhile our train had been slowing, groaning and screeching, again and again, into stations, into sidings, into the crowds, the *besprizorniki*, the soldiers with guns. I don't remember crying and being frightened, all that has gone, but not the rough feel of the dressing gown on my cheek as I sat on my father's good knee and saw the hungry faces at the window, peering in. But I was safe in his arms.

A small girl sits on the train seat with her teddy and the tiny cardboard suitcase that has teddy's clothes in it. She takes the teddy's clothes off, folds them just so, takes another set of clothes from the case, dresses the teddy, tells it to be good and sit quietly, takes this set of clothes off the teddy, folds them, takes a third set of trousers and jacket out, puts the taken-off clothes back in the case, folded perfectly, dresses the teddy. Over and over again, ordering the world, keeping control of events. There, you're a good teddy, nice and clean.

From Moscow comes the most powerful of all my early memories. I am in a hotel corridor, outside a door whose handle is high above my head. The ceiling is very far away up there, and the great tall shiny doors go all along the corridor, and behind every door is a frightening strangeness, strange people, who appear suddenly out of a doorway or walk fast past all the shut doors, and disappear, or arrive at the turn of the corridor and then vanish into a door. I bang my fists against our door, and cry and scream. No one comes. No one comes for what seems like for ever, but that cannot have been so, the door must have soon opened, but the nightmare is of being shut out, locked out, and the implacable tall shiny door. This shut door is in a thousand tales, legends, myths, the door to which you do not have the key, the door which is the way to – but that is the point, I suppose. Probably it is in our genes, I wouldn't be surprised, this shut door, and it is in my memory for ever, while I reach up, like Alice, trying to touch the handle.

And now we are in England. One might ask why none of the 'nice' memories, like snapshots, of pretty England, hollyhocks, cottage gardens, a thatched cottage, rocky seaside pools, are as powerful as the memories of dismal England – ganglia of black wet railway lines, rain streaming down cold windows, dead pale fish on slabs held right out into the street, the bleeding carcases on their great steel hooks in the butchers' shops. I met my step-grandmother, so they say, and there is a photograph of me on her knee, but not even a deduced truth emerges. I met my father's father, whose wife Caroline May died that year, and who was about to marry his thirty-seven-year-old bride: probably like all those women, she had lost her love in the Trenches, and marrying an old man was the only chance she had of a husband.

All kinds of visitings and little trips went on, but children are taken around like parcels. A Miss Steele helped with the children, and it is she who provides the sharpest memory of that six months. A room in a hotel. Again it is crammed with furniture, enormous, difficult to make one's way around and through. Two large beds, one mine, and a large cot. The flame on the wall, which is gas, is dangerous, and must be watched, like a candle, although it cannot be overturned like a candle, and it makes a striated light in the room, full of air that seems greyish brown. Dark rain streams down dirty panes. It is cold. The damp woollen bundle that is my little brother snuffles drearily in his cot. Miss Steele has ordered us not to watch her while she is dressing. Miss Steele is so tall she seems to reach the ceiling, and she has floods of dark hair about her shoulders, over her front, and down her back. She has on bright pink stays, and pale flesh bulges out showing through the hair, and below it around her thighs. I see my little brother's bright curious eyes, then he squeezes them shut, pretending to be asleep, then they gleam again. Miss Steele lifts her arms to slip a white camisole over her bushes of hair. Under her arms are silky black beards. I feel sick with curiosity and disgust. There is a smell of dirt and the unwashed smell of Miss Steele, sour and metallic, the smell of wet wool from my brother, and my own dry and warm smell that rises in waves when I lift the grimy blankets and take a sniff. The smells of England, the smells of wet, dirty, dark and graceless England, the smells of the English. I was sickening for Persia and the clean dry sunlight, but did not know what was wrong with me, for small children are so immersed in what surrounds them, their attention demanded all the time by keeping themselves upright and doing the right thing, they have not yet learned that particular nostalgia for place. Or so I think it must be. Or perhaps I was sickening for my lost love, the old cat. Long afterwards, I stood in Granada in Spain and saw the circling snow-topped mountains, and smelled the clean sunny air, and Kermanshah came back, in a rush: this was what it had been like.

But the question surely has at least to be put: why not remember just as intensely the jolly picnics in the hayfield, or the salubrious sandcastles, or the kindly arms of Aunt Betty and Uncle Harry Lott?

A sharp, indeed lurid, little memory is different from all the other English memories. A newspaper comic strip, about the adventures of Pip, Squeak and Wilfski must have been among the very first attempts at anti-Communist propaganda. Wilfski, a bewhiskered villain like a cockroach, was based on Trotsky. He always had a bomb in his hand, threatening to blow something or somebody up. He was designed to inspire fear, horror, and that is what he did.

When we left England for Africa, my father's father, the widower, stood in his thick tweedy clothes in a dark hall with a grandfather clock ticking just behind him, and he wept, and on his long white beard was a string of snot. This was what the child had to see, for the first years of children are devoted to subduing and ordering the physical, snot, shit, pee, a prison they struggle to get out of, and will not enter again until they are old. The old man wept, his heart was broken, he had not seen his son and his son's wife for five years, and he had only just met his grandchildren, but now they were off to Africa where the missionaries his church raised funds for converted savages who might even be cannibals. They talked airily of returning in another five years. He wept and wept, and his granddaughter felt sick at the sight of him and would not let herself be kissed. And perhaps he wept, too, because the family did not approve of him marrying Marian Wolfe, 'a girl half his age'.

The last weeks before leaving England were a rush of buying the things my mother needed for the life she thought she was going to lead. She was guided by leaflets and information from the Empire Exhibition, at whose instigation they were going to Southern Rhodesia, where they would be rich in five years growing maize. For my father, this was a chance to become what he had always wanted to be, ever since his country childhood with the farmers' sons around Colchester. And there had been farmers in his family. But he had never had the capital to farm. Clearly, the more Exhibitions a nation has, the better. That Empire Exhibition of 1924, which lured my father out to Africa – how often have I come on it in memoirs, novels, diaries. It changed my parents' lives and set the course of mine and my brother's. Like wars and famines and earthquakes, Exhibitions shape futures.

Apart from shopping at Harrods, Liberty's and the Army & Navy Stores, they both had all their teeth out. The dentist and the doctor said so. Teeth were the cause of innumerable ills and woes, they were of no use to anyone, and besides, there would not be any good dentists in Southern Rhodesia. (Untrue.) This savage self-mutilation was common at that time. 'We continue to burn candles in churches and consult doctors' – Proust.

The family stood on the deck of the German ship and watched the chalky shores of England recede. My mother wept. The desolation of separation was settling on my heart, but it cannot have been England I wept for, since I hated it. My father's eyes were wet, but he put his arm around her shoulders and said, 'Now come on, old thing!' And turned her away from the disappearing cliffs to go inside.

There was also on deck, apart from my little brother, Biddy O'Halloran, who was to be our governess. What I know about her is mostly what I was told. She was twenty-one. She was Irish. She was 'fast', a 'flapper', a Bright Young Thing. She was definitely no better than she ought to be. Why? She had shingled hair, used make-up and smoked, and was too interested in men. Much later my mother was remorseful, because she had given Biddy a hard time. This was when she, too, smoked, cut her hair, and used some lipstick. 'And I wonder what ever happened to her' – for Biddy clearly found the experience so appalling she never wrote to us. Later she married an Honourable and was in society newspapers.

But she was just one of the many people who had already appeared in my life and disappeared. Acquaintances, lovers, friends, intimates – off they go. Goodbye. Till next time. *A bientôt. Poka.*

Tot siens. Arrivederci. Hasta la vista. Auf wiedersehen. Do svidania. The way we live now.

It was a long voyage, weeks and weeks. A slow boat. Why a German boat? Perhaps my father was putting into practice his feeling of comradeship with the German soldiers who had been sold down the river by their government, just like the English tommies, and the French poilus.

My father was sick nearly all the way to Cape Town, and then Beira. My mother loved every second. This must have been the last time in her life she enjoyed herself in the way of deck games or bridge, dressing up and dancing and concerts – very much her way, her style.

On this boat I disgraced myself. I was miserable. First there was the Captain, my mother's chum, for she was up on deck with him when everyone else was in their bunks being sick in a Force 9 gale, and this established them in a teasing good-fellows' friendship. Joking, joshing, baiting, pulling each other's legs. 'Ribbing.' (Does this word come from the torture of tickling, great hands squeezing small ribs?) It was a most hearty jollity, and he was full of practical jokes. When I was dressed up in my party dress, he invited me to sit on a cushion where he had placed an egg, swearing it wouldn't break. Since it was obvious it would break, I did not want to sit. My mother said I must be a good sport. I sat on the egg and it sploshed under me and spoiled my dress and the Captain roared and rolled about. I was not only angry but felt betrayed. My father was disturbed, but to be a good sport, he must have felt, was the main thing. When we crossed the Line, I was thrown in, though I could not swim, and was fished out by a sailor. This kind of thing went on, and I was permanently angry and had nightmares. I think my mother was having such a good time that her normal obsessive care for her offspring was taking a holiday, for she was not one to take nightmares lightly – if she had been told of them. Besides, was not Biddy there to look after us?

It occurs to me that when my mother became such friends with the German captain two tributaries of a river met. The joshing, ribbing, teasing and ragging came from the English public schools she so much admired, and they were originally inspired by the Prussian elite schools where cruelty was practised on children. The Captain was hardly likely to have been a member of the Prussian elite but then, these examples of good living filter down. And was my mother cruel? Absolutely not. But we can all do whatever it is that is the done thing. Well, nearly all.

In the evenings she put on her beautiful evening dresses and went up to dinner at the Captain's table, to the parties, the dances, the treasure hunts. So did Biddy O'Halloran. We children were shut in the cabin and told to be good. My brother, as ever obedient, slept. I wanted to be where the fun was. But my mother said the evenings were for grown-ups and I would not enjoy it. But I knew I would enjoy it, and she knew I would enjoy it. I hated her. It was no good, the door was locked. I climbed up on to the dressing table and found nail scissors and cut holes in an evening dress. Small hands, the nail scissors were small, and it was hard manipulating them in the thick slippery material. I could not have done much damage, but it is the thought that counts. I was weeping and howling with rage. No, I certainly was not punished. But I was held on her knee through one of those scenes, her voice low, throbbing with reproach, intimate, while she talked about behaving well and about love – hers – and being good for the sake of being good.

And yet, while all these betrayals and injustices went on, the business of education went on too, for this was, after all, my mother's main business. Tiny children were held up in their parents' arms and instructed to watch flying fish, porpoises, the colours of sunsets, the trajectories of other ships whose funnels trailed smoke smudges across fair skies, the birds sitting on the rigging and on the rails, seagulls flying low after the ship to catch the scraps flung out to them by sailors, the phosphorescence on the waves at night, moonlight, and lifeboat drill – this last being far from an academic exercise, since her great love the young doctor had drowned for lack of a lifeboat. And, as a special favour from the Captain, we were taken down, down, through the world of bright corridors. And then, suddenly, we were in another world of oily metal stairways and big black pipes running and bending on steel walls. My brother and I clutched each other and stood looking down from what seemed a tiny platform, only part of a walkway into the bottom of the ship, where dirty half-naked men shovelled coal into the mouths of furnaces, one, two, three, four – more, we could not count them, and the flames reared up and flung red light on to naked sweaty torsos. These men looked up and saw two small clean children, the privileged, peering down at them with horror on their faces, and behind them the parents in their good clean clothes, and the Captain himself in this part of the ship where they did not expect to see him. And they swung their bodies hard in the rhythm of the work, while arcs of black coal reached from them to the flames, and then they looked up, and their white teeth showed in grimed faces. It was like the *besprizorniki* on the Russian railway platforms, it was the other world, where people had holes in their clothes and bones showed on their faces. I was afraid, looking down at the men who shovelled coal while the sweat poured off them, just as I had been looking out of the dirty cracked train windows.

In Walvis Bay I met death for the first time, on the beach, a sea ebbing from sands where tiny fish lay dying in a sea-puddle. They wriggled and writhed and gasped, and then I saw that drifts of dead little fish lay all over the sands. 'Are they dead?' I asked, wanting confirmation, wanting the word to fit what I saw: my father and mother understood the gravity of the moment, and my father said, 'Yes, I am afraid so,' and my mother said, 'Well, never mind.' A howlingly beautiful sunset filled the sky and I understood: this is how things are and there is nothing to be done about it.

Somewhere in the Cape, ostriches ran high-stepping across scrubby sands with blue mountains far behind them. Distance. The empty distances of Africa. But the family went on in the ship around the coast to Beira, of which nothing remains in my mind, not the railway journey up to Salisbury, nor Salisbury itself, which was then a little town you could stroll across in twenty minutes, nor the twenty miles' journey to Lilfordia, where we were to lodge while the farm was chosen.

Why ostriches, and not the ox wagons that still used the Salisbury streets, built wide so that the wagons could turn in them? Why the train in Russia but not the train Beira-to-Salisbury, surely equally exotic? Why remember this and not that? If I had decided to remember only the unpleasant, then why the ostriches, which were pure delight?

Lilfordia was the home of the Lilford family, later to be famous in the Bush War (the War of Liberation), because of Boss Lilford and his services to the white cause. Then it consisted of many rondaavels, solid and well-built thatched brick huts, scattered among shrubs which, we were at once warned, should not be approached incautiously, because of snakes. From the grown-ups' voices – the Lilfords' – it was clear these were no more of a danger than knocking a candle or a lamp over when playing too roughly, only something to look out for.

My father left us there and went off to look for a farm, I think, on a horse. This was when the white government was selling land to ex-servicemen for practically nothing, and when the Land Bank supported struggling white farmers on long-term loans. He would start farming on a loan. My parents had £1,000 and my father would have a pension because of his cut-off leg. He was also entitled to free repairs to his wooden leg, and, too, a spare one. This was well before the miracle legs of now, which can dance, climb, jump – do everything a normal leg does.

He chose the district of Lomagundi because it was a maize-growing area. It was in the north-east of Southern Rhodesia, very wild and with few people in it, and it stretched all the way up to the Zambesi escarpment. Banket, a large part of Lomagundi, not only grew good maize but had its name because it was full of quartz reefs similar to the rock formations called 'banket' on the Rand down south. So there were gold mines too. He and my mother must have realized by now that the enticements of the Empire Exhibition had little to do with reality. Fortunes had been made out of maize during the war, but were not being made now. But maize was what he wanted to grow. And that area was still being 'opened up for settlement'. It would not have occurred to them that the land belonged to the blacks. Civilization was being brought to savages, was how they saw it, because the British Empire was a boon and a benefit to the whole world. I do not think it can be said too often that it is a mistake to exclaim over past wrong-thinking before at least wondering how our present thinking will seem to posterity. There was another reason why my parents' view of themselves was similar to that of the English settlers on the eastern coast of America: they were colonizing an almost empty land. When the whites arrived in Southern Rhodesia thirty-four years before, there were, it is now believed, a quarter of a million black people in that land, roughly the size of Spain. When my parents arrived in 1924 there were half a million.*

My father was away some time and returned with the news that he had found a farm, or rather land that would be a farm – un-stumped bush, quite undeveloped, nothing on it at all, not a house

* See Note p. 421

or a well or a road. My mother went off with him to look at it. They were driven by someone from the Land Department. Meanwhile we children were left with Biddy O'Halloran at Lilfordia. There it was that I reached the summit of childish wickedness. The hut where my brother and I were lodged also held Biddy. What must it have been like to share air and space with two little children, both of whom spent so much time on the pot? – for toilet training remained a sovereign prescription for good character. In the hut were two low beds, made after the fashion of the time. Into the hard mud floor were inserted short forked sticks. Into these forks were laid poles. On this square framework were laced strips of ox-hide. The lattices supported mattresses. There was a large metal cot for Harry. It goes without saying that Biddy liked my brother, sweet, obedient, delightful, the ideal little child; I would have preferred him too. There were two Lilford girls, to me big girls, ten or eleven, sunburned, bare-limbed, bare-footed, athletic and lean, unlike any children I had seen. They included little Harry in their games, but not me. I thought them sharp and sly and cruel. Their accent made them hard to understand. I was afraid of them. I longed to be included in their games. 'Just now,' they said. 'Just now.' Meaning perhaps – sometime – never. The sharp pain of exclusion.

Now I began to steal, ridiculous things like pots of rouge, ribbons, scissors, and money too. I lied about everything. There were storms of miserable hot rage, like being burned alive by hatred. When my parents came back and asked, But why scissors? I said I wanted to kill Biddy. They knew what I needed was a regular nursery routine, an ordered life, but how and when? Before that could happen, there must be a home, and it wasn't built yet. We set off in an ox wagon on the road north. The road was then a track, and it was January, the rainy season, so the track was mud. The wagon was drawn by sixteen trek oxen. Into it went three adults and two children, and necessities, but the trunks of smart clothes, curtain materials from Liberty's, heavy table silver, Persian carpets, a copper jug and basin, books, pictures and the piano, would come on later, by train. We were five days and nights in the wagon, because of swollen rivers and the bad road, but there is only one memory, not of unhappiness and anger, but the beginnings of a different landscape; a hurricane lamp swings, swings, at the open back of the wagon, the dark bush on either side of the road, the starry sky. It was a covered wagon, like the ones in

American films, like those used by the Afrikaners in South Africa on their treks away from the British, north, to freedom.

We were again lodged with strangers, settler-fashion, paying our way, this time at a small gold mine, a couple of miles from the hill where the house would be built. It was managed by people called Whitehead, and owned by Lonrho. Nearly everything was, then. Lonrho was the successor to the British South Africa Company, which had helped Rhodes annex Southern Rhodesia, and for a long time it was referred to as 'The Company', and certainly not with affection. Again, there were many rondaavels, and a shack that was the central house. Beyond pale mine dumps stood up the grasshopper-like mine machinery. Beyond that was the mine store and then the compound of crowding thatched huts. Pawpaw trees, guava trees, plantains, marigolds, cosmos, cannas, moonflowers and poinsettias: these were the plants that then marked white occupancy.

Before farming could begin, at least a hundred acres of trees must be cleared, and the tree stumps dragged or burned out of the soil. Farm machinery and cattle must be bought. The house must be built, and the kraals for the cattle and sheds for the machinery.

The farm was a thousand-odd acres of bush, but there was some arrangement that enabled my father to use adjacent, non-allocated government land for grazing, and this land in our time was not settled, so 'our' land went on indefinitely to the Ayreshire Hills. There was no one at all living on that land, black or white.

Only one incident remains from that time that went on for months, later to be described by my parents, looking at each other with the awed, incredulous faces that accompany such moments of recognition, 'God, that was an awful time, awful, awful!' How did we live through it? – is the unspoken message that goes with the words. The small children, my brother and I and two others, were being settled for the night in a rondaavel on beds that had mosquito nets tucked tight down around us. An older girl came in with a candle, and set it down on an up-ended petrol box, so the flame was not more than a few inches from a net. My mother came in to check for the night, saw the candle, and shot across the room, clutching at her heart with one hand while she reached for the candle. She said in a voice hushed by urgency. 'What are you doing? What can you be thinking of?' It was true. If I had thrust a leg or an arm out the net would have reached the candle, and the hut would have gone up in flames – it was thatch on pole and mud walls. My mother stood there, the candlestick shaking in her hand, the flame trembling, candle grease scattering. Meanwhile the culprit wept, only now imagining possibilities. 'Why?' my mother went on in a low appalled voice. 'How could anyone in their senses do such a thing?' I have never forgotten her incredulity. Capable people do not understand incapacity; clever people do not understand stupidity.

My parents did not understand the Whiteheads, found them shifty and unsatisfactory, though soon they would become familiar with people who farmed, went broke, mined, succeeded, part-succeeded or went broke, farmed again, owned mine-stores – did anything that came to hand. Inside this same hand-to-mouth, hit-and-run pattern some people made fortunes. Others died of drink. The Whiteheads were not in any sense educated. They knew nothing but this settlers' life. My mother disliked them, and they must have found her more than a trial. As for my father, he was doing the books for the mine, and would for a couple of years after he was on his own. Already we were worried about money. There was an unpleasantness about the books. Mr Whitehead was either careless or dishonest, and he blamed my father. I have described this, humorously, in *In Pursuit of the English*, but for my parents it was the chief horror of '*God, that was an awful time.*' There was nothing funny in the living of it.

My father rode over every day to supervise the beginning of the farm, for already there was a 'bossboy', Old Smoke, from Nyasaland, who had brought his relatives with him, and a good part of each morning was spent in long, meditative consultations between the two men, who usually sat at either end of a fallen log, watching the labourers at work. Both men smoked, my father his pipe, and Old Smoke dagga, or pot. That was why he was called Old Smoke. My mother usually walked over for at least part of the day, and took us with her, so we could watch the cutting of the trees, the stumping of the lands, the new cattle in their kraals, the digging of the wells. Two wells were dug, according to the findings of the water diviners – everyone used diviners then for wells and, later, for boreholes. Above all, we watched the building of the house. The grass for the thatch of the house was still green in the vleis, but the pole and mud walls of the house could go up, and they did. This process I described in *Going Home*, the making of a house from what grew in the bush, and no house could ever have for me the intimate charm of that one. In London you live in houses where other people have lived, and others again will live there when you have moved or died. A house put together from the plants and earth of the bush is rather like a coat or dress, soon to be discarded, for it probably will have returned to the bush, from fire, insects, or heavy rains, long before you die. The minute the grass was ready, the roof went on, for the priority was to get away from the Whiteheads.

My parents had chosen a site which the neighbours all warned would give them trouble, on top of a hill, which meant dragging everything up and down the steep slopes by oxen. It was the beauty of the place, that was why my father chose it, and then my mother approved it. From the front of the house you looked north to the Ayreshire Hills, over minor ridges, vleis and two rivers, the Muneni and the Mukwadzi. To the east, a wide sweep of land ended with the Umvukves, or the Great Dyke, where crystalline blues, pinks, purples, mauves, changed with the light all day. The sun went down over the long low ranges of the Huniyani Mountains. In the rainy season it was extravagantly, lushly beautiful, mostly virgin bush, but even where it had been cut for mine furnaces the bush had grown up fresh and new. Everywhere among the trees the soil was broken by ridges and reefs of quartz, for this was a gold district, and on every reef of protruding rock you could see the marks of a prospector's hammer that had exposed a crust of fool's gold – pyrites – or the little glitter of mica.

Weeks before the house was finished, when it was still a skeleton of poles stuck in the ground, then poles covered with a skin of mud, then a roughly thatched house, with holes that would be windows, my parents were sitting on petrol boxes in front of it (where soon they would be in deck chairs), and they watched the mountains, or the sunset, or cloud shadows, or rain marching around and across the landscape. I sat on my father's good leg and watched too.

When the house was done, perched on the top of the hill, the bush was cleared not more than thirty yards in front, and on either side. At the back where the garage and store huts were, trees had been cut for a hundred yards or so. The real bush, the living, working, animal-and-bird-full bush, remained for twenty years, not much affected by us in our house, and right until my parents left it in the middle of the Second World War, you might startle a duiker or a wild cat or a porcupine only a few yards down from the cleared space. Two rough tracks led down from the house to the fields in front, and a steep path through thick trees and bush to the well. Down the hill in front of the house was a big mawonga tree, its pale trunk scarred by lightning, an old tree full of bees and honey. What impresses me now is not how much effect our occupancy had on the landscape of the farm, but how little. Below the hill on one side was the big field, the hundred acres, and there were smaller fields here and there. Cattle kraals, tobacco barns – and the house on the hill. The farm labourers' village on a lower hill merged into the bush, as our house did.

5

THE HOUSE ON THE HILL was not different from most first houses built by settlers who, when they arrived in the colony, were nearly always poor. Usually they were brick and corrugated-iron shacks, one room, or two. The most attractive houses of those early days were like the Africans'. An African family had a group of huts, each hut for a different purpose, and early settler houses were often half a dozen thatched huts, or brick or pole-and-mud, sometimes joined together by pergolas covered with golden shower or bougainvillea. The floors were of brick or red cement, more often of stamped dung and mud. The African huts had no windows, but the white huts always did, sometimes French windows, gauzed in, so it seemed like an aviary. The floors had on them reed mats or animal skins. The first beds could be strips of ox hide on poles. The furniture stores were miles away in Salisbury, and wagons brought the furniture out; even when they were brought by train, the tables and chairs would have to be trekked from the station to farms over bad roads. Farm sales, as farmers went bankrupt, which they so often did, recycled furniture among the farms. Furniture was often improvised from bush timber by any black man who showed he had an eye for it, and sometimes from petrol or paraffin boxes. In those days petrol and paraffin came in four-gallon tins, two in a box. A settee could be made from them. Sideboards, writing tables, dressing tables, were made with two or four boxes on their ends, with a board across, and boxes set horizontally on them. These exercises in spare living were civilized by curtains made from flour sacks. The flour came to the farms in thick white sacks which, when washed, went soft and silky and took dye well. Or curtains were of embroidered hessian.

If you were really hard-up, all that had to be bought was a Carron Dover wood-burning stove. Every farmhouse had one – or nearly every farmhouse: a just-arrived young settler might live for a season in a mud hut, and his kitchen was an open fire under a corrugated-iron roof.

The houses outgrew themselves, were demolished, to be replaced by the solid brick, ceilinged houses that announced success, or remained as the core of a spreading farmhouse, full of rooms.

The talent for invention, for improvisation, was never lost. Even in a house owned by 'a cheque book farmer' (I heard the old envious phrase in 1988 about a black farmer, by one who did not yet have a cheque book), there might be hessian curtains or hangings embroidered red and orange and black with wools, or appliquéd in the geometric patterns fashionable at the time because of 'the jazz age'. Or the white flour sack curtains, dyed. I have seen a farmhouse full of antiques – real ones from England and Scotland – with bedrooms where at the windows hung glazed chintz, the beds valanced in chintz, but with a fire screen of embroidered hessian, and bookcases filling whole walls made of painted petrol boxes.

Our house was different from these first houses only in its shape, built long, sliced across for rooms. A photograph of Mother Patrick's and her nurses' hospital in the early 1890s – the Dominican nuns were the first women in the Colony – is almost identical with that of our house, before it sprouted verandahs and porches and then another room joined to the house by a pergola. Inside it was better furnished than most: for instance, the living room where the dining table, made from bush timber, was set so we could look over to the hills as we ate. The pale grey mud of the walls had been left unwhitewashed, because it looked so nice with the Liberty curtains. The chairs, a settee, bookcases had been bought from a farm sale. The writing table was of stained petrol boxes, and John William McVeagh and his second wife, the daughter of the dissenting minister, looked through mosquito gauze at the verandah and the rows of petrol and paraffin tins painted green that held pelargoniums. The next room, my parents', had proper beds and mattresses, the curtains were Liberty's, the rugs were from Persia, the copper washbasin and jug stood on a petrol box washstand. Next door, at first my brother's and my room, then mine, there were reed mats on the floor, the bedspreads were of flour sacks dyed orange, the washstand and dressing table of petrol boxes, painted black. The little room at the very end had reed mats, and a petrol box wash table and dressing table. It was in this room that Biddy O'Halloran lived for a year. She embroidered the white flour sack curtains all over in glowing silks that were still fresh twenty years later.

Nor was there anything remarkable about the oil lamps that had to be refilled every morning, for those early farms did not have electricity. Nor the water cart under its shed of thatch, with its two casks side by side whose taps were never allowed to drip, for even a cup of water was costed in terms of the energy of the oxen who three or four times a week pulled the heavy barrels up the hill. Nor the lavatory, twenty yards down the hill, a packing case with a hole in it over a twenty-foot hole, standing in its little hut, with a thatch screen in front of the open door. And not the food safe, either, double walls of chicken wire filled with charcoal where water trickled slowly from containers on the top, dripping all day and all night, the food kept cool because the safe was set to catch any wind that blew. When a farmhouse took a step forward into electricity, running water, or an indoor lavatory, neighbours were invited over to inspect the triumph, which was felt to represent and fulfil all of us.

My mother must have realized almost at once that nothing was going to happen as she had expected.

Not long ago I was sent the unpublished memoirs of a young English woman, with small children, who found herself in the bush of old Rhodesia, without a house, for it was still to be built, no fields ready – nothing. And particularly no money. She too had to make do and contrive, face snakes and wild animals and bush fires, learn to cook bread in antheaps or cakes in petrol tins over open fires. She hated every second, feared and loathed the black people, could not cope with anything at all. Reading this, I had to compare her with my mother, who would be incapable of placing a vegetable garden where a rising river might flood it, who never ran from a snake or got hysterics over a bad storm. Another manuscript, this time from Kenya, was the same: wails of misery and self-pity, and what seemed like an almost deliberate incompetence in everything. The two memoirs reminded me of what was worst for my mother. Hard to believe that the first thought in the minds of the two memoir writers, with everything they were being tested by wildness and hardship, was this: were they still middle-class people, 'nice people'? But so it was. Similarly, my mother was unhappy because her immediate neighbours were not from the English middle class. How was it that my father, who, after all, must have at least noticed her preferences, chose a district where all the 'nice people' were miles away, on the other side of the District? Is it conceivable he really never understood how important it was to her? Or, perhaps, finding the land was all he had strength for, and then he had to make a farm from nothing, and start a kind of farming he had not imagined. He had always wanted to be a farmer, but in his mind were the patterns of English farming he had seen all around him as a boy.

Both of them believed, and for years, that a change of luck would bring them success. She might not have seen at once that her crippled husband would not be able to dominate the bush, and that they would never make the fortunes promised by the Exhibition, but she did see that the life of dinner parties, musical evenings, tea parties, picnics was gone. That meant she felt checked in a deep part of her. Going to Persia she had taken all the necessities for a middle-class life. Coming to Africa, she had clothes for making calls and for 'entertaining', visiting cards, gloves, scarves, hats and feather fans. Her evening dresses were much more elegant than anything likely to be worn even to Government House then. She probably thought that was where she would be invited. She might have defied her father to be that common thing, a nurse, but she never had any intention of giving up the family's status as middle-class. Her children would fulfil her ambitions and do even better. So in that first year, when she took a good look at her circumstances and her neighbours, she only postponed her ambitions. The farm would shortly be successful, and then she could go home to England, put her children into good schools, and real life would begin.

Meanwhile, she could not have made more efficient, ingenious, energetic use of what she found around her in the bush, on the farm.

And now I come to the difficulty of reconciling child time and adult time. There was a stage of my life – I was already in England, and trying hard to make sense of my life through a strict use of memory – when I understood that a whole tract of time had disappeared. There was a gulf, a black hole. Years and years of it – so it seemed. And yet the record of outward events is clear. In January 1925, the family was in Lilfordia. Between January and June 1927, I was at Mrs Scott's. Yet I had already been at school for a whole term at Rumbavu Park. All these hazy, interminable memories had to be fitted into one year and nine months.

Impossible. I simply gave up. But later had to come back, and back . . . and was forced to concede that between my stamping around in the mud and water that would make the plaster of our house, and my going to school, was – less than two years. And even now I feel incredulous, it can't be so. But it was so. Between January 1925 and September 1926, the following things happened.

All of us, the whole family, had malaria twice, badly. The new lands of the farm were stumped, the farm furnished with its necessities, the house built, and we moved into it. Biddy O'Halloran left, good riddance on both sides. My mother had a breakdown and was in bed for months. Mrs Mitchell and her cruel twelve-year-old son came and then left. I learned to read and triumphantly entered the world of information through print on cigarette packets, grocery packaging, the big words on top of newspapers, the Army & Navy catalogue, words written under pictures . . . and then, books themselves. My brother and I did lessons from the correspondence courses organized for farmers' children by the government.

That was the pattern of events, and it has little to do with what I remember, the chronicle in child time.

Biddy O'Halloran is leaning on my father's shooting stick, and we are in the big field, below the hill, grasshoppers and butterflies everywhere. She has had her appendix out, and she is telling my little brother that if he doesn't keep his mouth shut a grasshopper will jump down into his appendix and claw its way out through his stomach. He is crying with terror. 'Of *course* it's not true,' cries my mother later, at bedtime, while my brother sobs. But years later my brother told me he had an irrational fear of grasshoppers; I was able to tell him why. 'Do you mean to say that's all it was?' he demanded, trying to laugh, but shocked that what had so influenced him and for so long had been so insignificant.

Biddy O'Halloran had fair skin, and in the 'vee' of her cotton dress it was flushed a gentle red. A child's close stare revealed it was a mottle of scarlet and cream. Two small children gravely discuss how red jelly and cream got under Biddy's skin. 'It was poured into a hole and then it spread.' We made excuses to get close, were rebuked for staring, came away to tell each other there wasn't a hole. So she must have spread the jelly on, and it got through the skin. 'Mummy, how did the red jelly and cream get under Biddy's skin?' 'What red jelly? What nonsense!' The mystery was discussed gravely, scientifically, as we sat together under the eaves of the thatch, the cats and dogs in attendance. 'But perhaps it isn't jelly, it's blood from the roast beef!' 'But what are the white bits then?'

Or long, thoughtful stares at an adult's fingernails, where there is a pale blob in the pink of a nail. 'Mummy, why didn't God finish your nail?' 'What do you mean, finish?' Look, there is a hole.' 'What hole? That isn't a hole!' The hairs on an adult's forearm, each golden stalk in its little pit of brown skin. The smells. Biddy had a sour smell, sharp and hurtful to the nose, when she splashed on cologne. My mother's smell was vigorous and salty. My father's male and stale and smoky.

We watched from the edge of the bush my mother showing Biddy how to put bloody rags into a petrol tin to soak, under the thatch of the house at the back. The look of dramatic secrecy on my mother's face, her lowered dramatic voice. The deliberately languid, irritated movements of Biddy. We knew the 'boys' were not supposed to see the contents of this tin. We crept up to the tin when the women had gone and speculated: Biddy had cut her finger or her foot – that must be it. But why didn't mummy want the boys to know? We were always cutting ourselves, or had bruises, and sometimes the 'boys' washed the blood off for us. Why then . . . ?

The adult world, with its disorder, its lack of sense, its mysteries, two small children trying to get things into their right place, call them by their right names . . .

I lie on my bed, reading Walter de la Mare's *The Three Royal Monkeys*. One of the monkey brothers eats an orange, which he thinks is conveniently divided into segments for pulling apart and eating. I cannot make sense of this. The orange segments I am eating as I read are too big for my mouth. Yet I am bigger than the little monkeys we see racing about in the trees just down the hill, and which sometimes come into the house and investigate the rafters before running back into the trees. Did the monkey in the book mean those tiny globules of orange juice, each in its little bag, which I burst on my tongue, flooding my palate with scent and taste? But surely that couldn't be it: globules aren't segments. I lie and wonder, read, and think . . . The Royal Monkeys must be much larger than the little bush babies we know. When they pull the orange skins apart their fur prevents them feeling the showers of sharp juice that come out. The spray lives in the pores of the orange skin. When a visitor comes who has rough-pored skin on her face and neck, I stare secretly at the pores where water is standing. If pulled apart, would that skin send out a spray of . . . ? 'What is that child staring at?' 'Doris, why are you staring? It's rude.' I turn away, run off, sit under a bush down the hill, pull a leaf off a bush, look at the veins on the leaf and the pores between them. I pull the leaf apart but there is no strong-smelling spray on my face and hands. On the bush is a chameleon. I watch it creep with its slow rocking motion up a branch. And then suddenly . . . I rush screaming up the hill to my mother, sitting in her chair, beside my father, looking out over the bush. 'What on earth is the matter?' 'Mummy, Mummy . . .' 'But what is wrong?' 'The chameleon,' I weep, hysterical, terrified, 'The chameleon . . .' 'What chameleon?' 'It was sick and all its insides came out.' I run back down the hill. Behind me come my mother and my little brother. The chameleon is sitting quietly a little further up the branch, its eyes swivelling about.

I am in shock, it is like a dream. I saw the chameleon's insides come out and . . . it happens again, and I scream. 'Shhh . . .' says my mother, holding me tight. 'It's all right. It is catching flies, can't you see?' I am shuddering with disgust and fear – but with curiosity too. I stand safe inside her firm grasp. 'Wait,' she says. The club-like tongue of the chameleon darts out, a thick fleshy root, and disappears back inside the chameleon. 'Do you see?' says my mother. 'It's just its way of feeding itself.' I collapse into sobs, and she carries me back up the hill. But I have acquired adult vision; when I see a chameleon, part of my knowledge of it will be that it darts out its enormous thick tongue, but I won't really see it, not really, ever again, not as I saw it the first time.

In 1992 I was standing, a couple of weeks after the first rains, in Banket, near a mafuti tree, a big one. The mafuti is a serious tree, its fronded leaves dark green, its trunk thick and safe. There is nothing frivolous about this tree. But growing at its root was an excrescence, like a sea creature, coral sheaths where protruded the tender and brilliant claws of new leaves, and these were like green velvet. You would never think they had anything in common with the sober leaves above them. And suddenly I remembered how I rushed up to the house, screaming that a monster was attacking the tree, it was a beetle the size of a cat.

I wake in the night. All round me, above me, is a rustling, creeping noise. I start up on my elbow, peer up through the white of the mosquito net. My heart is beating, but the rustling is louder. The square of the window lightens, once, twice. Wait, is that a car coming up the hill, the headlights . . . ? No, my parents' room is dark, they are in bed, too late for a car. It is as if the thatch is whispering. All at once, as I understand, my ears fill with the sound of the frogs and toads down in the vlei. It is raining. The sound is the dry thatch filling with water, swelling, and the frogs are exulting with the rain. Because I understand, everything falls into its proper place about me, the thatch of the roof soaking up its wet from the sky, the frogs sounding as loud as if they are down the hill, but they are a couple of miles off, the soft fall of the rain on the earth and the leaves, and the lightning, still far away. And then, confirming the order of the night, there is a sudden bang of thunder. I lie back, content, under the net, listening, and slowly sink back into a sleep full of the sounds of rain.

Or it is just after we have been put to bed, and from the end of the house come the sounds of grown-up voices, and my mother playing her piano. I and my little brother talk in low voices, knowing we should be asleep. I continue my mother's bedtime stories, of the animals in the bush, the mice in the storeroom. Then I try to frighten him with the dragon from St George and the Dragon. I frighten myself. The dragon is spread all over the thatch, fills the sky, claws spread out, fire rushing from its mouth. I know perfectly well there is no dragon, yet I am frightened. Similarly, when I have convinced myself there are wicked fairies in the corners of the room, I know I have invented them. When at last I start yelling for my mother to come, and she does, she says, soothingly, that there is no dragon, no fairies behind the curtains, I feel impatient, because that is not the point. I need to be scolded for preventing my little brother from sleeping, for 'making things up'. Similarly, by myself at the very bottom of the hill, just where the lands start, I stand by an old gnarled and knotted tree, like the ones in *Peter Pan in Kensington Gardens*, and imagine fairies so strongly I am not far off seeing them. When I populate the antheap with its curtains of Christmas fern and its spider lilies with fairies and goblins, what I create is an intense listening silence, and I know if I turn my head fast enough, when they don't expect it, I'll see them. Which does not mean I actually believe they are there. Just as I believe and do not believe in the tooth fairy. My disbelief in Father Christmas does not stop me from expecting reindeer and explaining to my brother they will come in through the window, since there is no chimney. Long earnest discussions in the hushed voices that go with the turned-low lamp and the shadows in the room, about reindeer and how fast they would have to fly from England to get here in time for Christmas, and if the reindeer would have to descend at intervals to feed, and what would they think of trees and grass, since what they like to eat is moss. When my brother tells my mother I believe reindeer will arrive for Christmas and they will eat musasa and mafuti leaves, it can be seen from her frown that she is working out how to balance reality and useful and necessary fantasy, and I at once hurriedly say that of course I don't believe in Christmas reindeer.

My mother decided she had a bad heart. All her life she knew she had a bad heart and might die at any moment. In the end she died at the respectable age of seventy-three, of a stroke. Even as a small girl I understood the psychological advantages of a bad heart, and believed she was inventing it to get sympathy. I believed too that my father was not convinced by this heart.

Now I understand why she went to bed. In that year she underwent that inner reconstruction which most of us have to do at least once in a life. You relinquish what you had believed you must have to live at all. Her bed was put into the front room, because of the windows and the view to the hills, under the stern gaze of her father, John William, and his cold dutiful wife. All around her were the signs and symbols of the respectable life she had believed was her right, her future, silver tea trays, English watercolours, Persian rugs, the classics in their red leather editions, the Liberty curtains. But she was living in what amounted to a mud hut, and all she could see from her high bed was the African bush, the farm 'compound' on its subsidiary hill.

The doctor came often from Sinoia. They did not know as much then about anxiety as they do now. He prescribed bed rest. Doctor Huggins, her real doctor in Salisbury, when she appealed to him in letters, said, Why ask him when she had already had a doctor telling her what to do? Doctor Huggins – later Lord Malvern – was a testy character who did not believe in the need for a bedside manner, as a doctor or a politician: he was shortly to become Prime Minister.

Several times a day she summoned me and Harry to the bedside where she said dramatically, 'Poor mummy, poor sick mummy.'

It is this memory that tells me how badly she had inwardly collapsed. 'Poor mummy' was simply not her style. As for me I was consumed with flames of rage. My little brother embraced her whenever he was asked to. I embraced her warmly, but then resented and repudiated the emotion. Soon I refused to go to the bedside when called there by the cook. 'Mummy's ill,' my father directed me, and I snapped, 'No, she isn't,' for the conflict was unbearable.

Meanwhile our education went on and I can only admire the self-discipline this must have needed. Standing by her bed, or sitting on it ('Don't tire your mother. Don't lean on her. Don't . . .'), we learned our multiplication tables and did baby sums, but the reading lessons were already much too easy. She told us stories and she read to us.

Then Mrs Mitchell arrived, with her son, to 'help' my mother. Harry was still sleeping in my parents' room. I shared a room with Mrs Mitchell. Her son was in the room at the end of the house.

I experienced her as cruel and her son as a bully. She drank. When she left – soon, after only a few weeks – caches of empty bottles were found under bushes, in cupboards. She always smelled of spirits. What was she really like? If she was being nursemaid and housekeeper for a sick woman, and had a boy, school age, she must have been desperate. Widowed? Deserted? In flight from a brutal husband? This was before the Slump, when women whose husbands were out of work took any jobs they could find.

All my childhood we were told how poor we were, how hard-up, how deprived of what was our right. I believed it. Then, at school, I met children from really poor families. There was a stratum of people, white, in old Southern Rhodesia, who lived just above hunger level, always in debt, in flight from debtors, with drink and brutality waiting to swallow them up. Recently a book was published called *Toe-rags* by Daphne Anderson, the story of a girl who survived a childhood at this level of poverty. Often it was the black servants who cared for her. She was exactly my age, and compared to hers my life was gentle and privileged. This book is not likely – yet – to be read by black people in Zimbabwe, where it is necessary still to believe that every white person is, and was, rich. White people have proved reluctant to read it, because they don't like to think the whites in British Southern Rhodesia ever lived so low and so fearfully. The grandiose myths of White

Supremacy are made to look sad and sick by this book, even though the beautiful author married well, as we say, when she was in her twenties, and lived happily ever after. I hope *Toe-rags* will soon find a place on the reading lists of history courses in Zimbabwe.

Mrs Mitchell came from this frightful level of poverty. She could not have shared my bedroom for more than a term, perhaps even a school holiday. It was an endless misery, endless fear. I lay in the stuffy dark under the mosquito net. She was under the other mosquito net. I heard the sounds that meant she was drinking. I heard the bottle slide down between the edge of the bed and the net, and thump on the matting. She snored. She thrashed about in her bed. Next door the boy shouted in his sleep. Once she quarrelled so loudly with the boy my mother appeared in the doorway, a candle in her hand, and her hair flowing about her to stop the two yelling at each other, and saw the candle sloping in Mrs Mitchell's hand, the candle grease spattering, the flame lengthening and dipping and smoking an inch or two from the mosquito net.

Both Mrs Mitchell and her son shouted and screamed at the black servants. When my father remonstrated she shouted at him that he understood nothing about the country: perhaps it was the first time I heard all the white clichés: You don't understand our problems. They only understand the stick. They are nothing but savages. They are just down from the trees. You have to keep them in their place. (Just like Dr Truby King's infants.)

I was afraid to go anywhere near Mrs Mitchell's son. He was perhaps twelve but seemed to me as powerful as a grown-up. He tormented and teased the black child who was piccanin for the household. He chased and teased and tormented the dogs and cats. His catapult he used not only on birds, but to aim stones at the bare feet of any black person who came near.

Nothing I can do, no cajolements or enticements of memory, can bring back more than this: no incident or event bad enough to explain my dread of that woman. And probably there was no actual cruelty or blow, but only the foul angry voice, and the high scolding vituperation of the black-hater.

I cannot even begin to imagine what that year was like for my father. His wife was bedridden, and if she had 'a heart' there was no reason why she should ever get up. They had so little money, yet whenever she was worse the doctor would arrive from Sinoia. Two little children, one still not six, the other four. They needed tender care, but what they were getting was Mrs Mitchell and her bully of a son. My father was still trying to get lands stumped, bush cleared, fields made. He had to be down all day on the lands, for until there were fields there would be no crops. Meanwhile the debt to the Land Bank grew.

For some months he had an assistant, a Dutchman with many children. The story *The Second Hut* was written from memories of that year.

It was as early as that when we, the children, began to go with him down to the lands. The horse had died: that part of the District was not good for horses – they got diseases. It was on the sandveld at the other side of the District that horses thrived and people went in for racing. We bought two donkeys and my father rode one. We were put on the other. Later we got a car, an Overland, already third- or fourth-hand. We, the children, the two dogs, Lion and Tiger, cheerful mongrels, bottles of cold tea and packets of store biscuits went down to the land with our father, and played in the bush while my mother was in bed, being tended by Mrs Mitchell.

She left. Then came someone who was not paid, but helping out of kindness, Mrs Taylor, a Danish woman. Since she had a life of her own she did not move in, but might stay a few days, leave and come back, and soon we forgot the nightmare of Mrs Mitchell. She was a large, calm, good-looking woman, and my father liked her very much. My father liked women. Women liked him. He had a gentle, courtly, considerate way with him, and the undertones of regret and wistfulness were not anything a child could understand. All I knew was that throughout my growing up there was always this woman – the wife of a neighbour, or a visitor to the District – with whom he might sit and talk in this particular way, as if the time they were in, the two of them, was in another range of being altogether, something larger and tenderer than quotidian life, and where they shared, too, a rakish and amused recognition, never to be put into words. Mrs Taylor was not around for long – she was on the move somewhere. People were always moving about the country, farm to farm, from either to the town, or off 'up north' – meaning Nyasaland or Northern Rhodesia – or back to England, because they found the life disappointing. 'Not everyone can take the life, you know.' Women most especially often could not take the life.

If my father always enjoyed tender and, of course, platonic

friendships (now you have to spell out what would have been taken for granted then), my mother also had admirers who knew she was fitting her remarkable capacities into too small a space. One of them was a George Laws, who was a brother of Miss Laws, a teacher in Sinoia and some sort of a cousin of my father's. Mr Laws owned a timber concession in the government land between the rivers. It was he who made my mother a fitment that enabled her to read while she was ill, a bed rest, piano stool, couches and chairs of slatted wood, and 'occasional' tables so heavy they could scarcely be moved even when not buried under books, newspapers, magazines.

Then my mother got out of bed. She had to. She said her weight of hair was giving her headaches, and she cut it all off and appeared with a nude shorn nape. A 'shingle'. My brother wept. I wept. We sat in the pillows and billows of her brown hair and wrapped it around us and bawled while she sat and ironically watched us. She said *Right! That's that!*, and she wrapped her hair up in paper and threw it into the rubbish pit.

Correspondence courses still arrived by every post, but she wondered what she was paying money for when she could do better herself. She taught us geography by sloshing water into our sand pit and making continents, isthmuses, estuaries, islands. Being taught to see land masses and oceans like this repeats that stage of human knowledge when the world was flat. Then she ordered a little globe from Salisbury which arrived on the train, and with it we entered the mind of Copernicus. She sat my father in his folding chair on the sharp slope down outside the house, summoned the cook and the piccanin from the kitchen. My father was the sun. The two servants were the heavy planets, Jupiter and Saturn. Stones stood for Pluto, for Mars. I was Mercury and my brother Venus, running around my father, while she was the earth, moving slowly. 'You have to imagine the stars are moving at different rates, everything moving, all the time.' And then she abolished this system of cosmic order with an impatient wave of her hand. My father was now the earth, and my brother and I by turns the moon. 'Of course you have to imagine that . . .'

We interminably chanted the multiplication tables. We learned English trees and flowers from little books. One was *French Without Tears*. The inspectors came out from Salisbury to check on the farmers' children, and said yes, we were doing well. Yes, we were in advance of our ages. But we had to go to school. It was the law. Besides, children have to learn to be social beings.

For a time, my mother wondered about starting a little school there, on the farm. There were children of various ages on the near farms. But if this would be easy now, with good roads, then the question was, how to get those children every day, two, three, four, five, seven miles, to school and back again? Besides, this woman who had a genius for teaching small children was not qualified. And that was *that*.

I WAS NOW IN THE ROOM the third along from the front, which would be mine until I left the farm for good. It was a large, square, high-thatched room, whitewashed, full of light. From my bed I saw the sun spring up behind the chrome mountain and pass rapidly up out of sight, I saw the moon rise, soar up and away. I used to prop the door open with a stone, so that what went on in the bush was always visible to me – it was only a few paces away down the steep slope. I fought with my mother to have this door open. 'Snakes,' she cried, 'scorpions . . . mosquitoes . . . I won't have it!' But I kept the door open knowing I was safe inside the mosquito net. Besides, we took all that quinine for the months of the rainy season. Snakes did come into the house, and more than once my mother had to shoot one. The fact is, I was brought up in one of the most heavily snake-infested areas in the world. They were all poisonous, some deadly. For years I was in the bush with bare legs and often bare feet, and I was never bitten. Clearly they fear us more than we fear them. Impossible not to remember the threat of snakes dinned into us always. Remember to watch where you tread, never put your hand on a branch without looking, never climb a tree carelessly, puff adders like to lie out on hot paths and roads and they move slowly . . . remember, remember, remember. But my fear was for insects, so many, so varied, so large and black and horned or slim and jittery and invasive, spiders hanging in front of your face on webs spun in the night, lurking in your veldschoen, watching you from holes in the earth when you squatted to pee. It is a testament to the irrationality of humankind that when I look back at that time I think of those lethal but beautiful snakes with admiration and even affection, whereas the memory of harmless insects makes me shiver.

But I was under the mosquito net, so that was all right.

In the mornings I woke because the light had come, and the

sunlight a warmth on my face. I checked the net for spiders and beetles, then jumped up and tied it into its daytime knot. I flung myself down on my back, and lay spread out, the sheet kicked off, sniffing all the delightful smells in that room. First, my own body, its different parts, each with its own chummy odour. The thatch was damply fragrant or straw-dry, according to whether it had rained. The creosote the rafters were painted with was tar-strong, like soap. The linoleum, already wearing into holes, released oily odours, but faint, like the oilcloth on the washstand. The enamel pail under the washstand might have pee in it, but I learned to sneak out with the pail and pour it down the hill into the earth, where it bubbled up yellow, sank, and dried almost at once. The toothpaste was clean and strong. My shoes – veldschoen – smelled of hide, like karosses. But I refused ever to have a kaross on my bed, for a kaross was too close to the beast it came from, and anyway, the rough reek of kaross made me think of Mrs Scott and I never, ever, wanted to think of that place again.

I heard the 'boy' take tea into my parents, knew they were getting up, slid into my clothes before I could be fussed into them. I wore little cotton knickers, a cotton dress, sometimes made out of an embroidered flour sack, and a liberty bodice. The Army & Navy catalogue regulated our lives, as it did those of middle-class children anywhere in the colonies. Well brought-up children wore liberty bodices with their tabs for suspenders and stockings in cold weather. Worn without stockings they wriggled up and left red marks on your stomach. There was a day when I said no, I was not going to wear one, not ever again. And in winning the battle for me I won it for my brother too. He was still wearing the tight binders that were supposed to prevent chills on the liver, while I had long ago refused to wear one. We were meant to wear cotton hats, lined with red aertex, with red aertex pieces hanging down our backs to keep the sun off our spines. But no, no, no, I would not. 'No one wears a hat!' I shouted – and it was true, the farmers and their wives did not cover their heads though the women might wear a hat for visiting. My mother's pleas went for nothing: you will get deathly chills without your binders, bad posture without your liberty bodice, sunstroke without hats that have red linings. About the hats, it seems my mother and the Army & Navy Stores were right all the time. Recently (1992) I was at a skin specialist's in London, and he said most of his income comes from white sun-worshippers in Australia, South Africa, Zimbabwe.

I dressed myself as I had done since I was able to, whereas my little brother, now getting on for six, was still being dressed. He was supposed to be delicate, and often got bronchitis and was in bed with a towel over his head enclosing a basin of hot water that emitted the fumes of wintergreen, and friar's balsam. Not for another two years would he refuse to be called Baby, refuse to be delicate.

When I went into my parents' bedroom my father was putting on his wooden leg with its heavy leather straps, its bucket for his stump; my mother, in her flowered silk wrapper from Harrods, was dressing Baby. The Liberty curtains were still fresh. The whitewash glittered. The thatch above was yellow and smelled new. Years ahead was the gentle squalor that house at last subsided into.

We had breakfast in the room that overlooked the bush that stretched to the Ayreshire Hills. Mother in her fresh cotton dress, father in his farm khaki, the two healthy little children. The breakfast was the full English breakfast, porridge, bacon, eggs, sausages, fried bread, fried tomatoes, toast, butter, marmalade, tea. Also pawpaw in its season, and oranges.

That we should eat enough was my mother's chief worry. Now I cannot believe how much we all ate. And when a bit of white egg slime or a burnt bit of toast was left my father demanded with anguish that we should think of the starving children in India. If children were starving in Africa, or hungry or malnourished down in the farm compound visible from the windows, then that it seemed was not our responsibility.

But one of the difficulties of this record is how to convey the contradictions of white attitudes. My mother agonized over the bad diet of the farm labourers, tried to get them to eat vegetables from our garden, lectured them on vitamins. They would not eat cabbage, lettuce, spinach, tomatoes – now eaten by all the black people. They pulled relishes from the bush, leaves of this and that, and they brewed beer once a week, known to be full of goodness. But an ox was killed for them only once a month. Mostly, they ate the mealiemeal of that time, unrefined, yellow, wonderful stuff like polenta, and peanuts and beans. In fact, that diet was one that would be applauded by nutritionists now, but was regarded as bad then, because of its lack of meat.

There is a sharp little memory from then, and there were similar incidents throughout my childhood. My brother, or I, doing what we had seen others do, called the houseboy to bring us our shoes – which were in the same room. My father went into a shouting, raging temper – most unusual for him. He said how dare my mother allow the children to be ruined, how dare she let us call a grown-up man 'Boy'. Did she not care that we would get soft and spoiled being waited on? He wouldn't have it. He wouldn't allow it. Usually my father didn't lay down the law. But over this he did. Throughout my childhood he remonstrated with my mother, more in sorrow than in anger, about the folly of expecting a man just out of a hut in the bush to understand the importance of laying a place at table with silver in its exact order, or how to arrange brushes and mirrors on a dressing table. For very early my mother's voice had risen into the high desperation of the white missus, whose idea of herself, her family, depended on middle-class standards at Home. 'For God's sake, old thing,' he would urge, his voice softening as he saw the distress on her angry face. 'Can't you see? It's simply ridiculous.' 'Well, it's their job, isn't it?'

After breakfast, I might go back into my room to read. Or go with my mother to learn – well, something or other. For if her wonderful lessons stopped when we went to school, she never ever lost an opportunity for instruction, and now I am grateful and wish I could tell her so.

My brother always went down on the lands with my father, and I often did too. My father sat himself on a log or a big stone, and watched the gang of 'boys' hoeing a field, or wrenching the maize cobs off the plants, or pulling up peanuts, or cutting down the great flat sunflower heads full of shiny black seeds. Most wore rags of some kind, many loincloths, or perhaps a ragged singlet and shorts that might easily be laced across a rent with pink under-bark torn from a musasa tree. As they hoed, they conversed, laughing and making jokes, and sometimes sang, if threshing peanuts from their shells with big sticks, or smashing the sunflower heads to release showers of seeds. When the bossboy, Old Smoke, came to sit with my father, his two attendant young men always standing respectfully behind him, the two men might talk half a morning. For when they had finished with the mombies, the probabilities of the rain, the need for a new cow kraal, or a new ditch to carry water from the compound, or the deficiencies of the Dutch farm assistant

– but he only lasted a short time, because the Africans hated him so much – then they philosophized. At the African pace, slow talk, with long pauses, punctuated by 'Yes . . .', and from Smoke . . . 'Ja . . .' Then another slow exchange, and 'Ja . . .' from Old Smoke. 'Yes, that's it,' from my father. Smoke might sit on a log or on his haunches, with one forearm over his knees for balance – when my brother and I tried it was no good, our limbs had already set into European stiffness. My father sat with his wooden leg out in front of him, his old hat over his eyes for the glare. They talked about Life and about Death and, often, about the Big Boss Pezulu (the Big Boss above, or God) and His probable intentions.

Meanwhile my brother and I were watching birds, chameleons, lizards, ants, making little houses of grass, or racing up and down antheaps where often we startled a buck lying up through the hot hours under a bush.

Hours went by. Years . . . A bottle full of tepid sweetened tea would be produced with cake, biscuits, scones. Old Smoke would share this with us. More hours passed – years. Then the sound of the gong from the house. Men who had been at work since six or seven in the morning had an hour off, twelve till one. The gong was a ploughshare hit with a big bolt from the wagon. Then we drove up to the house where my mother had been working all morning, sewing mostly, clothes for her husband, her children, herself – she was always smart. Or she had cooked. She made jams, bottled fruits, invented crystallized fruit from the flesh of the gourds that fed cattle, filled rows of petrol tins with the sweet yeasty gingery water that would make dozens of bottles of ginger beer. And, like all the farmers' wives, she invented recipes from mealies, which were not called sweetcorn then. Because we were all poor, or at least frugal, saving money when we could, the women were proud of what they could do with what they grew. Not till I went to Argentina, which grows the same crops as Southern Africa – pumpkins and maize, beans and potatoes, tomatoes, peppers, onions – did I find anything like the same inventiveness. We ate the green mealies cut off the cobs and cooked in cheese sauces, or in fritters and milk puddings or in soups with potatoes and pumpkin. The maize meal was made into cakes and pancakes, as well as different kinds of porridge, or added to bread. There were a dozen ways of cooking pumpkin. Young peanuts found their way into stews, peanut butter made all kinds of sauces and breads.

We ate . . . how we did eat. Lunch was a big affair, meat, always meat, for this was before anybody but a real crank gave up meat. We ate roast beef and potatoes, or steak and kidney pies, or stews, or shepherd's pie, and potatoes and half a dozen vegetables from the garden down the hill near the well. Then heavy puddings, and cheese.

Then, it was time to lie down.

'But I'm not sleepy, Mummy, Mummy, I'm not sleepy.'

It was no good. In this climate, or on this altitude – and either might be cited as evidence against me – little children must lie down in the afternoons. I begged, I pleaded, even wept, not to be forced to lie down while my mother's voice got increasingly incredulous. 'What nonsense! What's the fuss about?' She did not know I was facing eternities: she looked forward to a few minutes snatched from the responsibilities of child-rearing, to write a letter Home. The orange curtains were drawn across the green gauze of the window, and the stone that propped the door put aside. 'Look, here is the watch,' and she arranged it on the candlestick by my bed. I had learned to tell the time because of the agonies of afternoon naps. My dress was pulled up over my head. She stood holding the coverlet back. I slid in. She turned away, her mind already on her letter. Now I was glad she had forgotten me. She shut the door into their bedroom where my little brother was already asleep. At once I nipped out of bed and pulled the curtains back again for I hated that stuffy ruddy gloom.

I lay flat on my back looking up. The cool spaces under the thatch welcomed me. Yes, and there will be an end to it, just as there was yesterday, and the day before. A lost bee buzzed about, tumbled to the floor, buzzed loudly, and I had an excuse to get up again to let it out of the door, but I did not dare replace the stone, set the door ajar. On my back, arms stretched, I took possession of my cool body, that thudded, pulsed and trickled with sounds. I flexed my feet. I tested my fingers, one by one, all present, all correct, my friends, my friend, my body. I sniffed my fingers where smells of roast beef and carrots lingered. The golden syrup of the steamed pudding sent intense sweetness into my brain, and made my nostrils flare. My forearm smelled of sun. The minute golden hairs flattened as I blew on them, like wind on the long grasses along the ditches. Silence. The dead, full, contented silence of midday in the bush. A dove calls. Another answers. For a moment the world is full of doves, and down the hill wings break in a flutter of noise, and the black shape of a bird speeds across the square of my window. My stomach gurgles. I put down a forefinger to prod the gurgle but it has moved downwards towards . . . but I had already gained full possession of my bladder, and had learned to ignore the anxious queries it sent up: should you take me to the lavatory? My hands slid, like a doctor's, down over my thighs to my knees. There was a spot there somewhere, if you prodded it, then just behind the shoulder there would be an answering tweak of sensation. The two places were linked. There were other twinned patches of flesh, or skin. I kept discovering new ones, then forgot where they were, rediscovered them. Just above the ankle . . . I lay on my back with my legs in the air and pushed my forefinger into the flesh all around the ankle bone – there it was, yes, and miles away, under my ribs, there was a reply, a sensation not far off pain; it would become pain if I continued to press, but I had already moved on, mapping my body and its secret consonances. Did I dare look at the watch? Surely the half hour must be nearly up? I had been lying there for ever. I sneaked a look – no, impossible! The hand must have got stuck, I snatched up the watch, shook it. No, it was alive, all right, and only three minutes had passed. A howl of protest, hushed at once; had she heard, would she come in? I shut my eyes, lying rigid, pretending to be asleep. But dangers lurked in the pretence, for one could easily drop off, and *I was not sleepy*. I lay listening with my whole body, my whole life . . . from the other bed I heard a sound like the disturbance of air when a small trapped moth flutters. My friend the cat was there. I jumped up and leaned over her, she was lying curled, and her grey silky fur moved with her breath; she was, like me, enclosed in her own time, in the time of her breath. I was convinced she understood the anguish of afternoon sleep, the half hour which never passed. I touched her little grey paw with my finger, and it tightened as I slid my finger inside it. The claws, like tiny slivers of moon, dug into my flesh and went loose. She made the little sound which meant, I am asleep, so I left her and flung myself down on the bed so hard the springs twanged.

But I could see her there, I had company, if I woke her she would come and join me, her soft weight on my shoulder. But that meant I would have to lie still . . . outside on the woodpile the houseboy was cutting wood, and the slow sound of the axe was like strokes of a clock. The doves were quiet, I could feel heaviness sit on my lids. I woke myself by drinking mouthfuls of the heavy sweet tepid water from the glass that had bubbles clinging inside it. Each bubble was a little world, and I picked up a straw that had fallen from the thatch and chased the silvery bubbles about inside the glass until they went out, one by one, like birthday candles.

The watchface said five minutes had passed. Misery seized me – dread. The eternities of Mrs Scott's were described as 'But it was only two terms, that's all', while my parents looked at me, as they so often did, with amusement and with incredulity. Ahead lay the convent and another exile from home . . . Eternity. My mother read us the New Testament from a child's version. Eternity: *time that never ended.* Lying flat on my back, arms flung out, eyes fixed on the cool under the yellow grass that seemed so high above me, I thought of time that never ended. Never ending, never ending . . . I was holding my breath with concentration. It never ends, never . . . my brain seemed to rock, my head was full of slowed time, time that has no end. For seconds, for a flash, I seemed to reach it – yes, that's it, I got it then . . . I was suddenly exhausted. Surely it must be time to get up? The watch said only ten minutes had gone past. Without meaning to, I let out a great yell of outrage, then slapped both palms over my mouth, but it was no good, my mother had heard and came bursting in. 'What's the matter? What's wrong?' 'The watch is wrong,' I wept. 'It's not working.'

She stepped efficiently to the watch, and checked. She had just had time to lay out her Croxley writing pad and envelopes, and sit, letting herself slow down, to assemble scenes from this life of hers, find words that would convey its improbability to her friend, Daisy Lane, who was an examiner for nurses in London. 'It's completely wild out here,' she might have decided to write. 'We have to bring all the water up the hill in a scotchcart several times a week, and we have to use oil lamps! I wonder what you'd say if you saw this house! But of course, it is only temporary. We're putting in tobacco this season, and you can make a pretty penny on that!'

She stood frowning at this difficult child, who was squatting on the bed, face streaked with tears, eyes imploring. The mother was uneasy. While the little boy, the good child, slept uncomplaining next door, this child looked as if she were being tortured. But it was with brisk humour she demanded. 'Now what *is* all this nonsense?' and pushed the child down with one hand while she flicked up the bedcover. 'If you thrash about like that you'll only get overheated.'

'But I want to get up, can't I get up?'

'No, you can't. You've not been there a quarter of an hour yet.' And she marched out.

'For ever . . . for ever . . .' The child was walking with Jesus and his disciples along a dusty road, and it was not the track along the bottom of the hill, where dust lay in thick drifts, soft, red, and where the tracks of beetles or centipedes or buck slowly eroded as the breezes lifted the grains of sand away. It was a rocky yellowish road in – well, it was Palestine, since that was where Jesus was, but the rough dry road was from Persia. The smell in her nostrils now was not Africa, but that other place, where sunlight smelled old, full of stories from hundreds of years ago, Khosrhu and his armies marching across a rockface, but that was before Jesus, thousands of years ago, and then Jesus walked with men in striped headdresses along a dusty track where they stubbed their bare toes on big hot stones and Jesus said, I am the Way, the Truth and the Life . . . what did he mean, what did they mean, hundreds of years ago . . . ? she would never grow up, never, why even to the end of the day and to bedtime was so long, long time, time was long, long . . . long time was not eternity, eternity was longer, it was unending, *it never ended.* From the bed next to hers, under its bundled mosquito net, came a small chattering sound. The cat was dreaming. Her teeth were making that funny sound. She was dreaming of chasing something? Like the dogs who would lie stretched out yelping and yapping with excitement as they chased a buck or a rabbit in a dream. Where was Lion? Where was Tiger? They were asleep in the shade under the verandah. Harry was asleep next door, the good baby. Daddy slept for a few minutes in his chair after lunch. The houseboy still sleepily measured time with his axe. And Mummy was writing to Aunt Daisy, who often wrote to me, from England, sending me presents, and often books about Jesus because she was my godmother. It was she who had sent me the stories about Jesus walking with the men in striped headdresses through the yellow dust . . . hundreds of years ago, hundreds.

Indignation had gone, a melancholy had seized her whole body. Sweat ran from her armpits. Her hair was damp. She felt her cheeks dragging with wet. She leapt up, but before she reached the other bed, controlled the impetuous movement, becoming as stealthy as a cat as she curled herself around the little grey cat, who let out her protesting sound, Let me sleep. But the child strokes and strokes, her cheek on the cat's side, the cat purrs, noblesse oblige, the child's face lifts and falls with the purr, the child's eyes close, the cat's purr stops, starts again, stops . . . outside two doves conduct their colloquy, Croo, croo, cr-croo, the axe thuds down, slow, slow, slow . . .

The woman writing to England sits with her pen suspended, smiling, for she is not here at all, she is dreaming of a winter's evening in London, crowded noisy streets outside, and she is with her good friend Daisy Lane, the little, wry, brisk woman who had not married, for she was one of the girls whose men had been killed in the Trenches. She thinks guiltily that she has never enjoyed anything as much in her life, talking with her friend Daisy in front of a good fire, eating chocolate, or chestnuts roasted in the embers.

Good Lord, it is already three o'clock. The children must be woken or they'll never sleep tonight. Not that Doris is likely to have slept, and she always gets so fretful and weepy, but perhaps she has dropped off. The woman felt surrounded by sleepers, safe in a time of her own, without anyone observing her. Her husband was lost to the world in his deckchair, snoring lightly, regularly. The dogs were stretched out. An assortment of cats, one curled up against the dog Tiger's stomach, all asleep. In the bedroom little Harry, her heart's consolation and delight, was asleep, like a baby, his fists curled near his head. Before gently waking him, she bent over him, adoring him. She loved the way he woke, whimpering a little, small and sweet in her arms, his face in her neck, nestling, as if with his whole body he was trying to get back inside her body. She took a long time waking him, gentling him into consciousness, then slid him into his little pants and shirt. 'You go and wake Daddy,' she told him. She went into the bedroom next door and stopped her hand at her mouth. Where was the child? Had she run away? She always said she would – a joke, of course. No, there she was, arms around the grey cat, fast asleep. 'There,' thought the mother, having the last word, 'you were tired, I knew you were, all the time.' She stood quietly there looking down at the little girl's tear-dirtied face. She always felt guilty, seeing the child with this cat, because of the cat left behind in Tehran, but what could she have done? After all, they couldn't have travelled for months and months with the cat, and anyway, it was such an ugly old thing. Never had there been such storms of tears as when the family left the cat, it was ridiculous, it was out of all proportion.

The mother did not touch the child but said briskly, in tones that sounded full of regret, a complex apology for what she was thinking, 'Up you get now, you've been asleep a good half hour.'

The child opened her eyes and looked past her mother at the room as if she had no idea where she was. Then she felt the cat against her face, and smiled. She looked up at her mother and sat up, and with a shake of her head, clearing her face of the sweat-sticky hair, 'I wasn't asleep.'

'Oh yes you were,' said the mother triumphantly.

'I wasn't. I wasn't.'

'Wash your face. Then we'll have tea.'

Tea was the family sitting in the hot shade under the verandah thatch, gingerbread, shortbread, little cakes, big cakes, scones, butter, jam. 'You can't have cake until you've eaten a scone.' Discipline and self-restraint, this was called. The dogs lay with their noses pointing towards the food. The cats gathered around saucers full of milk. The little girl carefully carried through the house a saucer of milk to her special friend, the grey cat. She sat on the floor watching the cat lap, pink tongue curling around the mouthfuls of milk. The cat mewed, Thank you, and sat licking herself a little, to wake herself up. Then she stepped out to join the other cats, the dogs, the family.

Afternoons were full of events, chosen by my mother to educate or in some way to improve and uplift. There was a treehouse, platforms of planks in the musasa tree just behind the house. 'Come up to our house, come up,' we shouted at Daddy, as he manoeuvred his great clumsy leg so that he got himself on to the first platform. Then up came Mummy, and she told us about life in England, and her voice was sad, so sad that he rebuked her, 'Don't sound such a misery, old girl. England wasn't all roses, you know.' And then he might tell us of another England, the beggars, the out-of-work ex-soldiers selling matches, and the silly Bright Young Things dancing and jazzing; they didn't care about the dead soldiers or the ones that couldn't get work. Or told us of his good times before the war, when he went to the races or danced all night.

Or we would be taken to see the man who made the rimpis for the farm. On a flat place down near the new barns were trees where ox hides hung to dry in the shapes of oxen, without their bodies. Or new hide, just lifted off the carcass, was being cut into strips, and then dunked into petrol tins full of brine. Soon they were hauled out, hung over branches, and then a couple of little black boys pulled and worked the strips so they remained supple and could be used for the many purposes of the farm – tying the yokes of oxen around their necks, or tying yokes to the great central beam of the wagon or the cart, making beds and couches, or dried to be wound into great balls like small boulders and kept in a hut till they were wanted. Or the little boys would be rubbing fat and salt on to the insides of new hides, manipulating them, moving them, rubbing them so they would be soft and good for karosses or floor mats.

Or the place where bricks were made. The earth was taken from the towering termite-heaps. It was piled on a flat place, sand added, and then water poured on, and again small black boys stamped around in it, and we, the white children, stamped and danced too, our mother encouraging us, because small children should play with mud and water, Montessori said so. In fact I did not like it. These occasions were like many others, when I was playing a role to please her. I did not like the mud on my feet, and splashing on my legs, but I went on with it, together with my brother and the black children. Then the piles of mud were ready, like poo, as my brother and I giggled, but never telling our mother why. Then the brick boy came with his moulds and one man filled them with mud while another carried the moulds to turn them out in rows over straw. There the sun soon dried them. Then they were built into kilns, and fires lit in the holes like ovens inside them. Soon there they were, the piled-up bricks, red, or yellow, and there we children climbed and balanced, feeling the hot roughness of the bricks on our soles, and we jumped off, and climbed up, again and again, while my mother watched, pleased we were having this experience.

Aeons later, eternities later, the sun slid down the sky into the spectacular sunset we took for granted. But I remember standing there by myself, my whole heart and soul going out and up into those flaming skies, knowing that was where I belonged, in that splendour, which was so sad, so sorrowful, I was not here at all, or I wouldn't be for long, I would get away from here soon. *Soon* – how, when a day took for ever and for ever?

Round about then I wrote a 'prose poem' about a sunset, a paragraph long, and my mother sent it into the *Rhodesia Herald*. My first printed effort. The complex of feelings about this were the same as now: I was proud that there I was in print, uneasy that impulses so private and intimate had led to words that others would read, would take possession of. I was wriggling with pride and resentment mixed when mother said Mrs Larter had said how clever I was to have a piece in the paper. And she's so young too. But I made a private oath that next time I was taken with a 'prose poem', it would remain my secret.

At sunset, the farm became loud with the lowings of the herd of oxen being hurried back from somewhere in the bush to the safety of the kraal. In the early days there were still leopards as near as Koodoo Hill, a couple of miles away. Until the family left the farm, there were leopards in the Ayreshire Hills. Sometimes a farmer would telephone to report that a leopard had taken a beast. And there were pythons, who liked calves. The oxen, though they were wild, unsubdued beasts and nothing like the comfortable tamed animals of England, had to be fenced at night. Besides, in the mornings the cows had to be milked. One cow was not enough – not of those thin, rangy Afrikander cattle. Five or six gave enough milk for our purposes. We were told of the wonderful beasts of England with udders that touched the ground and each one holding enough milk for several households. All that talk of abundant paradises . . . there are ways of listening to travellers' tales that keep you safe from them. That England they talked about, all that green grass and spring flowers and cows as friendly as cats – what had all that to do with me?

Then the children had supper. Eggs and bread and butter and a pudding. 'Eat up your food!' 'But I don't want it.' 'Of course you must eat it up.' 'I'm not hungry.' 'Of course you're hungry.'

By the time I went to my first school I had been reading for – well, how long? – when I am dealing with time as elastic as dream-time? I do know that from the moment I shouted triumph because I was spelling c-i-g-a-r-e-t-t-e from the packet, it was no time before I was reading the easier bits in the books in the heavy book-case. The classics. The classics of that time, all in dark red leather covers, with thin-as-skin pages, edged with gold. Scott. Stevenson. Kipling. Lamb's *Tales from Shakespeare*. Dickens. Curled in the corner of the storehouse verandah, on a bed of slippery grain sacks that smelled sweet from the maize meal, and rank from the presence of cats, I raced through *Plain Tales from the Hills*, skipping a good half, *The Jungle Book, Oliver Twist*, skipping, always skipping, and having found my parents weeping with laughter over *The Young Visiters*, read it with the respect due to an author two years older than I was, brooding over words like mousetache. Mouse-ache? Where did the 't' fit in and why did the mouse ache? To fit oneself to the mysterious order of the grown-up world was not an easy thing. 'There is a green hill far away *without* a city wall.' Why should the hymn need to specify the lack of that wall? Puzzles and enigmas, but above all, the delight of discoveries, the pleasure, the sheer pleasure of books which has never ever failed me. And not only grown-up books. Children's books arrived from London, and children's newspapers. If some enterprising publisher should now produce a magazine on the level of the *Merry-Go-Round*, with writers like Walter de la Mare, Laurence Binyon, Eleanor Farjeon, would it at once fail? 'It's television, you see . . .' The *Children's Newspaper*, with reports from Egypt and Mesopotamia of the archaeological discoveries from the tombs of Tut-an-Khamun and Nefertiti? But all this is on children's television. Then, just like now, children were supposed to be protected from horrors and, just as now, we weren't, for all the time, every day, those voices went on and on, about the Trenches, bombs, star-shells, shrapnel, shell holes, men drowning in shell holes and the mud that could swallow horses, let alone men. The wounded in the Royal Free, the men with their lungs full of gas, the death by drowning of my mother's young doctor, barbed wire, No-man's-land, the Angels of Mons, the field hospitals, the men shot for 'cowardice', on and on and on, my father's voice, my mother's, and, too, the voices of many of our visitors. What is the use of keeping the *Children's Newspaper* and *Merry-Go-Round* sweet and sane, when the News tells the truth about what is going on and the grown-ups talk, talk, talk about what will always be the most important thing in their lives – war. Whenever a male visitor came, the talk would soon be of the Trenches. No, it is not violence or even pornography and sadism that is the difference between then and now, it is that children were not patronized, much more was expected of them. I do not remember my parents ever saying, 'That is too difficult for you.' No, only pleased congratulation that I was tackling *The Talisman* or whatever it was. You would have to contrast the

Merry-Go-Round with the banal jokiness of a children's programme on television to see how much lower we all stand now.

Before I reached the big school, the Convent, there were two intermediate schools. The first was Rumbavu Park, just outside Salisbury, owned by a family called Peach. I, just seven, and my brother, four, were taken there together and I was instructed to look after him. But if I adored my little brother, so did everyone else. He was always in the care of the big girls, nine or ten, who took him about with them like a doll. This was a gentle place, run by gentle folk – gentlefolk. I use this word because the matron, Mrs James, did – constantly. Like Russians of the intelligentsia who talk now of being gentlefolk, with contemptuous dismissal of their decades of revolution and egalitarianism – 'my family are gentlefolk' – Mrs James made this claim, it seemed in every sentence. Here was another member of the English middle class threatened by rough colonial manners but, unlike most of them, who mean only that they are superior in some ineffable and indefinable way, Mrs James meant what the Russians mean: they are the inheritors of literary, musical and artistic culture. She was a large swarthy gypsy-like woman, with straight black hair, like Augustus John's Dorelia, an earth mother long before the word, and she was kind. When I wrote baby pieces about flowers and birds, she told me I was wonderful, and showed them around. She brushed my hair, and made me wash under my arms and between my legs for she was afflicted by a horror of natural processes, and she held me on her large lap and sighed and mourned the crudeness of the world and her sad fate, to be matron in a school. When my parents came to visit, Mrs James presented me and my brother to them as her achievements. Far from being unhappy there, I was full of the excitements and delights of discovery. The wonderful gardens spread all over a couple of hillsides – and still do. Terraces and fountains and pools and trees and flowers: it was a show place, and at weekends people drove out from Salisbury to admire it.

I was at school in Rumbavu Park for a term. It was an aeon. A forever. When sorting out the time-segments of those two years, I had to concede that it was only a term. I have to. Impossible, but so it was. If only I could have stayed there, but the Peaches went bust, hard luck not only for them but for the children at their school. Just before I left there was an incident that illustrates a theme of these memoirs which is: why is it we expect what we do?

Sybil Thorndike was on tour in Southern Rhodesia, and playing Lady Macbeth. The older children were to be taken to see her. I would go if Mary Peach did not return in time from England, where she was on holiday. She came back that afternoon, so I could not go. She came to me, a big girl, twelve or so, to say nicely she was sorry I was going to be disappointed. I remember stammering that of course it was all right, while inside I was the embodiment of all the insulted and injured of the world. Why was it that Mary Peach, who was rich and had just come back from England where I could not go – for the theme of the absolute out-of-reachness of England was already established in my mind – had the right to see Sybil Thorndike? Unfairness . . . injustice . . . the bitterness of it. But what I would like to know is, where did the violence of that sense of injustice come from? I was seven years old. This was not only the child's sense of injustice which we describe as 'innate': a child's betrayal of justice is, must be, love betrayed, and what I was feeling was social injustice. I can think of nothing in my life more cruel than that disappointment, as if it were the sum of the world's indifference. Surely it had to come from my parents, particularly from my father's voice murmuring through my days and through my sleep, too, of the war, the betrayal of the soldiers, the wicked stupidities and corruption of government, just expectation and faith betrayed.

My mother decided we should go to board with a Mrs Scott who took in the children of farmers so they might attend school in Avondale, a suburb of Salisbury, then on the very edge of the town. I was put in the class for my age, but at once put up, I think, two. In that class I discovered the pleasures of achievement, for the reading pieces were at first too difficult for me, and I was not able to skip as I liked. One, in particular, an abridged grown-up story of a man sucked into a sea whirlpool, nearly drowned, but then cast up by the sea, had words like 'maelstrom' and 'vortex', 'inundate' and 'regurgitate'. I stared at them, oppressed by failure, but was saved by context – and in no time this difficult story was mine. Is there any delight as great as the child's discovering ability? But if the classroom was all pleasure Mrs Scott's was all cold misery. Very far from gentleness was Mrs Scott. There was a Mr Scott, employed by the Mr Laws who had the timber concession. My mother had sent her two little children to the lumber camp to stay a few days in the bush with Biddy, for she never missed the opportunity to give them useful experience. We were in a tent, for the first time, surrounded by majestic trees full of cicadas, being felled one after the other, and destined to burn in tobacco barns and mine furnaces.

Already a social being, ready to please one set of people with agreeable information about others, I said to Mrs Scott that Mr Scott, her husband, had said goodnight to Biddy when she had on only a petticoat. The voice I used was my parents' – worldly and disapproving. I had no idea what I was saying. If Mr Scott had his arms about Biddy, his whiskers, scented with Pears soap, pressed against her ear, then this was only a sign of a general loving kindness I yearned for. Mrs Scott at once hated the messenger who had brought bad news, and made a loud and noisy scene with her husband.

I hated her. She was a large ugly woman smelling of stale sweat. He was large and smelly. There was no way of getting away from them day or night. Their bed was on the verandah just outside where my bed stood under a window. I did not like getting into my bed. The cover was a kaross, a fur blanket, made of wild cat skin. Everyone had karosses, which were cheap, costing only the price of a bullet, and the labour of the man who cured the skins in salt and wind. A kaross always smelled a little, especially in the rainy season. The kaross on my bed was badly cured and smelled stuffy. I lay in bed trying to keep my face in the air from outside, while outside Mrs Scott wept and said he didn't love her, and he soothed and reassured and said he did, it was only the word of a child. At this point I ought to be able to record listening to the sounds of sex and a resulting trauma, but no, it was the injustice of it, for I had described what I had seen. Mrs Scott never spoke to me in anything but a cold and sarcastic voice. There were other children, but I remember only her daughter Nancy, who bullied me in minor ways. Then she told her mother that at school I used to go round the backs of the lavatory blocks and look up at the shitty backsides. Such a crime had never occurred to me. Mrs Scott was not allowed to hit me – my mother did not hold with it – but she slapped and hit her own daughter, just as Mr Scott did. I was afraid she would hit me, for she did not believe me when I said it was untrue. She told my parents who came hastening, if that is the word for their dawdling progress, into town. Had I done this thing? No, I had not. Remember, it was wicked to lie. 'A lie is much worse than being naughty.' They believed me. My little brother giggled. Funny that I remember so little of my adored little brother except 'standing up' for him against unkind Nancy.

January to June 1927. My seventh year. I was homesick and miserable. But compared with what goes on in schools now, and the ugliness of the bullying, physical and verbal, Mrs Scott's unkindness and Nancy's malice were nothing. I listen to young friends' accounts of what goes on in well-reputed schools and cannot believe it. Not that children are cruel – for most are monsters, unchecked. No, that teachers seem unable to stop it. Perhaps they are not unable, but even like the idea? After all, Prince Charles reports that in the elite school, Gordonstoun, his head was held in the lavatory bowl while the flush was pulled. If that is what is prescribed for the highest in the land, lesser mortals need not expect better. We are a barbarous people.

For a long time, driving past that house, long since demolished, with its big garden, I felt ill and turned my head away not to see it. Avondale School, where I did so well, is still there, unchanged.

Among the reading matter provided by my mother was a series of improving tales for children about saints, like Elizabeth of Hungary, who earned from heaven chaplets of roses to shame her husband when he criticized her charities. An intense hunger for goodness took me over, and in the patch of empty ground behind Mrs Scott's, I built a cathedral of sunflower stalks. The pleasure of it, the accomplishment, planning the building while the tales of saintly women rising above all persecution saturated my whole being. I was handling the light dry stalks, three times my height, while in imagination I was creating a great church that God himself would congratulate me for, listening for voices which surely I would hear if I tried hard enough, all assuring me of fellowship with the saints. But Mrs Scott did not see the point of these stalks, dragged out of their piles where they were stacked for burning. If you fill children's heads with saintly tales they will build cathedrals and expect chaplets of roses and chanting choirs. This is as powerful a memory as any.

Why was I left at Mrs Scott's for two terms? Probably that child's taboo against telling tales out of school was already operating. Besides, all the time, there was the pressure of We are so poor, We are having such a bad time – meaning, We can't help it. I read 'Baa,

Baa, Black Sheep', and identified with Kipling as a small boy, just as my mother had done before me, but my mother didn't embrace me weeping, Oh my little child, my poor little child – as I raised my arm to ward off a blow. There were no blows, only cold sarcastic verbal bullying. And I was already reading *Stalky and Co.*, full of information about school brutality. Literature provides more complex news about the world than *It isn't fair*, but this lives in a different part of the brain.

I had begun, in short, to colour in the map of the world with the hues and tints of literature. Which does two things (at least). One is to refine your knowledge of your fellow human beings. The other is to tell you about societies, countries, classes, ways of living. A bad book cannot tell you about people – only about the author. A bad book does not know much about love, hate, death or so on. But a bad book can tell you a good deal about a certain time or place – about history. Facts. Mores. Customs. A good book does both.

But bad books were still in the future. Meanwhile, then and for three or four years, what came to the farm from London was an astonishing variety and number of books. They had to be written for, and the order took a month or so to get there. The books had to come by sea, three weeks, then a train to Salisbury from the coast, and another train to Banket, and then they had to be fetched from the station.

Here are some of what I remember. John Bunyan. *Bible Tales for Children. English History for Children.* The Crusades – with Saladin presented like an English gentleman. The battles of Crécy, Agincourt, Waterloo, the Crimea, biographies of Napoleon, Benjamin Franklin, Jefferson, Lincoln, Brunel, Cecil Rhodes. Children's novels like *John Halifax, Gentleman*; *Robinson Crusoe*; *The Swiss Family Robinson*; *Lobo, the Wolf* (from America, Ernest Seton Thompson). *Alice in Wonderland* and *Through the Looking Glass*, Christopher Robin, *Black Beauty*, Stevenson's *Verses for Children*, *Jock of the Bushveld*, Florence Nightingale, and – not least – *Biffel, a Trek Ox*, the story of a beast that died in the rinderpest epidemic of, I think, 1896, unforgettable by any child who reads it at the right time. *The Secret Garden. The Forest Lovers.* A whole range of little tales, purporting to be the lives of children in Iceland, India, France, Germany – everywhere – jolly little tales, jolly little lives, the equivalent of the readers that went, 'John and Betty had such fun playing with Spot' – but I suppose the information that in Norway they ski and in Switzerland they yodel is not unuseful.

It was mid-year of 1927 when I returned finally from Mrs Scott's, for I was due to go to the Convent, for which I was already being primed by warnings that they – the RCs – would try to 'get me' and I must be on my guard. The Convent, which always had more Protestant than Catholic pupils, was used to assuring anxious parents that the souls of their offspring were safe with them. The Convent, like convents in Britain, was supposed to be more genteel than the High School. I am always meeting women now who were sent to convents for the same reason. That the convent in Salisbury had this reputation was because of false comparisons with Home. I had been given a bursary. Into the third year on the farm it was evident things were going badly and not likely soon to get better. My father was building tobacco barns, because maize was no longer where fortunes were being made. And was he, with his wooden leg, his limited mobility, planning to get up several times during the night to check the temperatures in a barn a good mile away?

We could not have afforded unaided my uniform for the convent, piled on chairs and beds everywhere through the house. Pleated tunics in heavy brown serge and alpaca, with light orange cotton blouses, springy brown girdles that surely could never stay knotted, white panama hats with brown and orange ribbons, the brown blazer, piles of heavy brown knickers, and many vests and brown socks. Even to look at this stuff was oppressive, but luckily it was still the beginning of the holidays and time stretched endlessly ahead.

Just about then the family became characters in A. A. Milne, just as if we had never left England. My father was Eeyore, my brother Roo, my mother – what else? – Kanga. I was the fat and bouncy Tigger. I remained Tigger until I left Rhodesia, for nothing would stop friends and comrades using it. Nicknames are potent ways of cutting people down to size. I was Tigger Tayler, Tigger Wisdom, then Tigger Lessing, the last fitting me even less than the others. Also Comrade Tigger. This personality was expected to be brash, jokey, clumsy, and always ready to be a good sport, that is, to laugh at herself, apologize, clown, confess inability. An extrovert. In that it was a protection for the person I really was, 'Tigger' was an aspect of the Hostess. There was a lot of energy in 'Tigger' – that healthy bouncy beast.

But it was not Tigger that went off to the Convent, but a frightened and miserable little girl.

Mother Patrick came riding into the colony with her five Sisters just a year after the Pioneer Column in 1890 and they at once set up their hospital and became, but really, sisters of Mercy, because contemporary accounts speak of them like this. It was Mother Patrick who established the Dominican convent and she was, when I got there, a revered figure, spoken of in awed tones, like the other pioneer sisters. Sister Constantia and Sister Bonaventura were, I think, still alive, as silently influential as the statues everywhere of the Virgin. They had been lively and adventurous young women, and the administrative nuns that came after them were a different kind.

The Convent was a central mass with projecting wings, embedded in granite chips. When adults walk over stone chips, the pebbles are not very comfortable underfoot, but for a small child it is like toiling over the big sharp stones on a beach, each a hazard. The staircase to the small girls' dormitory was steep, every step at thigh-level. The little ones clambered up on hands and knees; going down meant jumping from step to step for the handrail was high above our heads. The day I found myself actually able to step down, the day I ran across the gravel, were markers on the road to being grown-up. This dormitory (which was over the gym), the refectory, the classrooms, the sickroom, were what the pupils knew of the Convent: most of the building was out of bounds to the children, and seemed a ghost story place of vast shadowy rooms full of nuns in their black and white robes floating like shadows. The nuns slept in dormitories too, but we knew white curtains separated their beds, making tight box-like cubicles. The 'little ones' dorm' was a long high-ceilinged room, and in it were three rows of beds, lined up head to foot, twenty-four beds. There were rows of high windows on either side. This large room, or rather, hall, was in daylight well lit and fresh, but at night a different place. A small table that projected into the room, making it necessary always to walk around it, held an assortment of holy objects, small statues like icing-sugar figures on cakes; and above it was a large picture where a man from whose head shot rays like those from behind a storm cloud pointed authoritatively to his swollen heart dripping blood. On the wall facing this altar was a tall picture of a man on whose head was clamped an enormous wreath of Christ-thorn, like that which grew on the kopje, with black spikes an inch or two inches long, and blood ran down his face from the spikes. Other pictures showed a man full of arrows that stuck out like porcupine quills, each in a bloody wound, and a woman holding a plate on which were two pink blancmanges in red jam sauce, but these turned out to be her cut-off breasts. In another a woman stood smiling while being burned to death by flames that curled around her like long witches' fingers.

When I recently drove through the countryside near Munich I kept coming upon horrific statues of tortured Christ. They were beside or in a pretty stream, or in a wood, a field, a garden. They reminded me of the pictures provided for the instruction of us children in the convent, all relish in blood and torture. The nuns in this convent were nearly all from South Germany, which was Hitler's country. Stalin the sadist came from a seminary. I was reminded of these feasts of blood when I was in Peshawar at the time when Shiah Muslims celebrate the murder of Hassan and Hussain, the Prophet Muhammad's grandchildren, more than 1,500 years ago. Young men ran or staggered in hordes through the streets lacerating themselves with heavy chains or whips, eyes blank or shocked with pain, till they fell, to be gathered up in ambulances that were patrolling the streets for just this purpose. Forgive me for the banality of this reflection, but there is something very wrong with the human race.

The youngest children in the great torture room were five or six, and the oldest were ten and eleven.

When we were in our rows of beds, the light was turned off, but the red light that burned always in front of the Sacred Heart and its bloody gouts lit the room with red. The nun in charge of us little ones came to stand in the doorway, the light behind her. In heavy German accents she said, 'You little children believe you are safe in your beds, you think that do you? Well you think wrong, you think the holy God cannot see you when you lie under the sheet. But you must think again. God knows what you are thinking, God knows the evil in your hearts. You are wicked children, disobedient to God and to the good Sisters who look after you for the glory of God. If you die tonight you will go to hell, and there you will burn in the flames of hell, yes I tell you so, and you must believe me. And the worms will eat you and there will never be an end, it will never ever end.' She would go on like this for a good ten minutes or so. Then, having cursed us to hell and back, she shut the door and left us to it.

Storms of sobs, and soft shrieks of terror. The older girls crept to the beds of the little ones, to comfort them. 'It's only Catholic,' they would say, 'we don't believe all that.' For most of us were Protestants. The Catholic little girls were protected by rosaries, holy pictures and bottles of holy water under their pillows.

When my parents warned me that the Catholics would try to 'get' me they had not foreseen anything like this, I knew that. I knew they would be appalled. This armoured me, and besides, one may believe and not believe at the same time. I do not know for how many years these horrific sermons went on: the impression the first term made on me was so strong I have forgotten the rest. I remember only lying in bed to watch the blood dripping from the big heart like a lump of fresh steak, making myself see that it moved, believing that I could actually see the blood trickling, while I knew perfectly well it didn't. The tiny children – I was already, at eight, in the middle range – used to cry out in their sleep. Sometimes one would go wandering around among the beds in her sleep, and an older child gently led her back to her own bed. One sleepwalking little girl persistently tried to get into the bed parallel to hers, because there was a kindly older child in it, who quietly made the swap when the small one was asleep at last, the nuns never knowing. In the morning there were dirty stains of urine in many beds. The nuns scolded and punished: for the Catholic girls the repetition of Hail Marys, for us, admonishment and threats.

The nun whose talent was for hellfire and the undying worm used a ruler on our palms when we were naughty. There were a thousand petty rules, and I have forgotten them, but remember the secret scorn they endangered: we protected ourselves by despising these dormitory nuns, making fun of their accents, telling each other if they weren't stupid they would be teaching nuns. Most of the rules were to do with washing. Not that we should wash, but that we should not. Cleanliness for these women was an invitation to the devil. We were told to wash our hands only to the wrists, keeping sleeves rolled down. Only our faces, with a washcloth soaped thick: if our eyes stung, we must offer the pain to God. We might bathe only once a week. The nuns told us that good children would agree to wear the wooden board that stood always against the bathroom wall, when we bathed. The board had a hole in it for the head, and was designed to rest on the sides of the bath, making it impossible to see our bodies. But no one would. We were allowed to change our underclothes once a week. We smelled. All our letters were read by the nuns and when I told my mother about the bath rules, the nun said I was disloyal and wicked and made me write the letter again. But at half term I 'told' on the nuns, and my mother was furious, protested – and thereafter we were all allowed to bathe twice a week and change our underclothes twice. We continued to smell. We had to put on smelly knickers and dirty socks. 'Vanity' said Sister Amelia, or Brünnhilde or whoever. 'All is vanity. You should not think about your body.'

There went on the usual school mythology about the slaps administered to our palms with rulers. We giggled, as is prescribed, advised each other how to soap our palms, recounted tales about a former pupil who was beaten till her hand fell off, and now she had an artificial hand. All this was as it always is, at this type of school. But if the rulers left hot red marks on palms, that was all, the nuns were not allowed to hit us anywhere else. It could all have been much worse. And I don't remember bullying, on the contrary, the older children were tender with the little ones, remembering their own misery.

The atmosphere in the Convent, in short, can only be described as unwholesome, a favourite word of mother's. How much did she know about all this? If it was within the code to 'tell' about the lack of baths, why not about the viciously slashing rulers, why not about those hellfire sermons? When 'Tigger' reported on them, she made a joke about it all. And certainly my mother knew about the sadistic pictures in the room we slept in, for she inspected the Convent thoroughly. But after all, she herself had had a strict, punishing upbringing.

The nuns never made any attempt to 'get' the Protestant girls. They did not need to. The atmosphere of magic and mystery was enough. Antonia White's *Frost in May* describes the allurements of the forbidden, though her convent was on a somewhat higher social level. Most of us at some time wanted to be Catholics, simply to be like the Catholic girls, who dipped their fingers into the holy water stoups beside every door, who crossed themselves and curtsied as they passed statues of Christ or the Virgin, who carried holy pictures in their pockets and rosaries wound around their wrists. They were always going off to special events in the cathedral. Bells rang from the cathedral a block away, several times a day, for Angelus, and for Mass. Bells tinkled from the nuns' chapel. The Virgin, a pleasant and beneficent figure, was often carried about the grounds on litters draped with coloured paper. Above all, there was the mystery of the part of the convent we were not allowed into. We believed there were hundreds of nuns, but perhaps there were not more than fifty. Most of them we never met. They worked in the kitchens, cooked our food and theirs, kept the convent and its grounds clean – there were no black servants. Some were taken out every day in lorries to the vegetable gardens. They all got up very early in the morning, four o'clock, some earlier. If you woke at night you could hear the sweet high chanting voices from the chapel. There were often funerals. If we begged hard enough the Protestant girls were allowed to go in the lorries to the cemetery with the Catholics, where we stared in a romantic trance at the coffin, violin shaped, bright white and pink, like a cake, with messages in gold script, Sister Harmonia, Bride of Christ, RIP. She was very young to die, said the other sisters. Knowing that eighteen, twenty, was thought young shocked us with our small sum of years, for it was hard to believe we would ever be as old as this dead woman.

Now I think these girls died of broken hearts. Nearly all were poor peasant girls from Germany. The Convent in Salisbury, Southern Rhodesia, was an extension of the economic conditions in Europe. Germany had not recovered from the First World War and reparations. As had always happened in the poor families of Europe, one or two girls in a family became nuns, to save their families the burden of feeding them. They found themselves thousands of miles from home, in this exotic country, doing hard physical work, as they had all their lives, but in the heat, and with no prospect of seeing their families again. Their only consolation can have been that their loneliness and exile made things easier at home. Once, when I was in the sickroom, a nun came to sit on my bed (against the rules) while the Angelus rang its call to prayer and the sky flamed red, and she wept, and crossed herself, crossed herself and wept, saying she longed for her mother. Then up she jumped, asked the Holy Virgin to forgive her, told me to forget what she had said, and ran out. She was eighteen.

Our speculations about the nuns' secret lives were innocent. Now children of five or six would probably talk knowledgeably about lesbianism. Their bathing arrangements part consoled us for ours. They took baths once a week, wearing a white shroud, and kept the board around their necks. They never saw themselves in a mirror. Their heads were shaved. They seldom changed their undergarments. We knew what they wore, for we could see acres of white garments on the washing lines. There were layers of vests and knickers and petticoats under the heavy white serge robes we could see, that had over them the black robe, the crimped wimple, and the two veils, white and black. The nuns smelled horrible.

The nuns who taught us were educated women. One at least was a Nazi – so says Muriel Spark who writes about the same convent in her autobiography. Sister Margaret taught music, was kind to the little girl whose mother kept insisting she was a musical prodigy. She knew my mother could have had a career in music, listened gently to her tales about thwarted ambition, and for four years taught me scales and apprentice pieces, told me about great musicians and the obstacles they overcame. She never even hinted I had no particular talent. There was a Sister Patrick who, the nuns said, was a real lady, from Ireland, but she had given all that up for the love of God. She was a tall thin woman, with a fine elegant face, and she was dry and witty and sometimes unkind. She might quote from French or Latin and then say, 'But you will not have heard of him I suppose,' and sigh.

I was clever, that was my attribute, clever little Tigger Tayler. School lessons were never difficult, exams pleasurable. But being clever was not something I was prepared to go along with, for from the start I was quietly sliding out, not knowing what I did. My cleverness was a continuation of my mother's, like my musical talents, insisted on, held up to other people for admiration, boasted about to the farmers' wives, used as a means to get bursaries and special privileges.

What was my own, where I belonged, was the world of books, but I had to fight for it as soon as I arrived at the Convent. The school library was several rooms full to the ceiling with books neatly covered in brown paper, the titles and authors written on their spines in ink. I felt as if I had walked into a treasure cave, but the library nuns did not believe a child of eight had read *Oliver Twist* and *Vanity Fair*. They insisted I must have the permission of my parents to read such unsuitable books. My weekly letter home read, 'I am very well. I hope you are very well. How are Lion and

Tiger? Sister Perpetua says I must have your permission to read books. It is only four weeks and three days and seven hours to the holidays. Love to Harry.' While waiting for permission, the library nuns urged on me improving literature, which filled two long shelves. The word 'unwholesome' is hardly adequate to describe the moral climate these novels came from. The plots were all the same. A pure young man or girl met, apparently by chance, a worldly person, usually a woman, well dressed, older, but whose every smile or glance promised enticing initiations. The neophyte was invited to a country house, full of cosmopolitan older people, who all had the same air of mystery. The bemused one found herself, himself, attending seances, table-turnings, and ambiguous services in ruined chapels and sylvan glades. And then – the choice! The left-hand path into Satanism, the right-hand path into tedious virtue, which was fit only for the stupid or the timid. I did not find anything like this mix of eroticism and black magic until the TV series *Twin Peaks* from the States a couple of years ago, but the convent novels had nothing in them of that grotesque wit.

These novels were not as compelling as the library nuns would have liked. I had never heard of seances or Satan. For the four years I was at the Convent I was being urged to read them. Now, when I ask Catholic friends, they know nothing of these books or anything like them. Perhaps some pious library at Home was pruning itself, and thought: 'Pity to waste them. I've got it, they'll do for those heathen natives in Africa!'

I was at the Convent for four years. Or for eternity. I used to wake up in the morning with the clang of the bell and not believe I would live through that interminable day until the night. And, after this endless day would be another. Then another. I was in the grip of a homesickness like an illness. It is an illness. When I was in my late sixties and succumbed to grief, I thought, My God, that's what I went through as a child, and I've forgotten how very terrible it was. What did I long for? Home. I wanted to be home. I wanted my mother, my father and my little brother, who until he was eight was still at home. I wanted my dogs and my cat. I wanted to be near the birds and animals of the bush. I wanted . . . I yearned . . . I craved, for this anguish to be over. I did not believe it would ever end. I have exchanged recollections with men who were sent to schools in England aged seven, and some remember this weight of misery. There must be by now hundreds of memoirs, autobiographies, testifying to the misery of small children sent too young to school. It is a terrible thing to send small children to boarding school. We all know it. Yet people who remember very well how they suffered, sent from home aged seven or eight, do the same to their children. This says something pretty important about human nature. Or about the British.

I could not conceivably have lived through four years continuously in the grip of that pain, but whenever I take out my mental snapshots of the Convent, I am immersed in grief.

When I went home for the holidays, the end of them seemed so far away it was like a reprieve. Six weeks. Even four weeks. When every day was endless, then even a week was an ocean of time.

For two years my little brother was at home being taught by correspondence course and slowly he fought his way out of being Baby, or Roo, insisted on being called Harry, and took firm hold of his birthright, which was physical excellence. If my early memories of Baby are all of a cuddlesome complacency, on someone's lap, usually mine, then later they are of him in energetic movement, flying down the hill on his scooter, then his bicycle, brakes off, or at the top of some fearsome tree, or hitting sixes over the roof of the house while he ran like a duiker. He was like all the other white boys of the District, a lean, tough, sunburned child, his knees always scarred, his shorts torn, and his eyes inflamed by the sun, for he was out in it from sunrise till sundown. My mother read us *Peter Pan* too often, and her voice broke when Peter returned, found the window shut and went flying off again. 'Come on old girl,' urged my father, 'it's not as bad as all that.'

But for her it was. Nothing she had wanted for herself was going to happen. All her energies were in her children, and particularly her darling little boy. But he – and quite suddenly – did not seem to be aware of her. Interesting, the different ways children rebel, preserve themselves. I cannot remember a time when I did not fight my mother. Later, I fought my father too. But my brother never fought. He would smile, quite politely, as my mother tried to make him eat this, wear that, think this or that, see the children on the other farms as common, or see 'this second-rate country' as a place he would not stay in. But, if he did as he liked, it was within the limits of what she chose. He went to Ruzawi, a prep school modelled on English lines, and later, into the Navy, though he did not want to. It was not until he married that he made a big choice for himself. Now I see it as an instinctive passive resistance.

I begged my mother to have another baby. She was a maternal woman all right, and it must have been painful, when that little girl's pleadings reinforced suppressed instincts. 'Please, Mummy, please, I'll help to look after it.' 'But we can't afford it,' she said, over and over again. And then, already, and so early, 'Besides, Daddy is not very strong.' The strength of my yearning for that infant mingled with my homesickness; I am sure yearnings of this intensity are for some other good lost perhaps when we are born. But when I mourned that there would not be another baby, I learned how much 'Baby' had been my baby as much as my mother's. After that, if there was a baby or a small child anywhere in the District, I adored it, could not be separated from it, begged to be allowed to bring it home. This passion became quite a joke in the District – a kindly one. 'Your little girl, she's a funny one for babies.'

In the paraffin box bookcase beside my mother's bed, behind the Liberty cretonnes that were beginning to lose freshness, was a book about the process of giving birth, the manual on obstetrics from the Royal Free. I lay on my mother's bed, studied the stages of the foetus's growth, pored over the enlarging slopes of the stomach, and, in imagination, went into labour and gave birth. So strong was my identification that I *almost* believed that yes, there would be the baby, lying there on the bed. This fantasy was also erotic, but in flavour, not in physical fact. Who was the male? One of the little boys in the District with whom I was in love and with whom I was making a family.

The holidays were crammed with incidents and events. My mother made sure they would be. Not only did our instruction continue, stories from history, geography, exploration, but there were visits to and from the other farms. When the families arrived and the children were sent off to play, it was not play at all. We stalked animals and hid to watch them, watched birds, learned how to distinguish tracks in the dust of the roads, searched reefs for gold-bearing rocks. My brother was given his first airgun, and he shot every bird he saw. The guns divided the gang of children into boys and girls – the boys shooting, the girls playing family. But when I was alone with my brother, we went together into the bush.

My mother's genius for social life showed itself in picnics, either with other families or when we were on our own. The car was piled with food, and we went off to some clear place in the bush and made a fire and cooked sausages and eggs, and lay under the trees watching the moon rise, or naming stars. If there were other children we sang jolly songs, like 'Campdown Races' and sad ones, like 'Shenandoah'. We sang American, not English songs.

Several times a day, through the holidays, I, or my brother, or both of us, would be summoned to learn something. My mother, or father, had found a skull or skeleton in the bush, or a lump of gold-bearing rock. She boiled the skulls and skeletons of birds and small animals until the flesh fell off so we could learn the structure of bones. She blew birds' eggs, and dismantled birds' nests. She cut open termite nests to show us their gardens, their nurseries, their roads, their galleries. She showed us cast snakes' skins and the eggs of spiders and snakes. She pulled flowers and leaves apart, and made us draw their parts.

Meanwhile, all the time, it seemed day and night, talk of the war went on. Sometimes it seemed as if the house on the hill was full of men in uniforms, but they were dead, just as in all the houses of the District were photographs of dead soldiers. And, too, cripples from the war. There was Mr Livingstone with a wooden leg, like my father – but he did much less with it. Mr McAuley had a steel plate over his stomach, to keep his intestines in – so they said. In the Murrays' house, a sad, stoical, woman mourned the death of a husband and four sons in the Trenches. There was one son still alive, to take the place of all of them. In the Shattocks' house was the picture of a beautiful little boy who, when a boat was sunk by a torpedo in the war, was sucked into the funnel to drown. Sometimes, when the talk of the war began again – and again, I wriggled away, tried to get out of the room, and if my father caught me he would shout, 'That's right, it's only the Great Unmentionable. It's only the Great War, that's all!'

There is a question: there has to be. Four years at the Convent, but also four years of holidays, weeks of holidays that seemed when they began as if they would never end. There were a hundred kinds of experiences, good times, picnics, family outings, the dogs, the cats, cuddling babies, or walking all day with my brother in the bush, sitting up at night to watch the stars. But the dark times, the miseries are stronger than the good times. Why is that?

'Give me a child until it is seven,' they say the Jesuits say. The talk of war was probably the first thing I ever heard. So perhaps if there had never been the Convent with its bloody and tortured people everywhere, its tortured but smiling saints, it would have been the same. Suppose the Convent had nothing but sunny pictures of woods and fields and kind faces, would then the talk of war have proved stronger? Or is there something inherent in our composition that disposes us to grief and memories of grief, so that days or even weeks of good times prove less inviting than pain? This question has a rather more than personal relevance.

I had not been at the Convent a year when I escaped into the sickroom. First, I was really ill, with something then generally called B. Coli. A kidney infection, with high temperatures. And thereafter I was always reporting to the sickroom, with vague symptoms, and being kept in bed. My mother saw this as a sign of being 'delicate'. I knew I was homesick, but did not know that what took me to the sickroom was Sister Antonia, a kindly and affectionate woman, who mothered me and all her charges. These imaginary illnesses had a double face. First, being delicate removed me from my mother's insistence that I should be clever, 'Just like I was,' and continually being shown off to neighbours who, I knew, would be derisive as soon as the telephone was silent, or our car had driven off. 'Who does she think she is?' But worse than the neighbours was the pressure of that ferocious energy of hers, insisting I must be clever, that if I got 70 in my maths exams it could be 100, that I would soon get a scholarship, and go to school in England. But illness also delivered me to her, helpless: doctors, illness, medicine. It is like looking back into something like the cold fogs that, sometimes, my father said, lay over No-man's-land, or even clouds of poison gas. Illness permeated everything. Why was it doctors always did what my mother said? For one thing, she demanded the right to be considered a colleague. 'I am a sister from the Royal Free in London.' She knew as much and more than the nurses. I was always being taken to Doctor Huggins for tests and checkups, some of them involving catheters. Now I know I had cystitis, but the most minor inflammation was seen as a symptom of something serious. I used to scream at even the idea of a catheter, so they chloroformed me.

There were powerful reasons for my mother's obsession. First, one has to ask what need in a woman is fulfilled by making a baby 'clean' from the age of a few days, by spending hours in every day 'holding out' a baby over the pot – or making someone else do it? When the infant becomes 'clean', the mother's occupation is gone. When I took command of my own bladder it was a moment of exhilarating freedom. I shouted, 'No, no. I *won't* use the pot, I'm going to use the grown-up seat.' Meaning, I won't have you examining my products several times a day. And 'Baby' very soon made the same step towards independence, mostly because I helped him insist on it. More importantly, we really were threatened by diseases. After all, the whole family did have malaria twice in the first rainy season. People did die of blackwater fever, then supposed to be the result of malaria, and early signs of it were in the urine. Bilharzia was a threat throughout my childhood: again, one of the signs was blood in the urine. All kinds of sickness, now only a question of taking a few pills, threatened long and perhaps fatal illnesses. My mother's imagination was always flaring into disasters: how can I not understand it, when I am the same myself?

All those skills, how she must have longed to use them, as she watched her children running around in the bush, full of snakes. 'Where are your shoes? The dust is full of filthy diseases. Where is your hat? You'll get sunstroke. Have you taken your quinine? Do you want to get malaria again? You won't listen to me, but one of these days you'll be sorry.'

We all took quinine night and morning through the rainy season, October to April. The pills were then large bright pink tablets, and they made your ears ring.

In the sickroom in hospital, I lay in a neat white bed in a bright light room, where the bells rang loudly for Angelus, where on the walls there was only one small pretty statue of the Virgin, and where I read. And read. And read. If there was some symptom worse than usual, my mother would make my grumbling father drive in, and having examined me, in the sickroom, she insisted on the doctor coming. Visits to the hospital, visits from the doctor, nearly always a false alarm, punctuated life in and out of school, and if I feared them, I also shared in the drama of it, the lift of excitement in my mother's voice, the threat of death, of hospitals, of invalidism.

The threat of illness for my father ran as an undercurrent to everything during those years. His wooden leg was only a small part of it; he was often sick, and when the doctor came from Sinoia

I would hear my mother and him talk in low voices: 'You can't expect a man who was in bed for a whole year just to get over it like that.' And, 'He had shell shock, did you say?'

My father most valiantly fought the disadvantages of a wooden leg. It was when I saw Captain Livingstone, who only walked on safe flat places, that I understood how much my father did with his. He went down mine shafts, deep ones, in a bucket, the stiff leg sticking out, knocking against the shaft wall, and making the bucket spin, while he shouted up from the depths: 'Wait, hold it,' and the windlass was held still, the spin slowed, and he was able to go on down. He dragged himself through enormous clods of a newly ploughed field. He drove the old car everywhere, through the grass and bush and rough places where there was no road. When he later got gold fever he walked miles with the prospecting hammer and the divining rods. That is, he did until he became too ill.

When I was in love with one of the little boys, I forget which, and dreaming of him, I broke a small branch or twig, near the platform in the treehouse, and bandaged it, then wetted the bandage, murmuring his name. This confession, I know, is of the deepest psychological importance, but more interesting to me now than my needing a wounded lover, like my father, is that this was magic, an act of magic-making, when I knew nothing about it, was the child of parents committed to being rational. Instinct was instructing me how to cajole circumstances, how to manipulate the outside world by means perhaps millions of years old. Any shaman would have understood what I was doing.

Something else took me into the sickroom, when I could persuade kind Sister Antonia. It was that I did not fit in with the other girls, not only because, as it seemed everyone complained, I was too old for my age. A scene: ten or twelve little girls sit around a refectory table, darning lumpy brown stockings and lumpy brown knickers, watched over by the sewing nun, Sister Theresa. They were boasting about their holiday experiences, trips to Durban, trips to Cape Town, even a trip to England. I, offering them my dearest and best, spoke of going into the big field to pull maize cobs, and afterwards stripping the seeds to make a cheese dish. The girls exchanged scornful looks, and the nun at once congratulated me on my proper humility and frugality, while I simpered and hung my head, hating myself. After that I kept quiet. Boarding school life and home life are always kept sharply divided, but there can never have been a home life so jealously kept from school friends. The boarders were nearly all farmers' daughters, but I soon learned they had more in common with the town girls, who were day girls, than with me. They knew nothing about the bush, seemed afraid of it. They did not learn farm skills. By the time I left the Convent I knew how to set a hen, look after chickens and rabbits, worm dogs and cats, pan for gold, take samples from reefs, cook, sew, use the milk separator and churn butter, go down a mine shaft in a bucket, make cream cheese and ginger beer, paint stencilled patterns on materials, make papier mâché, walk on stilts made from poles cut in the bush, drive the car, shoot pigeons and guineafowl for the pot, preserve eggs – and a lot else. Doing these things I was truly happy. Few things in my life have given me greater pleasure. That is real happiness, a child's happiness: being enabled to do and to make, above all to know you are contributing to the family, you are valuable and valued.

WHEN I WAS TWELVE OR SO I wrote a little piece called 'The Treasure Trunk', making it a symbol for my parents' exile from good. The contents of an old cabin trunk were all magic and mystery. I forgot it was there for days, weeks, but then something my father said about England, or a sigh from my mother, sent me back to it. At first it was out of bounds, not to be touched. Then, as my parents understood how unlikely it was the trunk would ever be opened in a real house, where what was in it would find their right places, I was allowed to take things out, provided they were put back, exactly as they had been, in their layers of crisp white tissue paper, the soft black paper that protected gold and silver and bronze lace which would otherwise tarnish.

The top layers were baby clothes, the tucked, embroidered, flounced lawn and muslin dresses and petticoats and jackets once worn by my brother and by me. When I mourned over the baby my mother would not have, I laid out these little clothes and stroked them and wept. It was then that I was given my first real doll – and my last. My mother sent to the Army & Navy Stores for Christmas presents which, we were reminded and reminded – and reminded – cost more than they could afford. Both cost exactly one pound – a fortune. My brother's steam engine, which when you lit a fire in the boiler puffed a few paces, was a half-day wonder, but then he went back to the bush. My doll was baby-size, had blue eyes that opened and closed, and it made a sheep-like noise when tilted. For days I dressed it up and undressed it, and all the old baby things had a third life, and I hugged and rocked the cold unresponsive thing and sang it songs. And then I forgot it. Forgot the trunk. Quite soon, another passion drove out the first, and I painted the doll's head into a black shingle, with kiss curls, and her mouth scarlet, and her nails red. I had begun to fight my mother over my clothes. She made them all. She made them well. She made them too young for me.

The baby clothes lay about, became painting rags, disappeared: if they were exhibited now people would not believe an ordinary humdrum baby wore such exquisite garments.

Under the tray of baby clothes was the one where my mother's evening and afternoon dresses lay in scented tissue paper. Quite soon everyone would sneer at the 1920s clothes. 'How could they have worn those ugly things – no waist, no breasts.' Under the dresses were the underclothes to go with them, crepe de Chine camiknickers, and bands of stiff material designed to flatten the bust. I remember the exact moment when – obedient to some wind of change – opinion reversed. I was on the top of a bus, early 1950s, and saw a young woman walking down the Bayswater Road, in a grey silk waistless shift dress – it might easily have come out of her mother's cupboard. Oh, but that's beautiful, I thought, and saw the female passengers on the bus craning their necks: *Wow. Where did she buy that?*

The short smooth skirt of an evening dress in dull green silk was of bronze lace bands, finished in 'handkerchief points' of the silk. The lace had a strong metallic smell, as powerful as sea-smell, rocking the brain. I put my face into the dress and thought that there was a world where people wore dresses like these, and not for 'dressing up'. Soon these wonderful dresses were used for games and I wonder what my mother felt when she watched the little farm girls peacock about in the dresses she never would wear, using high excited affected voices in a parody of grown-up formality.

The third tray had in it my father's dinner jacket, his tails, the starched white shirts, his officer's dress uniform. No ambiguity about this. 'Thank God I'll never have to put those on again,' he would say looking defiantly at my mother, for he was saying, 'Thank God we won't be living the life you want to live.'

The fourth tray had an apron, a trowel, a lady's white dress gloves, and some books with strange designs on the covers. About these my father was contemptuous. He had become a Mason long ago, in the bank in England. 'What a charade. What a circus. But if you didn't belong you didn't get promoted. At least in Persia we were free of that. And the police too, so they say. In the bank if you weren't a member they'd give a little hint. "Were you thinking of that manager's job in Cirencester, Tayler?" The Royal Family too. The fish starts rotting from the head. Thank God I'm out of it all.'

The well of the trunk was filled with brocade shoes, silver and gold shoes, black satin shoes with rhinestone buckles; evening bags; a brown satin winged hat trimmed with beige ostrich feathers; a beige ostrich feather boa; a fox stole with black beady little eyes caught nose to tail with a gilt chain; a tray of my father's war medals; kid, leather, and silk gloves, each in watered silk envelopes; fly-away silk scarves, just the thing for dressing-up. My father's army gloves. Packets of old photographs, wrapped in oiled paper against fish moths. But fish moths are not so easily deterred, and the photos were like lace. Insects, insects – it was insects that in the end brought the old house to its knees.

Most children announce they are really orphans, foundlings, even of royal or noble birth. I did not want to discard my father, but decided my mother was really the Persian gardener. (Who was the Persian gardener, what was he that I chose him for my mother years after he tended the water channels of the gardens at Kermanshah?) That this was impossible I knew with one half of my mind, for was I not familiar with the book on obstetrics, and with Marie Stopes? Yet the Persian gardener it had to be. The announcement of who my real mother was, and similar dramatic presentations, were made by Tigger, the joker, the zany fat and clumsy animal, and when I announced my mother was the Persian gardener, it was only Tigger's fun. Just as Tigger joked that the nuns would only eat a sweet if they whirled around on their heels to pop it into their mouths, or wouldn't let you change a gymslip that had soup spilled on it, and were so silly about hellfire.

It was also Tigger who ran away, and joked about it afterwards.

But it was no joking matter, the intermittent rages of hatred against my mother that decided I must get to Beira, on to a ship and away to – well, anywhere. And not an easy business, running away from the farm, in the bush. I worked it all out. Into a pillow-case went cheese, a tin of bully beef, sardines. I stole ten shillings from my mother's handbag. I had birthday and Christmas money saved up. Harry did not want to run away. He did not understand why I did. But my calm and my ruthlessness overcame him. 'Of course you do,' I kept saying. It would be easy, nothing to be frightened of. But he kept snivelling, and burst out, 'But I don't want to!' 'Yes, you do,' I informed him. As soon as we were put to bed we would steal away down the hill, and walk along the track to Banket. At that time we had not walked that road, but knew the Africans did, all the time. The train would arrive early in the morning and then . . . the thing is, I hoped it would, but did not know. In short, here was the hazy mental place where most of the world's brave plans dissolve like sugar. I wanted it so badly, and so it must happen. Once we were in Salisbury we would . . . but there the haze became a deep fog. Somehow we would get hold of some money and get on a train to Beira and then . . .

I kept my little brother awake by telling him stories, and then, already chilly in the night air, we sneaked from the house, crept past the lit windows where our parents sat reading, and ran off down the road in the dark, the pillowcase with its loose tins banging against my legs. Harry was crying loudly now. The bush was not then the domestic bush it has become. It was full of dangerous noises, owls and nightjars, the crashing leaps of a disturbed buck, but above all, mysterious presences from our fairy tales, released from the pages of books, and at large here all around us, in the trees, behind bushes, running silently beside us in the road. And then something happened, the two dogs arrived, to lick our hands and whine and jump around us. What are you doing? Where are you going? But we had not remembered the dogs. We told each other we could not run away because the dogs would come with us, and we ran back up the dark track in the bush to the house on the hill, the dogs making a game of it, bounding and barking . . . we fled into our bedroom and into bed. We giggled and laughed and shrieked with relief, and the dogs went quietly back to lie in their places in the lamplight. Next day I told my mother we had run away, got scared stiff and come back, but 'Tigger' told the story, it was funny enough, and she did not believe me.

But I was not really laughing, I was burning with shame. I had planned it all so carefully, but not remembered the dogs. The trouble was being a child, that was it. I must grow up fast.

The incident with the burned hut told me the same thing. My father was obsessed with fear about the thatch catching fire, with reason. When veld fires swept too close about the landscape the air was full of sparks and black bits of grass that sometimes had red hearts to them. At such times the labourers were summoned from the fields by the ploughshare gang, and they balanced on ladders, or on the treehouse tree, or they swarmed up poles, and petrol tins full of water were handed up, and soon the thatch was soaked and safe. Then the water cart stood empty, and, too, the great water tank that was full of rainwater, if there had been rain. One day my father called me to him and said that on no account should I play with matches, for it would be so easy to set the whole house on fire. It had not occurred to me to play with matches, but now I could think of nothing else. The occasion is clear in my mind. There he sat in his folding chair at the back of the house, watching as the thatch was patched with new long grass, yellow on the old elephant grey thatch. I had to set fire to something. I had to. And with what cleverness and cunning did I work it all out. I would set fire to the little shelter built for the dogs – not that they used it. I knew it had taken an hour or so to put up: it was of poles with a topping of thatch. To rebuild it would take only an hour. Not much money would have been wasted. When my parents went off to the front of the house and the thatchers came down from the roof, I lit a match and set fire to the dogs' shelter. The trouble was, it was built under the projecting verandah of the storehut, also thatched, and not more than a few paces from our house. I saw the little house flare up, and at once the flames swept into the storehut, which was a roaring fire, sending sparks up everywhere. This I had not foreseen. Panic. Terror. I ran away into the bush sobbing. I thought our house would be burnt down and that would be the end of our family, for we had nothing except what was in the old house.

I sat with my back to a tree, and heard the clanging of the ploughshare, heard the shouts of the men as they came running up through the bush. I could not keep away. I crept to the edge of the bush where I could see the house, its roof covered with men pouring water on it from tins, and beyond it, the burning storehut. The dogs roamed around barking, or stood with their tails between their legs. And where were the cats? Then the flames sank into the blackened mess that had been the hut, and the air was all smoke and bits of burning grass. My dress had black smudges on it. My mother came down the hill to find me: they had a good idea who was the culprit. It would be better if I told the truth, they said, and I did. The crime was so enormous no punishment they believed in could fit it. They kept saying, Did I not know how poor they were, how little they could afford the money to replace what had been in the storehut – the bins of flour and sugar, the mealiemeal, the groceries, the sacks full of chickenfeed, the waterglass eggs, the sides of bacon? It would cost them at least a hundred pounds to replace everything. But of course I knew how poor we were, money, money, money, talk of it went on always. I could not make them see that my intention had been to burn the dog's shelter, not the whole hut: they simply could not believe I had been so stupid. The punishment was that I should not be allowed to eat cake for a month. My father said irritably it was absurd, how could I understand the seriousness of what I had done, if it was matched against tea-time cake. I agreed with him about the cake: there was an anticlimax about it, something silly, and in fact long before the month was up, the punishment was forgotten. But I never forgot how my father's obsession, his terror of fire, had transferred itself to me with a few words, took a different form, possessed me. Words indeed have wings.

And again the message was that a child could not encompass all the possibilities, she was bound to be tripped up by something unforeseen. I must, I must grow up, grow up soon . . . and yet how far away it was, the condition of being grown-up and free, for I was still in the state where the end of a day could hardly be glimpsed from its start.

The main reason, the real one, why an autobiography must be untrue is the subjective experience of time. The book is written, chapter one to the end, in regular progress through the years. Even if you go in for sleights of hand like flashbacks or *Tristram Shandy*, there is no way of conveying in words the difference between child time and grown-up time – and the different pace of time in the different stages of an adult's life. A year before you are thirty is a very different year from the sixty-year-old's year.

When scientists try to get us to understand the real importance of the human race, they say something like, 'If the story of the earth is twenty-four hours long, then humanity's part in it occupies the last minute of that day.' Similarly, in the story of a life, if it is being told true to time as actually experienced, then I'd say seventy per cent of the book would take you to age ten. At eighty per cent you would have reached fifteen. At ninety-five per cent, you get to about thirty. The rest is a rush – towards eternity.

It occurs to me to wonder about the infant in the womb. The foetus repeats evolution: fish, bird, beast, then human. Does it experience the *time* of evolution? Is it possible that poor creature is submerged in near eternities? A nightmare. This is so terrible a thought it can scarcely be borne.

My brother was eight when he went to Ruzawi, the prep school modelled on English lines. Again, mother got him a bursary. Our lives at our schools could not have been more different, but the moment the holidays began, nothing had changed. We were often up with the sun and off into the bush, and by breakfast time might have travelled miles, just as my father, up with the sun, had been a couple of hours with 'the boys', or with Old Smoke, down on the lands. Sometimes we two went off for a whole day, by ourselves, with a packet of sandwiches. We drank out of rivers, but did not tell the parents this, because of their fear of bilharzia. Or Harry might go off with my father, for he helped in all kinds of ways. For one thing, he learned to drive when he was eight, standing to reach clutch and brake. I might retreat to the bedroom and read. And read. I lay on my bed and read. Often I ate while I read, usually oranges. We could order a whole sack of oranges from the Sinoia Citrus Estates for the price of a large joint of beef. This combination of reading and eating is common among child-readers. You ingest images through your eyes, calories by mouth. I was still reading my way through the books in the bookcase. The families in the District who read exchanged books, most of them war memoirs, war histories. By now my picture of the First World War was no longer only made by my father's voice telling about the Trenches. 'And then – it was just before Passchendaele . . . I felt so awful, it was just like being enveloped in a black cloud, and I wrote to the old people and said I was going to be killed. But I wasn't killed, the shrapnel got my leg.' Here were other tales of the war, notably *All Quiet on the Western Front*, fiercely approved of by my father, whose identification with the German soldiers, betrayed like the English tommies by their generals, was sometimes queried by my mother. 'But they were our enemies,' she might protest, in a troubled voice, for she was as instinctively obedient to authority as he rebelled against it. And he: 'Shouldn't have happened, any of it. The ordinary people didn't want the war, any more than we did. I never knew a tommy who was bloodthirsty. No, it was the armament manufacturers. The war suited them all right.'

There was an account of English nurses working with the wounded Serbs. (As I write this the Serbs are very far from the pathetic victims who needed the loving care of our VADs.) Another was of a Russian girl who fought with the soldiers on the German front, but they never knew she was not a man. I identified most passionately with these women, and dreamed of being a nurse, but certainly not in a London hospital, no, on savage battlefields, or retreating with the Russian soldiers when their lines collapsed and they deserted back to their villages. For years, for decades, in the West, our picture of Russia at the time of its collapse was of a vast, shambling, lice-ridden demoralized hungry country ready for revolution. There is a little book by Bulgakov, his first, *A Country Doctor's Notebook*, describing his work in a village not far from Moscow, but his battles were against peasant superstition and ignorance, and the reader only knows this was wartime when a soldier comes home from the front.

And still arrived, it seemed by every train, parcels of books from London, the children's books including many from America. The *Anne of Green Gables* series. *The Girl of the Limberlost* and its successors. *What Katy Did* and its sequels. The American experience, mostly in its Midwest, of a generation or so before me was nearer to the child in the bush than my mother's reminiscences of urban jollity, or my father's childhood with the farmers' sons in the green fields of Colchester.

The trouble with these books was their seductiveness, their ease, like sucking toffee. Sometimes I wonder about children's books. When no one suggests Dickens might be too difficult, or any adult book, you stumble and stagger through them, skipping when you have to, but soon they are your possession. Children's books make you lose the appetite for effort. I read those books for years, lulled myself with them, daydreamed over them. I lived almost entirely in daydreams. Except when with my brother in the bush, where you needed your wits.

We were not affectionate, my brother and I. English families do not teach their young loving ways. Or did not: perhaps things are different now. At the Convent I was learning the skills of the survivor, of loneliness, of exile. And if you send a little boy to boarding school at eight, then you can expect him to return with half his heart sealed off. But we were easy companions. Till I left home in my fifteenth year, whenever we were on the farm at the same time we at once took to the bush, as if we had never left. A few weeks ago – 1992 – I was in the bush not far from where I was brought up, as it happens a wet and trickling bush, for it had rained, and again, I might not ever have left it, this was where I belonged. The two children set off on deliberate excursions, perhaps to the Ayreshire Hills on bicycles, or wandering about without aim, the dogs panting beside us. We sat under a tree watching birds, ants, beetles, chameleons – always chameleons – but there are few left now. We observed, all day, the thousand dramas of the bush and at night went home to that high and airy house on its hill as if to prison. It was too confined for us. We usually took the .22, for that was a promise made to the parents. After all, there *might* be a leopard . . . but in all those years, and there were leopards in the kopjes, I saw only a tail disappearing into rocks. Like snakes, leopards know how to get out of the way. Usually Harry used the gun: there is sense in conceding priority to excellence. He was a very good shot, and this made my competence second-rate. Mostly he shot duiker, for the family to eat, or as a present to the compound. All the farming families ate as much bush meat as could be shot, to keep bills down, just as, in the compound, Africans set their dogs on duikers or larger buck, or brought them down with skilfully thrown stones, or catapults, or spears.

I might be asked by my mother to shoot her eight to ten pigeons for a stew or a pie: in certain parts of the farm there were hundreds of pigeons. Later, I shot her guineafowl, tough birds, which were first steamed and then roasted or braised. I never liked shooting buck.

Harry and I did not talk about our schools, or if we did we made giggling jokes about teachers. How funny that hellfire nun became, when Tigger described her. What good clean fun the 'ragging' that Harry had to endure as a new boy. But that was not the world we shared. We talked about the animals we lived among, the birds, our dogs. We talked about Mummy and Daddy. I was fiercely in need of support against my mother. I needed him. I had watched him refuse to be Baby, and to be ill all the time. It seemed he did not know he had done this, had got his own way simply by not noticing there was a battle. I believed he should be fighting her openly, her pressure, her insistence, her close jealous supervision, her curiosity – all the pathetic identifications of a woman whose gratification is only in her children. I believed he was being damaged by being blind. It was the principle of the thing! – that was the substance of my thought, even if I did not have the words. The truth! Facts! The right words to describe an action, a situation. He did not understand what I was talking about. Just as I rebelled by instinct, so he conformed. How can brother and sister be so different? But one asks that question only until one has had children, after which it is hard not to see the womb as a sort of nourishing conduit for the inexhaustible variety of humans, each one conceived different, an individual.

Did we talk about the Africans? – the blacks – the 'munts' – the 'kaffirs'? Not much. They were there, taken for granted. No white child learned Shona – considered to be a kaffir language. I don't propose to elaborate on white settler attitudes, there's nothing new to say about them. My *African Stories* describe the District – Southern Rhodesian farms – at that time. I would pick out *The New Man* – which describes how the white farmers operated as a community – *The Nuisance, No Witchcraft for Sale, The Old Chief Mshlanga*, which proved itself as 'politically correct' I feel (for it seems the new dogma is that whites cannot write about blacks), when a young Nigerian black sent that story in, under another title, but using his name, to a competition for short stories. There is also a tale called *Little Tembi* which ends with the words, 'What is it he was *wanting*, all this time?' What he – they – were wanting was a warm-hearted, generous, open sharing of the benefits of 'white' civilization, instead of doors shut in their faces, coldness, stinginess of the heart.

We did talk a good deal about Old Smoke, because we used to sit listening to the long, slow, conversations between him and my father, which had about them something of the philosophical hesitations and cadences of two doves in midday colloquy, on and on, then a long silence, you'd think the talk was ended but no, it starts again, There's another thing . . . Yes? . . . Another thing . . . What is it? . . . The business of the ox . . . Yes, that is so . . . What should we do? . . . I will talk with the herd-boy . . . A funny thing about mombies . . . They know what they want . . . The black cow needs some muti . . . Tell the herd-boy, Smoke . . . Yes, yes, I'll tell the herd-boy . . . Silence. A long silence. Nkosi? . . . What is it, Smoke? It's the business of Jonah's wife . . . No, come on, Smoke, I shouldn't have to deal with your marital affairs . . . That is so, but . . . Let's have it? . . . Her baby is sick and you must take it to the hospital . . . But how can I when she won't come up to the house? . . . You must tell Jonah to tell his wife to let you take the child to the hospital . . . But you should tell him, Smoke . . . I have told him, and now you must tell him . . . A long silence. Jonah is a man who has a skellum eating him from inside . . . What kind of skellum, Smoke? . . . That is why he is always so angry. It is why he hits his wife . . . Then how can we cure the skellum, Old Smoke? . . . Ah, Nkosi, and now this is a hard matter. Only the Nkoos Pezulu can cure Jonah's skellum . . .

And now, there they were, this is what they both liked, to philosophize. And the two men might sit there all morning, the doves' voices, like theirs, going on . . . on . . . on . . . Into the heat, into the bush silences, with sometimes the voices of other birds in counterpoint. All these conversations were in kitchen kaffir, the disgraceful lingo used by all the whites at that time, a mixture of Afrikaans, Shona and Ndebele, everything in the imperative, Do this, Bring this, Go there. But could have its gentler moments.

The easy good nature I and my brother enjoyed was only for when we were alone. When his schoolfellows appeared he at once became one with them. His particular friend was Dick Colborne, who lived across the District and was at Ruzawi. He might come to 'spend the day' as the custom was, riding over on his bicycle, or the Colbornes came to drink tea with us, or we went to them, and the children were always expected to go off together. Dick and my brother gave me a bad time, sneering at girls in general, pretending to throw stones or throwing stones which – since they were expert – were aimed to miss by an inch or so. They ran off into the bush to lose me. They enticed me up to high places in a kopje which I could not reach by myself, then went off, to enjoy watching me trying to get down, while they laughed raucously. They behaved towards me, a girl, just as schools of that kind prescribed, and this gang behaviour worsened through the years as they graduated to the older school. When I asked my brother before he died if he remembered what a bad time he and Dick gave me, he was surprised and upset. The point was, the betrayal: for a day, or days, for weeks or months, he and I were friends, never a suggestion of my being a girl or inferior, then the moment another boy appeared, that was it, he was my enemy. To him it was normal. None of this will come as a surprise to women married to or living with men who were at that kind of school.

There is a memory, a most particular and special memory . . . for years, knowing how little my brother and I agreed on, or shared, I would think: I wonder if he remembers that day, but as it turned out, he did not. It is an odd business, holding in your mind a memory where someone else is so alive and *there*, but then he doesn't remember it, not a thing.

We knew that buck like to spend the hot daytime hours in the shade of antheaps where there is thick cover. We went together silently, making sure we did not tread on twigs or dry leaves, to an antheap where it could be seen animals had been. Flattened grasses, a clean edge of stone where a hoof had chipped a bit off, fresh pellets of dung. We found a high place on a rock, shielded by branches. We were careful as we climbed up, since this was a place a snake would choose. We waited. It was about six in the morning, and the sun had just got up. Not easy for a nine-year-old, or ten-year-old – and we couldn't have been more – to sit absolutely still. My brother amused himself by making dove calls. They arrived in the branches just above us, and sat cocking their heads this way and that, peering at us – but not finding another dove, sped off again. We heard small sounds, and then there he was, a male koodoo, slowly picking its way up through the drapes of Christmas fern, the boulders. He stopped and looked nervously about. He knew something was wrong, turned his great spiralling horns, looked back over his shoulder where the sun was glistening on his fur. We could see liquid dark eyes, dark lashes . . . we sat scarcely breathing, stiff with the effort not to make a sound. The beast stood there, nervous, unhappy, for a good minute or two. We had never been so close to a living koodoo – dead carcases yes, brought to the house to skin and cut up. We were not making any mistakes, apart from being human children in the wrong place, and emitting, probably, signals we knew nothing about. Meanwhile the koodoo stood, and turned itself about to look down the way it had come, turned back again to face us. We were seeing how the beast experienced its life, the constant threat, always on the watch for enemies, always wary, listening, turning its head this way and that. Yet here it was, full-grown, it had survived, and it was not in danger from us that day since we did not carry a gun. A long time went by, or so it seemed, while we waited, and the beast listened and looked. Was he looking at us? Yes, but what did he see? His gaze moved on. And then – but what? – a breath of wind from a different quarter? Or, in spite of ourselves, had we made a sound? The beast turned and rushed off down the anthill, not in the panic of urgency, for we knew that crashing run, when terror seizes a buck's legs, but fast enough, to get away from this dangerous place, the anthill, where some kind of danger was, though he did not know what.

The koodoo was one day, one time, one memory, but some memories are composites of days, perhaps hundreds of days. Sound, the sounds of the time, taken for granted . . . my brother and I often went down to the place where the telephone lines marched down the ridge from the Mandora Mine, across a grass stretch to our big field and there up its edge, then up the hill, and to our house. We sat under the lines, our ears to the metal pole, and listened to the thrumming, drumming, deep-singing of the wind in the wires where we watched, as we listened, the birds, hundreds of birds, alighting, balancing, taking off again, big birds and little birds and plain birds and birds coloured like rainbows or sunsets, the most glamorous of them the rollers, mauve and grey and pink, like large kingfishers. We sat deep in the grass and listened, we sat concealed and looked. And heard, as our ears opened to sound, since we were silent so as not to frighten birds or any beasts around, how the farm's activities, the life on the farm, told us what went on everywhere. On the big land below the house a man was ploughing and shouting admonitions to the oxen who dragged the ploughshares squeaking through the heavy red soil. On the track to the station the loaded wagon with its team of sixteen oxen creaked and groaned while the driver cracked his whip that reached to the horns of the leader oxen and yelled on a note only they understood, for they put their shoulders into the pulling and lowed, again, again, to each other – to him? – a protest, there was no doubt of that. When the whip cracked, it was like lightning, as if the air were being torn, but the line of the whip snaked over the oxen's heads and did not touch them. The driver's voice, screaming, shouting, or conversational, like the oxen's with each other, lessened as they went off into the trees towards Banket station. On the telephone wires the birds twittered and sang, sometimes it seemed in competition with the droning metal poles, and from the far trees came the clinking of hidden guineafowl flocks. The wind sang not only in the wires, but through the grasses, and if it was blowing strongly, made the wires vibrate and twang, and then the flocks of birds took off into the sky, their wings fluttering or shrilling, and they sped off to trees, or came circling back to try

ABOVE: My mother, Emily Maude McVeagh, as a girl.

LEFT: My father, Alfred Cook Tayler, in 1915.

BELOW: My father, after he had had his leg cut off, in his room at the old Royal Free Hospital, with my mother, Sister McVeagh.

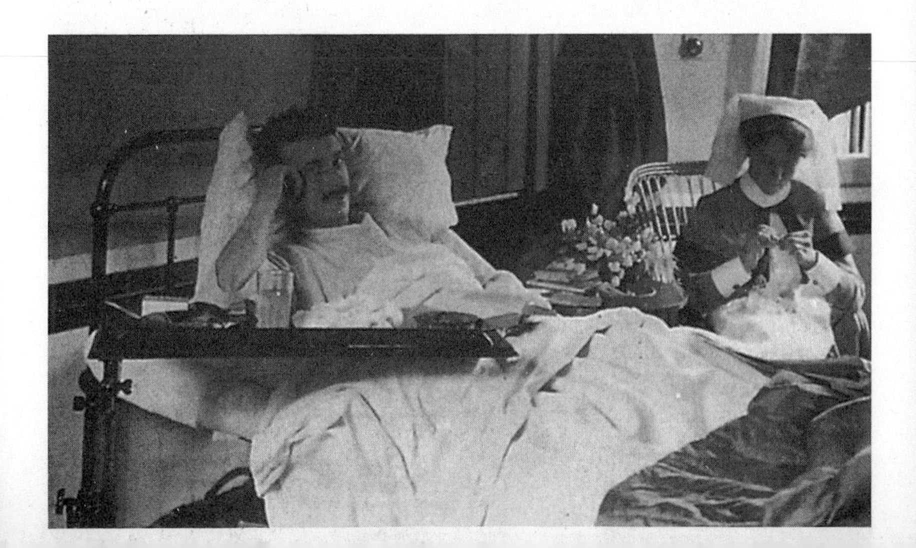

ABOVE: The house where I was born, in Kermanshah. The first and second windows from the right are those of the room I was born in.

INSET: The view from the verandah of the house in Kermanshah.

BELOW: A picnic on Boxing Day, 1919. The relief of the man on horseback with a spear depicts Khosru, the king and warrior. My father is on the extreme right.

RIGHT: My father and mother with me, aged three months.

BELOW: 'Old Marta', the Syrian nurse, with my brother and me, while Gerald Nelligan sits on the donkey.

Here I am with my teddy, aged three and three months.

Fancy dress Christmas party 1923, in Tehran. My little brother is the rabbit. I am Bo-peep.

Grandfather McVeagh and Stepgrandmother McVeagh, who was Maria Martyn, the daughter of a dissenting Minister. She was described by my mother as cold, correct and dutiful.

Stepgrandmother McVeagh with my brother Harry, and Grandfather Tayler with me. This was during the family's six months' sojourn in England, between Persia and Southern Rhodesia. Grandfather Tayler married Marian Wolf, aged 37, in this year, 1924.

Making the pole framework of the house.

ABOVE: The house, just before the thatching is finished. The storehouse, which I burned down, because of an error of judgement, is on the right. The watercart is just visible under the roof of the kitchen on the left. You can also see my brother and some of the men working on the house.

RIGHT: The newly completed house, before verandahs, porches and pergolas were built around it.

I am with my calf, Demi. He was left an orphan.

Ploughing a new-stumped field for the first time. A span of oxen for heavy work could be as many as sixteen, or eighteen.

My father, my brother with his dog, and I.

My father, my mother, my brother and I.

LEFT: About nine years old, with my mother, who is looking ill and unhappy. She had put a lot of face powder and rouge on to be photographed.

BELOW: With my dog, Lion.

This is Granny Fisher's house in the Vumba. No one now seems to know where it was. Decades later, my son John Wisdom lived in a house very like this one.

ABOVE: In Umtali.

RIGHT: On the top of a mountain near Umtali (now Mutare again), being blown off my feet. I was thirteen.

BELOW: Aged thirteen, I am in the navy blue uniform of the Girls' High School. The car is, I think, an Overland.

WISDOM — TAYLER. — At Salisbury, on Thursday, the 6th April, 1939, Doris May, only daughter of Capt. A. C. Tayler and Mrs. Tayler, of "Kermanshah," Banket, to Frank Charles, second son of Mrs. E. Wisdom, of Salisbury, and the late G. E. Wisdom. 5270xf14

ABOVE: Frank Wisdom and I, just married.

RIGHT: With the two babes, John and Jean, in 1942.

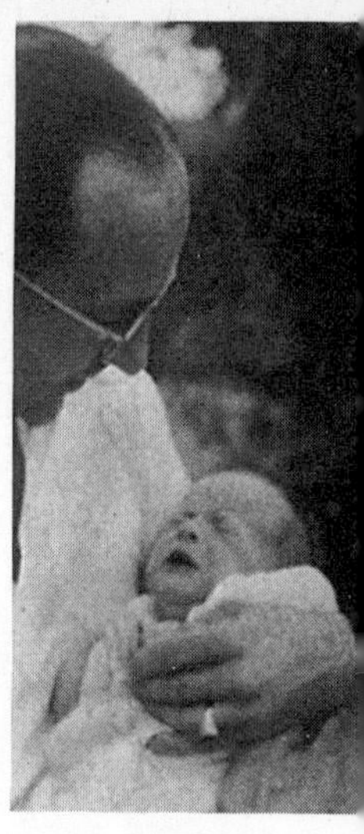

ABOVE LEFT: In 1946, with baby Peter.

ABOVE RIGHT: Gottfried Lessing with baby Peter.

LEFT: In 1949, just before I left for London.

again. Dogs barked from the compound. Our dogs lay in the grass near us, where ticks crawled onto them – we could see the ticks leaving the grass for the rough fur – and they panted, and sometimes lifted their muzzles to yawn with a sound like a whine. Their tails went flop flop, or their teeth clicked and chittered through their fur after the ticks and fleas. A car came slowly along the track to the station, a sound unlike any other, in its steady grill and grind. From the gangs of men at work across the track hoeing peanuts, came singing, a work song, that went on, and on, with a grunting refrain that made the men laugh and scuffle their hoes more vigorously among the weeds. From the compound the voices of women calling to each other, high and clear, and then a laugh. From the house on the hill the clang of the ploughshare, for it was lunch hour. The working men flung down their hoes and went off talking and laughing. The crickets sang in the grasses, and cicadas in the trees just across the track, and from somewhere in the trees came the sound of what we called a hand-piano, the mbira, for a man there was playing to himself as he walked along. Far far away, over the Huniyani mountains where clouds were banking grey and purple half-way up the sky came a small growl of thunder – but it was still midday, probably the rain would come rushing and swishing over the bush tonight. Meantime the heat shimmered on the track and over the iron roofs of the barn, which cracked and crackled.

A hawk detached itself from the circling company of hawks far overhead, and came floating down to look at something in the grass twenty yards away. The dogs raised their heads, the hawk saw them, swerved, and came speeding past under the wires, the wind shrilling in its wings. The dogs barked at it, for form's sake, and let their heads drop back on their paws.

Doves cooed from the big vlei, hundreds of them lived down there, a somnolent accompaniment to all the rest.

The gong had rung, it would be lunch soon, we wandered back to the house, birds singing around us as we went, the telephone wires humming.

In the late 1950s the BBC asked me for a radio play and I suggested one based on the sounds of the bush – on the farm, at the station at Banket. There the sounds of the train would make a background, a frame for the story, the shunting, the huffing and puffing, the long screams of greeting and farewell, the squeal of the brakes. The wagons creaked into the station, the oxen stood patient, lowing their thirst, and then they were outspanned two by two and taken to the troughs at the water tanks, where water dripped and splashed from their muzzles. From the verandahs of Dardagan's store and the kaffir store came the sounds of the mbira, and, too, the tinny wind-up gramophone playing Caruso singing 'The Volga Boatmen', or 'The Waltz of the Toreadors'. The story would be told through and by sound, about people for whom the train sounds were news of the outside world, the farm sounds and the station sounds the breath of their lives. Ideas for programmes, stories, plays, are always being born, and then die. This one I regret. We might now have a record of the sounds made by an Africa that no longer exists.

8

VERY EARLY, LONG BEFORE I can remember, began two themes or streams that have dominated my childhood.

One was the world of dreams, where I have always been at home. Until I was ten or so, they were most often nightmares. Nothing unusual in that, for small children. But I had rituals to avoid them or make them harmless. I had learned that often a nightmare has in it a germ of something everyday, a word, a sentence, a sound, a smell. If you allowed this excitatory moment – or substance – to creep into your sleep unexamined, then you were helpless. But you could disarm these enemies. Every night before going to sleep I went over the incidents of the day, those that seemed to have the stuff of potential nightmares. I ran emotion-loaded incidents again and again in my head, till they seemed tame, harmless. I think this is the technique used to accustom people to a fear of spiders or snakes. Spiders were sometimes in my bad dreams, but then they were in my life, everywhere and always. I would scream for my father to come and lift out of the room some black horned monster, or a great hunting spider, mottled grey and pale, pretending to be plaster, crouched, I was sure, to spring at me. My father always came, grumbling, demanding that I should grow up. But spiders and snakes were the least of it. My father was the worst, and I knew why. It was the 'tickling' of my earlier years. In these dreams great hands pressed and squeezed my ribs and I screamed and wriggled, and heartless brutal faces loomed down at me, not necessarily my father's face, for the Dream-worker took for its purpose any old material lying around – Mr Larter's face, or Mr Macdonald's from over the ridge. There was also the dream of the pit, which I cannot match with a memory, and a boy with a stick. The other nightmares were what everyone has, the staircase that levels out and lets you slide back, or ends suddenly in air, so that you step into emptiness – the great stone steps of the little ones' dormitory provided material for this dream.

But it was not the stories of these dreams – the plots – that interest me, rather how I learned to sanitize them, take the danger out of them, running through the events of a day in this monitoring way over and over, reducing the interminable day to something like a picture storybook whose pages you flick through faster and faster. And this not only sanitizes, but shortens. I was learning how to make short work of Time – no, of course the days still crawled, would for some years yet, but I could also reduce a day to a few incidents. Like everyone I have had nightmares occasionally throughout my life, but after I was grown up, very few. The worst years were when I was seven, eight, nine, ten. For all I know, earlier.

The other theme, or motif, or – no, these words have in them a sense of the continuous and what I am marking here are the special moments, when one is alive, and noticing, as if injected unexpectedly with some substance whose gift is that you should see clearly.

One of these 'moments' will have to do for them all.

I walk up out of the bush where I have been by myself, and stop when I see my parents sitting side by side, in two chairs, in front of the house. For some reason, perhaps because of my thoughts in the bush, I see them very clearly, but from a child's view, two old people, grey and tired. They are not yet fifty. Both of these old faces are anxious, tense, full of worry, almost certainly about money. They sit in clouds of cigarette smoke, and they draw in smoke and let it out slowly as if every breath is narcotic. There they are, together, *stuck together*, held there by poverty and – much worse – secret and inadmissible needs that come from deep in their two so different histories. They seem to me intolerable, pathetic, unbearable, it is their helplessness that I can't bear. I stand there, a fierce unforgiving adamant child, saying to myself: I won't. I will not. I will not be like that. I am never going to be like them. I shall never sit drawing disgusting smoke into my lungs holding cigarettes with fingers stained orange. Remember this moment. Remember it always. Don't let yourself forget it. *Don't be like them.*

Meaning, never let yourself be trapped. In other words, I was rejecting the human condition, which is to be trapped by circumstances.

There were moments like this all through my childhood, the most powerful influences of my life.

I will not. I will *not.*

And now, themes from the Convent. The first has to be food, for how can you write about boarding school and not describe what obsesses children? The food was in fact better than most school food, but for us it was horrible, heavy and greasy and full of unknown tastes. It was the German peasant food of that time. We ate thick floury soups flavoured with caraway seeds, and were often sick afterwards. We ate slabs of meat always crusted with fatty crumbs, and rich stews with dumplings. When the kitchen nuns made us treats for the many feast days, they were nearly always pancakes soaked in some kind of oil, and inside were little rolls of greasy paper hiding tiny holy objects, like rosaries and crosses. We never got fruit or salads and the vegetables were overcooked potatoes and cabbage, also tasting of caraway. According to current ideas, this diet represents the worst one possible. We survived. Meanwhile the nuns scolded us at every meal for our wastefulness, our heedlessness of God's good gifts. Now I know their often tearful voices were because of what that food represented to them. All had come from hungry families, and by the time I left the Convent, the Slump had made conditions in Germany, already very bad, worse. Everywhere in Germany were soup kitchens, food lines, riots about food, and here were these ungrateful and wicked little children . . . Tigger entertained her Mummy and Daddy to the point where they wept with laughter, with tales about this food.

In those years the Talkies arrived, and while the nuns hated doing it, parents insisted their children must have the best of everything, and so, in long crocodiles, in our brown alpaca tunics and yellow shirts, in our panama hats with the Convent yellow and brown ribbons, off we went to the Bioscope. We saw *Rio Rita.* We saw Al Jolson's *Sonny Boy.* From time to time I read how experts say children are not affected by what they see on screen. After *Rio Rita*, dozens of little girls were lost in amorous fantasy with John Boles and his moustache in their arms. After Al Jolson, sick and dying babies added an intense melancholy pleasure to the nuns' funerals. There were not then violent films of the kind children are brought up on now, but if we had seen anything like them doubtless we would have fantasized killing the hellfire nun. Given the opportunity – and the information on how to go about it – I can easily believe that I would have killed that woman, and with the feeling not that she 'deserved it', but that she already embodied cruelty and killing. I think it was Orwell who said there is no fool like an intellectual, meaning exactly this, a kind of clever stupidity, bred out of a line of logic in the head, nothing to do with experience.

Jazz arrived too. Jazz on a little portable wind-up gramophone set on the refectory table, is where my exposure to the second most powerful influence of our time began.

Blue skies smiling at me,
Nothing but blue skies, do I see.

So sad is the music that I thought it was Grey skies, for years.

Red sails in the sunset
Red sails on the sea

Sad, sad, sad.

Out of all this emotion washing around the refectory garden, I wrote a Shakespearian one-act play, full of warring kings and queens, which the nuns said was too old for me, and, too, a very small piece, no more than a paragraph, about Echo, who, when hearers went to investigate, was 'only a tired boy lying on the rocks'. It is the pleasurably intense sadness that I remember, the self-consciousness of that 'tired boy'. My mother said it was too old for me. But, one has to ask, if potted feudal epics and the 'tired boy' were too old, then why not Bebe Daniels in the arms of John Boles, and those yards-long tears on Al Jolson's cheeks?

And now I was ten years old and a big girl and the stage was set for all kinds of crises, not least an exam for which I was specially coached, having missed so much school. I was in the big girls' dormitory, and left behind that great room with its twenty-four beds and the bloody pictures. This was the old part of the convent with all the associations of the antique – at least thirty years old – and ghosts were the least of it. There were smallish rooms, and each bed was curtained, which meant that, like the nuns' quarters, long white presences stood by each bed. These curtains of white cotton were supposed to be pulled around every bed after we got into them. If we were Catholics, we should be sleeping with our hands crossed on our chests, and on our lips the words 'Holy

Mother, if it is thy will I die in my sleep, then take my soul . . .' and so on.

Long before we actually reached the dormitory, we were primed for fear. We set the stage for every kind of ghost, dead nuns, dead fellow pupils, and . . . what had the Ides of March to do with a convent in the middle of Africa? But it was the words, the squeaking and gibbering in shrouds, they were irresistible. It was in that place that for the first and last time in my life I walked in my sleep. The washrooms were down a steep flight from the dormitory rooms, and I woke as I tried to climb on to the basins, thinking they were a bed. I fled up the stairs in the dark into the darkness of the dormitory that glimmered with white shapes.

My mother was writing by every mail that I had to do well in the exam – the equivalent of the eleven-plus – and wrote and telephoned the tutoring woman. 'Tigger' was in control, and I clowned and was pert and 'clever' with Mrs (I think) Baxter. Her physical type was one that recurs in my life: she was a wholesome, blonde freckled woman, who I suppose must remind me of some early warmth, probably the Danish woman, Mrs Taylor. To my mother's letters, almost incoherent with anxiety, I returned jolly 'clever' letters.

And then, deliverance, we got ringworm, all of us. Our limbs were covered with itching inflamed circles. Then, we got lice in our hair – every one of the girls in the 'big' dormitory. In came the parents from all over Mashonaland, and girls were removed from school, I among them. Lice in her daughter's hair! – for my mother it was the ultimate in degradation. As a nurse she had had to deal with slum children who had lice in their hair. She knew that poverty and dirt caused lice. My scalp crawled with lice. When I looked at my hair in the mirror I could see it moving, and in the morning lice crawled on the pillow.

My lice were cured with paraffin, as my mother had used it in the slums of Whitechapel. Hair was combed through with paraffin, then held in a towel tied tight, and in no time the lice had gone. Paraffin makes the hair thick and shiny, too.

This crisis of dirt and insects came only just after a rather more serious one. I had a sudden conversion to Roman Catholicism. This was regarded by all the Protestant girls as bound to happen at some point: we knew that the English Church and its wholesome ways were no match for the ghostly nuns, the holy water, the bells, the

Virgin Mary, and the visits of the priests from St George's, the companion boys' school, who reduced the nuns to fluttering little girls, as unctuous male hands covered with rings were held out to be kissed. The nuns curtseyed and blushed, and we little girls – the Protestants too, who had begged to be allowed to join in – blushed and giggled afterwards.

My submission was sudden and total. Rather, it could be made to seem sudden, if you left out of account the ring-kissings and the glamour of incense and high chanting voices. And the loving kindness of Sister Antonia, for now it is clear that I was not moving from the English Church to the Roman Catholics, but from Sister Antonia to the Virgin. I spent every spare minute in the Roman Catholic church, just down the road, which then seemed to me a tall shadowy place full of mystery but, above all, there was a statue of the Virgin, and there I knelt beseeching her for recognition of my love for her, and trying to persuade myself I saw her smile. I did make myself believe it – almost. In my blazer pocket was a medicine bottle full of holy water, and I practised crossing myself and curtseying every time I passed the Sacred Heart or any picture or statue of the Virgin. The fever of this time was like being in love, a condition I knew well, but it was heaven that swam about me, making me vague and stupid. The nuns saw the holy water in my pocket, and were troubled. Please would I tell my parents, they begged, that no pressure had been put on me? This seemed irrelevant. If the Truth encompassed us, manifest in statues, pictures, holy water, then how could they say no pressure had been put?

I went home for the holidays and my mother saw the holy water and the rosary under my pillow and exploded into reproaches. This marked the beginning of a rejection of my mother, like a slamming of a door. She called me out to sit with her in front of the house, set her chair opposite mine, and began on a history of the crimes of Roman Catholicism. The Inquisition figured as the chief wrong, but others were cited, for instance the way Catholic missionaries converted the Africans they taught to their religion. By now I was listening, full of cold loathing for what I saw as illogic masquerading as virtue. I lost religion in a breath; Heaven fled from me on the wings of Reason, when I said that everything she said was true of the Protestants, who had burned Catholics at the stake, just as Catholics had burned Protestants. The brown wrapping-paper-covered books in the Convent library held not only cautionary tales about the dangers of black magic, but histories of Protestant wrongdoing. I had become an atheist, but what I really did was to put an end to the conflict of being a Protestant in a Catholic school, the interminable anxious queries from both parents about whether I was being 'got' by the RCs. And I could no longer stand my anguishing love for the Virgin, with her sweet and indifferent smile. Having announced I was an atheist, at once I found a thousand allies, for atheism then was all stern righteousness, not to say self-righteousness, just like religion. I was the heir to all the virtues of the Enlightenment – though I did not know that then – but just as if I did know it, began with a good conscience to despise the religious for weakness of mind and moral cowardice.

They would have removed me from the Convent then and there, but it was a question of the bursary and of the exam, and so back I went, only to be rescued by lice and the ringworm, for these were stronger arguments than the threat to my soul.

Goodbye Convent! Goodbye, nuns. Goodbye to that exam which I would not sit and could not pass. Goodbye, goodbye. The four long years rolled themselves up and were put on a shelf in my mind labelled 'Convent', a place not to be visited for years, except in dreams, for there waited only grief, pain, and – always – uneasy incredulity that those grey eternities could be described by the words, four years.

How neat, how convenient, if I could now say, That was it, so ended warm feelings for my parents, as I entered a premature adolescence. What happened then was dysentery, and those who have never had it will not believe the intensity of the intestinal gripes, like jabbing beaks or hands wringing. My little brother had it, and lay cowed and shrieking. My father had it, stoically, as an old soldier should. I had it, and so did my mother, but she was nursing us all and kept quiet about what she suffered. As I became convalescent, and weepy and enfeebled, I begged her, 'Come and cuddle me, come and cuddle me.' She had held the writhing child close and kindly, but now the exhausted woman was summoned every few minutes. She laid herself carefully down, put out an arm so I might put my head on it, and said, 'Dear, oh dear, what a fuss.' And fell asleep. I lay hardly breathing, thinking, Perhaps when she wakes . . . but, being only a few inches from her face, saw it tight with pain and worry. And then, when I next coaxed her, 'Come and cuddle me, come and cuddle me', and heard her wail, mimicking me in gentle mockery, 'Oh, oh, come and cuddle me,' Tigger echoed her, teasing the sick child and her importunities. 'Come and cuddle me' became part of the repertory of family jokes, and saved me from the embarrassment of remembering that I had so recently begged for love.

And now ahead of me stretched the long months until the High School, bush-months, farm-months – when I learned something new every day.

The best was looking after the sitting hens. I was shown how to choose only the largest eggs, assemble them in a box filled with straw in a safe place, wait until a hen fell broody. In hot weather there was a broody hen nearly every day. But you could induce broodiness with a little spoon of sherry, while the chosen hen was tempted up to the waiting eggs. When the hen was ready, she marched into the nest, her claws settling gently among the eggs. She dipped her beak into the tin of water set in a corner of the big box. She fluffed her feathers, and was at once ready to peck at your enquiring hand. Several times a day I checked her: did she have water, did she seem content? What was time like for a hen squatting there all day, all night, her fierce eyes on guard? Once a day she was lifted off the eggs, squawking, and encouraged to eat thrown-down grain, stretch her legs, empty great gobs of hen-dirt from her feathery backside. Meanwhile I would be sprinkling the eggs with tepid water, to make them easier for the chicks to break. She hustled back, pecking at me as she came. And so, day after day, while the eggs got heavier in the hand. When she turned the eggs, she might roll with her beak an unsatisfactory egg to the edge of the nest, and then I took it away and flung it into the bush where it imploded with the dull thud of a rotten egg. Fourteen eggs, fifteen eggs under the big Rhode Island hens, more under a black Australorp whose downy under-caverns seem spacious enough for as many eggs as you slid in there. Sometimes there was a wild stir and cackle from the back of the house, and I went running to find one of the dogs had gone too close, or a hawk sat on a nearby tree. It was not unknown for a rat to come sneaking around in the dark, or even a snake. Once a hen was found stretched dead, the eggs cold and scattered. A snake had taken off two or three. But the dogs who prowled all night, and the cats who seemed to know everything that went on, were a good warning system.

And then at last it was eighteen, nineteen, twenty days . . . I sat holding a hot egg in both my hands looking to see if there were signs of a chip, or held it to my ear. You could hear the chick turning and shifting, and then there was the minute crumbling blemish on the shell, and it became a tiny star, and the chick's beak, with its pallid hardened tip, showed in the hole. And, soon, the egg fell into two halves, and out flopped the pathetic ugly wet little chick, with a look about it of lizard – the sloping head, the big helpless claws – but within a few minutes it had dried, it had achieved its status of being adorable, nestling in the outer fence of its mother's feathers, cheep, cheep, while under the hen the still unhatched chicks knocked and tumbled about in their shells. Being adorable was a condition it kept for only a day or so, for it would be stringy and lumpy for the weeks of its growing up, but then become a handsome beast, like its mother, destined for a life of egg-laying and sitting, or, if a cock, with less good prospects, for most soon ended in the pot. Even a good-sized flock of fowls needed only a couple of cocks.

But what you were learning was the precise timings of nature. If the period of development was checked, if for some reason the eggs got chilled, then that was it, no chicks, only rotten eggs: it had to be three weeks, it had to be twenty-one days. Like cooking a cake, timing was everything.

Most precise and mysterious timings regulated the many existences parallel to ours in the bush. A moth came in from the dark every night soon after the sun went down and settled on the wire gauze over a window. I dipped my finger in golden syrup and put my fingers through a hole in the gauze and the moth crawled onto it, clinging with its soft legs, and feasted there – for minutes, for half an hour. A big soft, brown wonderful moth with antennae like feathers. Then it took off back into its life in the bush. Every night, a visitant that made my heart beat faster with love and gratitude. I used to wait for it, watching for the blur of wings in the yellow light that fell from the window. And then – that was it, no moth; its time was up for one reason or another.

Or we all sat out in front of the house, in the dark, beyond the reach of the lamplight, watching the life of the stars, waiting to catch the moment when falling stars streaked across the sky. Then the bush seemed to enclose us with its life, and the house shrank back and down, becoming small under that great sky. From the compound came the sound of drums and sometimes of voices. If it was the dry season, fires burned on the hills and ridges, crawling across the dark in long bright chains. If it was the wet season, the sky might blaze or flicker with lightning. There are single memories – koodoo, moth, or the cobra that crawled into the dining room and had to be hunted out from behind a bookcase – but there were years and years of nights. 'Shall we take our chairs out?' my father would enquire. 'Well, don't forget they have to go to bed,' said my mother. But once we were out there under the stars and the moon, the claims of bed lost force. My brother might fall asleep in his deckchair and have to be awakened, and go off rubbing his eyes and yawning. Sometimes when I had been sent off to bed, I crept around the side of the house in the dark, my hands on the dogs' heads, telling them to be quiet, and I stood looking at my parents in their low chairs, watching the night and smoking, their cigarettes glowing red and fading, burning up again. Their voices were kept low because of the bush and its animals, because of the nightjars and the owls. 'Lord, old thing,' my father might exclaim, in a passionate incredulous voice, 'this kind of thing, well, it makes everything worthwhile, doesn't it?'

'I suppose so,' she would agree, but sigh.

I learned to drive during those holidays, for my father was never happy driving with his awkward leg, and my brother was not always there. I was driving by eleven. All the boys on the farms drove but the girls did not. A boy as young as ten or eleven might drive to another farm with a plough part, a sack of meal, or a gift of fruit, and if a policeman from the camp at Banket happened to be on the road on his horse or a motorbike, then he took no notice. Everyone knew that the farmers' sons were often as good as the regular assistants to their fathers.

My father had still not yet become ill, and hope was keeping the family buoyant. If not maize or tobacco – for tobacco had turned out to be less than it promised – then prices might flare for one of the crops we grew, as they had in the War for maize. Sunflowers and peanuts, millet and cotton, surely one of these would come up trumps? Meanwhile my father grumbled that he was growing food for the birds and animals. When the sunflower heads were ripe, the big birds scrabbled out the black shiny seeds with their claws, and every one of the enormous tilted plates showed gaps in the tight rows of seeds. The little birds crowded up and down the millet plants. Pigs dug up the peanuts. But never mind, something would turn up. Sudden lucky chances, windfalls, the sweepstake, or a pan that showed an inch or so of trailing gold grains instead of the one or two we regularly saw; the gifts of fortune did not entirely dominate the family talk, though they soon would. I had already learned not to listen, and my brother didn't either. We might smile at each other when the talk was of 'getting Home to England' when a real life would begin. Meanwhile the Slump was getting worse there, and deepening here too, and desperately thin anxious young white men kept turning up on the farms, walking, to ask for jobs, or in the post arrived letters from women begging to be allowed to look after children or come for a week to do the sewing. These visitors and letters were all out of the 'toe-rag' level of poverty. How could we go back to England in the middle of the Slump? It was absurd, and we knew it. But that need not stop one daydreaming. They daydreamed aloud, about Home, and Harry and I smiled and escaped into the bush.

Poverty is supposed to mean that you cannot get credit. But all the farmers lived on loans from the Land Bank. When we were in Salisbury, we drove to the Land Bank where we children squirmed on the hot back seat for hours, for ever, while the parents went in to negotiate yet another loan. We continued to farm only because these loans were renewed, to be reduced every season, and then soon reached their former level. White farmers were, more than any other kind of settler, White Civilization, and they had to be very bad farmers not to get loans. Few of the settlers had capital. My father said his real capital had been the Trenches and his lost leg: the small amount of capital, £1,000, he brought with him would not have lasted long. But 'they' would think twice, he said, before making a cripple from the war a bankrupt.

Being poor meant that we were always in debt to Dardagan, at the store in Banket. Meant my father protesting when my mother bought something unnecessary, like English jam. Meant that at Christmas our 'present' from Dardagan was a tin of stale biscuits and a bottle of sherry, while the rich farmers like the Larters got a bottle of whisky, boxes of chocolates. Being poor meant my mother made everything we wore, including my father's bush shirts, and we sent for shoes from catalogues from Home, Lilley & Skinner, Dolcis, while the children wore veldschoen. Being poor meant that medicines (and the household consumed so many) were brought from a catalogue where common medicines like aspirin, or the then equivalent of vitamins, like Phospherine or Parish's Chemical Food, were made up under numbers, not the brand names that cost ten times as much. Being poor – well, it was the climate you lived in, but I already knew we were lucky. At the Convent, when I went out for a day with a friend, Mona, with dubious permission from my parents, her father, a sad blustering guilty man, kept leaving us outside the bars where he went for yet another drink, came back and said, 'Oh there you are, my girlies,' and went off again to return unsteady, jovial – and miserable. That was real poverty and I knew it. Mona kept saying he had lost his job, and she wished he wouldn't drink because they couldn't pay the food bills.

Poverty was the rapidly worsening shabbiness of the house, where the piano, the Persian rugs, the copper jug and basin, the silver, the pictures, already seemed to belong to another house, another world.

In the early 1930s all the talk on the farm verandahs was of the Slump, falling prices, bad times, bankruptcies. But not one of the farmers near us went bankrupt. The grandsons of some of them farm there now and in the same houses, though much extended and smartened, and they are known as 'the high-tech' farmers. Banket is 'the high-tech district'.

Throughout our childhood Harry and I went visiting around the District, walking or on our bicycles. The Matthews were the nearest, their house visible from ours. They were Big Bob Matthews, and Little Mrs Matthews, and Bobby Matthews, much older than us and already during the holidays at work with his father. In farming districts, where you do not choose who lives on the near farms, you cannot afford to dislike this one or that one, you have to get on, your lives closely meshed with give and take. But the fact was, my parents did not like Big Bob. Nor did we, the children: we liked Little Mrs Matthews, who would telephone to entice us to tea, a Scottish tea, with at least twenty different kinds of cakes, biscuits, scones, flapjacks, shortbreads, enough to feed a fête. She was always alone, her men down on the lands. Alone, and lonely – as so many farmers' wives were. But we did not see this, thought her begging us to stay a little longer, why must you go so soon, come tomorrow – was just politeness. Her soft Scots voice, chatting, chatting about a distant girlhood in Scotland, how she met

Big Bob, a policeman on the beat, how she married him – we listened to it, and I can hear it now, the charm of it, but what she spoke of was too remote from us. And soon, we had excuses when she rang us up.

The Whiteheads had long since gone from Mandora Mine. Where? Who knew? Who cared? People came and went. There was a new manager at the mine, which was a place Harry and I visited to see the mine stamps at work, the ore being sluiced through the slimes. A hot, dusty, acid-smelling place, with a mine office where a shelf held exemplary lumps of rock from reefs all over the District. Some of them Harry and I took there, to be told, 'No, that's just fool's gold, it's just pyrites.' Or, 'Not a bad bit of rock that. That's a grain of real gold there. Tell your Dad to take samples from that one.'

A short way past the mine was the MacDonalds' house, where there were three children, one of them Norah. We did girls' things together, tried on each other's dresses, and taught each other a new dish.

Old man McAuley ran the Ayreshire Mine. I put him in a story called *The Antheap*. About him nothing is made up, but the manager and his wife were brought in from another mine, another district, a story told to me of a drunken small-worker and the manager's wife bossing him around. Mr McAuley was lonely, though he had coloured children in the compound. He loved having us there. His house was a two-roomed shack under an iron roof and it was horribly hot. He lived on greasy boiled beef and potatoes. He was a very rich man. Beside his bed was *All Quiet on the Western Front* with the dirty bits underlined. My father disapproved of Mr McAuley utterly, because he cheated his workers, employed criminals from everywhere, and would not own his bastard children. This did not mean the two men did not sit for hours talking about the Trenches.

The other very rich man of the District was Mr Muirhead, an Australian plumber who came to Southern Rhodesia because he was a wandering man by nature, and bought the land near Salisbury which later found itself in the path of the Salisbury–Umtali railway. He bought a farm in Banket, and employed the MacDonalds to run it. He was a vegetarian, ate only bread and boiled vegetables. He read the Bible and health magazines. Aged ninety, he would climb on to the corrugated iron roof of his little brick shack to mend a gutter. He also got malaria, because he did not believe in taking medicines, and my mother walked across the bush to nurse him. Where she walked lived a pair of yellow cobras, for Harry and I had often seen them there. She put my father's army revolver, with which she despatched inconvenient snakes, into her cretonne carry-all with the tortoiseshell handles, and continued to walk through that bush. She found unopened envelopes with cheques in them in the wastepaper basket and reproved him, 'If you don't need this money, Mr Muirhead, I can give you a list of charities.' 'Filthy lucre, Mistress Tayler,' he shouted at her, 'it's filthy lucre, the Good Book says so.' It was Mrs MacDonald's task to keep an eye on Mr Muirhead, but she was lazy, and anyway the family grumbled day and night about the salary Mr Muirhead paid them. My mother suggested to Mrs MacDonald that if she were nice to Mr Muirhead he might be nice to her. But Mrs MacDonald was not going to lower herself, emptying his slops, not on what he paid them. Mr Muirhead did not employ a 'boy' – he believed in being independent. It was a deadlock. Meanwhile the MacDonalds said my mother fussed over Mr Muirhead because she was hoping to be remembered in the will. Why else would she be doing it? Mr Muirhead bought one of the first Packards, and he went roaring into Salisbury, a tiny gnome of a man, peering through the steering wheel. Sometimes my father asked for a lift, and remonstrated as the old man skidded at eighty miles an hour over the rough dirt tracks. 'I shall go to the good God when it is my time to go,' yelled Mr Muirhead, who was deaf. 'But is your time the same as my time?' yelled back my father.

Unlike Mr McAuley, Mr Muirhead did not much like us children turning up, waiting to be invited to sit at his dirty kitchen table to drink tea and chat. His little grey eyes peered suspiciously at us. 'What did Mistress Tayler send me, show me, show me!' For she sent him vegetables from her exuberant garden, and even flowers. But Mr Muirhead could not see the point of flowers.

The Dodds' farm was near Banket station, and there were four sons. Mrs Dodd was a little friendly woman, like Mrs Matthews, but Mr Dodd was a lean, scholarly looking benevolent man who went about his farm with his four boys, all his assistants from the age of six or so. Mrs Dodd had longed for a girl, but Mr Dodd had said at the fourth, 'Now, my dear, that's enough, we can't afford it.' Harry at once became part of an all-male gang when we visited together, but sometimes I went by myself, and sat all day with Mrs Dodd on the verandah, while she asked me about my doings, and smiled gently at me and sighed, and sighed – and pressed on me another little jam tart, just one, just another piece of cake.

The Colbornes were also near the station. Mrs Colborne was a lean, severe, dry, witty woman who, my father complained, put the fear of God into him. She certainly frightened her husband, a fat and foolish fellow who joked as a defence. My mother and Mrs Colborne were friends until my mother died, two clever women whose lives did not permit them to use their talents. Anne Colborne should have been my friend as Dick was my brother's, but she was a polite and subdued little girl. Probably the sad little girl who lived well hidden by the mask of 'Tigger', and the frightened little girl I am pretty sure lived behind all those obedient ways would have got on, if they had ever met.

We and the Colbornes exchanged books and magazines.

Two families were far enough away to need the ritual, spending the day. 'Come and spend the day and stay for supper.'

The Livingstones were Captain Livingstone, with his wooden leg, Mrs Livingstone, the Captain's wife, and Master Livingstone, the Captain's son. They lived in a big stone house overlooking a wildly beautiful view of the Umvukwes with its crystalline lights and distances. The three subdued, polite people, each lived inside an invisible membrane containing grey thin air, which dimmed and slowed them. English. These were the English 'nice' people my mother yearned for, but in fact, she found them all too much of a good thing. I used the man and the woman, not the son, in a story called *The De Wets Come to Kloof Grange*.

The Shattocks lived in a house consisting of half a dozen rondaavels laced together with bougainvillea-covered pergolas, on the edge of a deep-running rocky river where, when it rained upstream, the water came thundering down in a wall ten feet high. All through the house were pictures of the baby who had been drowned. The family had much more than its share of misfortune. There were two boys, Jim and Nick. Nick was, as his parents kept telling him, as beautiful as an angel, and just like the drowned baby. Later he was killed in the war. Leonard Shattock, a tall thin burned-brown man, all restless energy, died suddenly of bilharzia. Nancy Shattock went to England to become a nun in a contemplative order. With

Jim I was wildly in love for two years or so. Few experiences in my life have been more erotic than when, we four children having spent a morning climbing over and around the kopjes near the river, Jim stood directly in front of me, his large grey eyes looking steadily into mine, and gently, deliberately, and slowly, picked from the muslin over my still new and self-conscious left breast thick clusters of black-jack seeds, while I made myself look back at him, scarcely breathing.

After we had 'spent the day', came the drive home in the old Overland through the night-time bush, full of animals. Every few minutes we had to stop, because green or yellow eyes blazed back at us from the road and switched off like lights as the animal turned to run off into the bush – a duiker, a bushbuck, a koodoo, wild cat. Or a tree by the road would light up as if strung with green fairy lights, and bush babies clung to their branches and watched as we went by. Once I saw zebras. Zebras were supposed to have left the district and when I said, 'Stop, there's a zebra,' they said, 'Nonsense.' The thing is, they were zebras, their stripes dazzling as the car lights passed over them. They were probably just moving through, on their way north away from all the new farms. In 1992, I stood in the bush and there was elephant dung at my feet. 'Oh, yes, they were passing through here a little while ago. The drought you know.'

Within bicycling distance too were the Watkins. He was another of the lean, driven, burned-almost-black men fighting to survive. Mrs Watkins was thin, with fair straw-like hair and blue eyes alight with religious conviction, a Christian Scientist. She kept having miscarriages. Every time Mr Watkins rang up mother, frantic with anxiety, and she went over to find her patient dying from loss of blood and refusing medicine. Sister McVeagh stood no nonsense, but packed her into the car and went with her to the hospital at Sinoia. There, her life was saved, and she came home again, pale and anaemic. The four sat on our verandah. My mother was patient and ironic, my father irritable, Lyall Watkins miserable with worry, and Mrs Watkins bitterly accusing. 'You shouldn't have, you shouldn't, it's against my religion!' 'Well, my dear, you're alive, and you wouldn't have been,' says my mother, reasonably. 'Yes, you could have died and what about our baby?' says her husband. This happened more than once. I used her dry straw-hair for Mary Turner, in *The Grass is Singing*.

The neighbours I remember best were the Larters, four miles away – the Cyril Larters, for there was a brother. Cyril Larter was a short, strong, stubble-headed man with cold blue eyes and teeth always bared in a sarcastic grin. He scared me. He was, of all the farmers in the District, none of them famed for their gentleness, the most brutal with 'his' natives. And proud of it. He always had some incident to tell of his skill in his handling of his labour, while his cold eyes watched for reactions. It was he who tied up the houseboy all one day to make him confess to the theft of some soap, beat him, and then, when the man complained that by law only the police were supposed to beat a labourer, tied him to the car and made him run all the way to the police station in Sinoia. He would always produce some tale like this to shock my father, and then turn away, grinning. He carried a sjambok, the hide whip which already belonged to the past. Cyril Larter was a type – a kind – a genus. They were scattered about on farms and mines all over Southern Africa. They defined themselves by their toughness with their black workers. It never occurred to them to be ashamed. They might be just, in the sense that they paid wages regularly and stuck to the letter of the law, or they might be unjust. But they were brutal. Probably their own childhood experiences made them so. This was a European type, not just a British one.

Mrs Larter was a gentle kindly soul who, observing that my mother and I did not get on, invited me over to stay, for days at a time. She gave me a large room and never came into it. My mother did not understand that children need privacy. She would crash into my room, any time of the day or night, not necessarily because she was curious, but because she was on the way somewhere. Half-way across the room she would frown, something on her over-full agenda had come into her mind, and she stopped. Was there paraffin in the lamp on the wall, matches by the candle? Had I changed my knickers that day? She would pick up garments off the empty bed, examine them, focus critically on my neck or hands, say, 'You'll ruin your eyes reading by candlelight' and stride off again, pursued by fiends and furies of worry.

Mrs Larter and I sat on her verandah and chatted about what I was reading – wistfully, for she was not an educated woman, but would have liked to be. She made me dresses suitable for my age. When I spent hours floating on the great dam watching the water birds rock on the waves beside me, the hawks in the sky, she never said no, reassuring me with 'But bilharzia needs running water.' And did not tell my mother, who would have been frantic. She was infinitely kind. Children or young people who are unhappy can survive if there is even one person like Alice Larter to befriend them. I think of her often. Yes, she was lonely too: how could she not have been, a gentle soul married to a brutal one?

A mile or two from the Cyril Larters were the Lattys. She was a tall slender creature, willowy, like the models in the fashion books. She wore the new long floaty fragile dresses, which she made herself. Often her sleeves were 'puff' sleeves, half of the material of the dress, but the bottom half of organdie, whose transparent stiffness made a coloured bubble where could be tantalizingly glimpsed a beautiful arm. It was evident she found our farmers and their wives dull and old-fashioned, but she gave wonderful parties and served food we had never heard of. She had a small baby and that was why I could not keep away from her. I adored the baby, loved it to death, wanted only to cuddle and hold it, just as if he were – no, not my own, not at all – this was my baby brother again. From memories of this baby I am able to deduce the strength of my passion for baby Harry. All my life there have been times when my arms have ached, yearned to hold a baby, and they are the arms of a little girl wanting her baby brother. Grown up Harry did not remember our early fondness.

The Latty farm was also an unfortunate one. The baby's older brother, a beautiful little boy, got diabetes. 'He can die at any moment!' wept Mrs Latty as she telephoned one farm after another, as if expecting that someone, somewhere, might have an answer. This was just before they discovered insulin, and the child did die. And Mr Latty died soon after, I think of blackwater. Mrs Latty went home to England.

From this time in my life came the stories *Old John's Place, The New Man*, and *Getting off the Altitude*.

All these so different people, who had so little in common, regulated the affairs of the District from the verandahs, an 'informal activity' as the sociologists call it, for I do not remember a meeting ever being called to discuss a neighbour's delinquency over a fence, or damage done by a bush fire. They also discussed politics whenever they met, or on the telephone, and this meant either the Native Question, or the Company – Lonrho. The Company owned farms and mines, represented Big Business and was loathed by the citizens who would have been surprised to hear that they shared views with socialists. From one end of Southern Rhodesia to the other, at any given time one or other or both, these topics were being discussed. There were also frequent public political meetings, and for these the District turned out in force. The Banket Hall was not large enough, and so the visiting politician might stand on a platform made of petrol boxes set up on Dardagan's verandah, and address 200 or 300 people who stood to listen. At the back of this white crowd might be a few natives, but the meetings were not supposed to concern them.

The member of Parliament for Lomagundi was Major Lewis Hastings, who had been part of the colony's first cabinet, under Prime Minister Coghlan. He was famous for his oratory. He was famous for his love affairs, possibly because he wrote poems not unlike Rupert Brooke's and a good many were love poems. Very handsome he was, like a lion. He was a dandy, with a suggestion of military swagger, but this was used for dramatic effect. He would stand at ease on his box platform and entertain the farmers and their wives and their children with speeches that could have been printed, garnished with Latin and Greek. The crowd stood about in the red dust, the men in their khaki, the women in their best dresses, the children behind him on the verandah, while the ox wagons went groaning past on their way to railway tracks, and Major Hastings said – he was talking about Native Policy, but don't imagine that he disapproved, '*Volenti non fit injuria* – which means, as of course you all know, "No harm is done to him who consents".' And everyone laughed, but letting it be seen that they weren't going to be seduced into approval of government policy by this kind of thing. Major Hastings loved his audiences too much to despise them, but Prime Minister Dr Huggins, who came out less often, could be observed only just controlling his irritation. He loathed the business of hand-shaking and admiring infants. Major Hastings did it all with just a touch of parody, his smile inviting us all to share with him his style, his bravura. How could wives not fall in love with him? Not to mention daughters. There are men who – with not so much as one second's impropriety, with no more than a look – perhaps without even intending it, promise a half-grown girl that one day she, too, will be a member of the freemasonry of love. As for Dr Huggins, he stomped grumpily off when a meeting was over, and did not invite us to vote for him.

But who in fact went on as Prime Minister, year after year, and ended up Lord Malvern? And was not his Party always referred to, simply, as 'Huggins' party'?

It was in that year between schools that I began the business of leaving home, and how could that not be a conflict, when the farm was where I belonged? But I had to get away. I joked all the time about running away, and the jokes suggested a solution. I was rescued by a custom of the country which was even more of an institution than going to spend the day. People went to stay for a few days, a couple of weeks, a month, for the same reasons Jane Austen would have understood, or Tolstoy: it was the bad roads, the old rattling cars and, of course, that there were servants. My parents were happy to have me go, but for different reasons. My father was increasingly irritable because of our bickering, my mother and I, and she lay awake at night because of the narrowness of social opportunity for her young.

Once I was making a mental list of all the places I had lived in, having moved about so much, and soon concluded that the commonsense or factual approach leads to nothing but error. You may live in a place for months, even years, and it does not touch you, but a weekend or a night in another, and you feel as if your whole being has been sprayed with an equivalent of a cosmic wind. Finally I came up with a list of about seventy places, which included a verandah where I spent an afternoon, but left out a house where my occupancy was long enough to figure in a street registry. At various times in my life I got to know several houses in Marandellas, but only one do I remember well. It was not more than seven or eight miles from the place in the bush where we – the family – camped, near my brother's school, two or three times a year: I wrote about this in *African Laughter*. I did not think of this farmhouse as near this camping place, or in any way linked with it. Children can experience houses in the same street as if they come from different worlds, and even rooms in the same house, each with its own atmosphere – air, smells, texture of light. What a child notices, visiting a new house, creates a picture of it the adults living in it would not recognize. The loose green weave of a linen sleeve is like a smile, the curls hanging from the ear of a dog say, I love you, sweat filling the crease in a red neck is almost too much to bear, while the sprightly ranks of china vegetable dishes around a dinner table seem to be ordering you to behave in a certain way.

It would be nice – it might be useful – to report on this house more generally, in a district the rest of the country referred to as 'they are all cheque book farmers there', for it was a community quite different from Banket's where everyone had at least begun poor. These people tended to be well-off. Many had seen the writing on the wall in India, and settled here. They kept horses. There were two hunts in the district, their quarry – leopards. (My story *Leopard George* came from tales of these hunts.) I ought, too, to be able to relate informative conversations with the servants who in this district were not from Nyasaland, as they were in Banket, but Shona, or Manika people. They wore crisp white uniforms and white tennis shoes, and I had never before seen shoes on black feet. They seemed always to be cooking food or serving it, for the household's rituals centred on food, beginning with the early morning tea that came with Marie biscuits, at six. Breakfast . . . morning tea . . . lunch . . . afternoon tea, snacks with the drinks that arrived at six o'clock, the famous sundowners, dinner. Mealtimes were occasions for a good deal of proselytizing, for in this household vegetables were bad for you unless 'thoroughly' cooked, whereas my own home was in the grip of the conviction that only lightly steamed vegetables were wholesome. Both diets were presented by experts as definitive and for all time, just as every new diet is now: no argument was possible, or even discussion.

The house was the usual one, many huts linked by pergolas, but they were of fine brick and well-thatched, nothing like our already shambling and shabby house. On the highly-polished floor were rugs as fine as ours, and the same heavy silver loaded the dining table. Two large and authoritative women took charge of me. They wore clothes I already knew how to place, for they were 'English' clothes, that is tweeds, and linen and cotton dresses of a certain cut, with self-confident pleats and buttons. Living with them was a handsome upright man, ex-Indian Army, but he shared something with my father: he was an observer, he watched, he noticed everything. He was the husband of one of these women? A brother? It never occurred to me to wonder. They were lesbians, so I understood later. The man was like a guest in his house: that I did notice, and felt a kinship with him. But, really, my attention was on the books. There were thousands. Each hut was completely lined with carefully fitted bookshelves, a carpentering feat, since the huts were round: the walls seemed to be made of books. I took one look at them and lost my head and my heart. I was tired of the books at home, I had read them all to death, Dickens and Kipling, Shaw, Wells, Wilde and the rest, and the innumerable books, more and more of them by every mail, about the First World War. Those writers were all here somewhere, but subdued among bright paper covers with the names on them of writers I did not know. Seeing where I was staring, the two women took me in hand. 'There is only one real modern writer,' I was instructed. 'Ann Bridge. Compared to her – well, don't bother with the others, that's our advice.' The inside of my hut, the guest hut, was also lined with books, but I had to carry there the complete works of Ann Bridge and promise fidelity to them. I did not think then how improbable it was that in Marandellas, Southern Rhodesia, I should be commanded to-read the novels of a lady chronicling the delicate relationships of Embassy life in Peking, particularly love affairs, most of them unfortunate, but suffered with good taste. Meanwhile, the man, having watched with disapprobation this attempt at the indoctrination of a defenceless mind, came into the guest hut with a single book. He was also in the grip of the need to form the young in his mould, perhaps the strongest of the drives towards immortality.

'This is the only book worth reading that has been written since the war,' he pronounced. 'You don't need to read anything else, you can take my word for it.' And with this, he gave me a nod, a stiff but charming smile, and strode out, touching up his moustaches. The book was Stapledon's *Last and First Men*, and I read it with *Peking Picnic* and the works of Beverley Nichols, Aldous Huxley, Sholokov, Priestley and Dornford Yates, and a great many more, for in that place I read. That's all I did. I read all day lying on my stomach on my bed, and I read nearly all night, while the candles burned themselves out one after another, and I fitted new ones. I lit the candle instead of the oil lamp because it gave out less light, and I was afraid the two large ladies would appear in their curlers and navy blue tailored satin housecoats to order me to stop, particularly as I was reading so many authors who were not Ann Bridge. They did not come. When I peered out of my window across to their windows they were dark. But through the Major's window I could see him sitting in his dressing gown by an oil lamp, reading. He was there when the dawn chorus split the silky grey sky, silencing the owls and the nightjars; he was there when I guiltily blew out the candle and rubbed my burning eyes and slept until wakened for breakfast, a meal which could not be missed, because God only knew what would happen to a growing girl if she did not eat properly five times a day.

Now I wonder how *Last and First Men* got itself across the sea and on to that farm. Wells recognized Stapledon. Perhaps that was it.

I could not have been at home longer than a few days before I left again. I had to.

Again it was the eastern part of the country, not far from Rusape, on the Salisbury–Umtali road. They did not farm, they were living on money from England, but hardly on the same level as the 'cheque book farmers' of Marandellas. He had been in the war, was wounded, and now had his pension. She had a little money of her own. These so English people lived in Southern Rhodesia because they could live so cheaply. No one could have changed as little as the Watsons away from what they would have been in England. He was a tall, stooping, lean, silent man, with a long lean head, and strong sceptical lips that always manipulated a large curved pipe. His grey eyes were half-narrowed because of the smoke and his slow strong scrutiny of you, of his wife, of everything. The man from Norfolk, my father called him. They were distantly related. She was fat, foolish, yellow-haired, blue-eyed, good-natured, and was, said my father, your typical Saxon. She had trained as a nurse under my mother in the Royal Free Hospital during the war. She said my mother was a martinet, but fair. She would laugh, ebbingly, as she contemplated distant tickings-off. She might laugh, weakly, when she said that Bob, her husband from Norfolk, was a hard man, oh, a hard hard man, no one would believe it. Bob and Joan were yet another couple who kept me wondering and even frightened because of the mysterious demands of Nature who had ordained that these two people would spend a lifetime making each other miserable.

They had built the basic house which was the alternative to the collection of huts, linked together. It was brick, with three rooms side by side under corrugated iron, and a verandah the length of the house, which was where life went on among every kind of plant standing, or dangling or wrapping the brick pillars. Friendly and companionable dogs – just like everywhere else. Decorative cats, like everywhere else. She liked cats. He liked dogs. His were the wonderful Alsatian bitch, Stella, and her puppies. I wrote about them in *The Story of Two Dogs*. This was where I first saw my dog Bill hurtling around the space behind the house, drunk with the full moon and the joy of being a puppy. This space was between all kinds of different buildings appropriate to a real farm, including a cow shed where lived the house cow, milked by Joan. She would not allow her husband to milk this cow; his hands were too clumsy and the cow always got upset. When she said this – often, Bob said nothing, but only bared his teeth over the stem of his pipe and laughed silently.

The cow meant it was hard for them ever to leave the place. They had only one 'boy', a general servant, and they did not trust him with the cow. Besides they were poor, genteel poor, prisoners of their tiny income. He was out of the house nearly all day, walking through the bush with the dogs obedient behind him. She stayed in the house and often wept. Otherwise she cooked, cared for her plants, and sewed. She had arrived in the Colony with trunks full of Liberty prints. Just like my mother. Is there a thesis to be written on 'The Role of Liberty Prints in the History of the British Empire, Final Phase'? My mother's taste was for the striking and the definite, while Joan draped little bits of flowered stuff across straw hats, wrapped pretty rose and blue flowers around her neck, and went to bed, large and piglet pink and sighing, in sweet pea shades of Liberty lawn, trimmed with lace.

'They are all remnants,' she informed me, with pride. 'I went to all the sales. I never missed.' And she sat on her cretonne sofa, her legs apart to make a lap full of floral bits and pieces, a yard of this, three-quarters of that. She would pick up a remnant still in the folds it had been given by a rushed Liberty assistant ten years before, hold it up, survey it critically, like a rich woman examining diamonds she fully intends not to buy, then throw it with a slicing sideways movement, like a quoit, on to a chair. 'I can take them or leave them,' she would announce proudly. 'They're nothing to me. I'll give you a piece or two when you leave and you can run yourself up a nice little blouse.'

I was out of the house as much as Bob was, walking by myself, as I did on our own farm, for hours, not seeing anyone. The veld here was different from ours, few trees, mostly long rolling open downs. 'Like Kent,' said Joan, the tears at once filling her eyes, always reddened by exile or by Bob.

There was short springing grass, and overhead the birds tumbled and soared, singing fit to bust. Small animals started away as you walked, and I found a baby duiker, not much larger than a cat, crouched under a bush with a rock at its back for safety, left there by its mother. Perhaps she had bounded away when she saw me coming and was waiting for me to go. But I could not see her anywhere in the expanse of grass. And one day I found a spring high on a slope that reached down from a rocky headland where hawks built their nests. I did not at once see it. The short grass was damp . . . then it was wet and soggy . . . then it was giving squelchily under my feet . . . then I was standing up to my ankles in water. I stepped carefully through water till I saw on its surface a bubbling movement near a large flat piece of granite. I knelt on the hot granite and leaned over. There was a clear space among the grasses where clean white sand was arranged into patterns by water gently welling from under the rock. A spring. The grass for yards around was soaked. I had to walk in a circle ten yards across until I saw the water running secretly down the slope under grasses, and then there was a rill, where the grasses leaned over, meeting across the water, which was becoming a streamlet on a bed of small white pebbles, but soon it was a real stream, running fast and clear over stones and widening and deepening with every few inches of its descent. I followed the stream, the hot sun on my back, half drunk with the smell of wet grass, to where it fled into a patch of musasa trees to join with a rivulet coming from another high grassy place on the plain, and then it was a little river running fast through big rocks with an exuberant splashing noise – and so it would run, fast or slow, always joining with other rivers until it found the Indian Ocean.

When I told Joan she sighed and said, 'Are you sure you ought to be . . . ?' This meant, as it did when my mother remembered to say it, 'Are you sure a white girl should be risking rape by a kaffir by running around through the bush by herself?' I took no notice. No one was ever raped, I believe. I was recently told of a young woman in Matabeleland left widowed, with a small girl, who farmed by herself, in a house where the doors and windows were always open, day and night. When Bob heard Joan's plaintive protests, he said curtly, 'Of course she should. When I was a boy I was out by myself all the time. That was near Norwich,' he said to me – but did not add, as I longed for him to do, that he too had found the spring and knew how wonderful it was, that he would go with me to see it, and then take me with him on his long slow silent walks. But he never did invite me. It was not his way. He probably did not know how cold and contemptuous he sounded. Tears filled the foolish blue eyes turned in appeal to me . . . see what a brute he is?

She liked me to run errands for her, walk to a near farm to fetch a recipe or take a present of vegetables. She liked me to feed the dogs, clean out their kennels, help with the cow. But best of all was when she told me to go down to the vegetable garden.

I took two large baskets, smelling of herbs, from the hooks in the pantry, and walked off from the front of the house by myself: the dogs always went with Bob. It was a sandy winding path between musasa trees, a mile or so. Half-way to the garden, a hundred yards off in the trees, were granite boulders where the python lived. Anywhere near those rocks, when with Bob, the dogs were called to heel, and he had his rifle ready on the slope of his arm. 'A python can move as fast as a horse,' said he. 'A dog, that's what they like best. A python got poor Wolf, two years ago.' I always walked slowly past that ominous pile, looking hard for the python. I saw it once, a grey coil, motionless in dappled light, easy to mistake for granite. My feet took me off in a spasm of terror as fast as I could go down to the garden, though I would have to return this way and now the python knew I was there he would . . . delicious terror, because I did not believe the python was interested in me. There were pythons on our farm and we often saw them, and I had never been chased by a snake speeding through the grasses. They were always sliding off as fast as they could go.

I stopped before reaching the garden, and stood sniffing that air soaked with herbs, tomatoes, the clean smell of peas. The garden was a half acre fenced to keep the duiker out, but baboons sometimes got in and threw aubergines and green peppers around, and made holes where they dug up potatoes. The tomatoes sent out a smell so strong it made me giddy. A row of them, yards long, of plants as tall as a man, weighted with green tomatoes, yellow tomatoes, green tomatoes red-streaked which I sometimes had to pick for chutney – and so many ripe tomatoes there was no hope of ever picking even half of them. I filled the baskets with these dead ripe, heavy, aromatic, scarlet tomatoes, added bunches of thyme and parsley from beds crammed full of herbs, and went out, carefully fastening the gate. As I left the birds descended from the trees, and even the sky, where they had been waiting for me to leave, commenting in their various tongues on this interruption of their feasting. Some of the tomatoes had been hollowed out by their beaks, and peapods had been opened and bright green peas rolled about the paths. Joan said, 'We don't grow food for ourselves, we are a charitable institution,' and Bob said, 'Birds and animals have to live too.'

I walked slowly back up the long path, feeling the heat get to me, and the tomatoes dragged my arms down. I did not run now as I passed the python's territory, though I watched the grasses for a rippling movement that meant he was coming for me. Slowly I went on, listening to the birds, the birds, the birds of Africa, and particularly the doves, the slow sleepy sound seducing you into daydreams and longing.

I put the baskets side by side on the kitchen table, and drank glass after glass of tepid water from the filter. 'Just make us some soup for lunch,' Joan called from the verandah where she reposed on a long grass chair, another beside her loaded with cats. I filled the grate of the Carron Dover stove, the same as ours – the same as everyone's then – fitting the wood in so there were proper spaces for air, and soon the fire was going. From a hook over the stove I lifted down the enormous black iron pot that always smelled of herbs no matter how much it was washed. Into the pot I emptied baskets of tomatoes, twenty pounds or more. The pot was set over the flames, and I went to the back verandah and sat there, legs dangling, watching the wandering fowls, the dogs, if they were there, the cats, whose lives were parallel to the dogs, neither taking notice of the other. Cats had their own chairs, places, bushes, where they waited out the long heat of the day. Dogs flopped about on the verandah, but never in the house, which was Joan's territory and the cats'.

After an hour or so I took the pot off the stove. It was now filled with a gently bubbling red pulp. Stirring it with a wooden spoon in one hand, I fished out bits of skin with a silver spoon in the other. This was a slow pleasurable process. When all the tight little rolls of pink skin were out, in went salt, pepper, a handful of thyme and about a quart of yellow cream. This simmered for another hour.

Then, lunch. Platefuls of reddish scented brew, making your head reel with its smell. I did not eat it so much as absorb it, together with thoughts of the vegetable garden where by now hundreds of birds would be drinking from the water buckets, or fluffing themselves in the dust between the beds. The doves' long slow cooing, the tomato reek, the python – all this became part of the taste.

That's tomato soup. Never accept anything less.

Ah yes, fond memory, fond lying memory, picking out the high points of everything, in this case all pleasure, crystal springs, pythons, vegetable soup, the somnolence of doves, cats luxuriously rolling under my hand . . .

However, the truth compels me . . .

I did not remain aloof or indifferent to this marital battle, and for most of the time which I am making sound so idyllic (and it was), I was also in a state of hot and stupid emotion. And a very odd business that was, too. At home I identified with my father, who I had decided was the victim of my mother, but here I reversed identification and was allied with Joan against Bob. What a monster of cruelty! How badly he treated her! When she wept, and that was a good half of every day, I comforted her with many an exclamation about the thickheaded insensitivity of men, until I could not stand her or myself another second, and escaped into the bush, which is what I did at home. When Bob came in for meals, and sat silent, eating anything put in front of him, ignoring her signs and exclamations and complaints to me about him, I sat fixing him with accusing angry eyes. Then she and I sat together at the end of the verandah in an alliance of womanly sensitivity, while he sat at the other, smoking and reading all manner of journals from England, mostly to do with farming. We gave him long, slow, bitter sarcastic looks, and tittered and exclaimed in low voices about his rudeness. When he got fed up with us, and went out into the bush with the dogs, even though it was very hot, he left behind him a loud chorus of complaints and accusations.

He must have been delighted when I left and the marital war, evenly weighted, could continue at its usual level. I could not wait to tell my parents about their old friends, his callousness, her sufferings, the awfulness of it all. They were at first silent, and then pointed out that Bob and Joan had done all right during those years of marriage, and it was probably just their way.

I was the wrong age to listen to this philosophy. In those days people did not divorce, or, if they did, it was a scandal and an outrage. Bob and Joan could not divorce for they needed each other's small incomes. That one might need to sigh and suffer and be misunderstood, and the other to remain proudly isolated and misunderstood, was not anything I could bear. I wept hot angry tears in secret about the unhappiness of Joan and of Bob, the unhappiness of mismatched Cyril and Alice.

It was thoughts of the animals, the birds, the bush that I took with me to the new school, as shield and buckler. After – well, how long? The interval between the schools seemed forever, but already I was looking back into the endlessness of child-time with awe, with dread, with terror. How had I survived that? (How does any child survive it?) Worse – *what was to stop me sliding back into the quicksand*? This fear was with me for years. I was at the Girls' High for a year, and it was a long time indeed, but more like 'Was that *really* only a year?'

The place was indeed a shock after the Convent, for the atmosphere was brisk and cool. Religion, instead of being on every wall, in every corner, in the halls, in the rustle of the nuns' robes and their voices chanting from the chapel, became a church service on Sundays to which we were marched in pairs, long lines of us along the edges of the streets under the jacaranda trees, dying of heat in our navy blue serge tunics, white crisp shirts, white panama hats with, this time, blue and green ribbons. At the Convent every meal had begun with a long grace in Latin, and prayers punctuated the days.

The Girls' High, like the Convent, had day girls and boarders, but there did not seem so much of a divide between them. There were four Houses, for the boarders. Most had been sent to school first at seven from farms or small towns. You would think that there would be immediate evidence of a lack of gentility: after all, it was to escape the vulgarities of a government school that I was sent to the Convent. Neither then, nor with the eye of hindsight, did there seem much difference. Both schools had 'a better class' of girls. In both were girls whose parents, in trouble because of the Slump, had gone to government offices to beg, plead, weep, that if their daughters were not taken in free, they would not get an education at all. There were many bursaries: I was given one.

The House I was in was dominated by a group of girls whose

'tone' was bad. We knew this because the House Mother kept telling us so. She was recently from 'Home' and would go back there. They were common, said she, and when we were separated, sheep from goats, on leaving school, the common girls would be mere shop assistants and cinema attendants. I have always found it at least interesting that my mother's preoccupation with nice and less nice people, with commonness, had so little effect on me that the House Mother's exhortations seemed only absurd. The democratic spirit of the Colony was too strong for her. (Democratic, that is, for the whites.) For instance, it was usual at that time – and had been from the start of the Colony – for the better-off farmers to have an open day on Sundays where everyone was welcome, farm assistants, the mechanic from the garage, the policeman from the camp, the struggling farmers. The towns were already more snobby and would get worse.

These common girls, seven or eight or more of them, went about in a group, loud, insolent, self-assured and indifferent to everyone else, particularly the loudly contemptuous House Mother. Their exploits were admired, and these were all to do with Boys. We were forbidden to have anything to do with our brother school, Prince Edward's, full of long-legged, smelly, raw, gangly, raucous, jeering louts, just like my brother when with his peers. We were less impressed by the Boys than by tales of exploits to do with Boys. Our exemplars passed notes to Boys in church, or when the long crocodiles chanced to be in the same street at the same time. They also claimed to climb out of the windows at night and creep through the shrubs to meet the Boys, even go to the Bioscope. This last boast proved they lied, because in small-town Salisbury, any truant would be seen by someone. Never mind, this was like girls' schools anywhere. Fashion magazines circulated. Photographs of film stars were passed about like pornographic pictures. Feasts after lights-out had to be horrible, to appal the parents – sardines with condensed milk, or biltong with jam. In our boasts these midnight feasts were almost nightly affairs, but in fact were few, like the exploits with Boys. The dominant group, whose tone was so bad, were proud to be bottom of the class, but were not emulated. It was quite all right to do well in class.

There was a little bullying. Love affairs had the flavour of the cinema. A girl older than myself lured me to a balcony, I could not at first see what for, and behaved like a film lover, like Ronald

Colman. In a low thrilling voice she said I was beautiful, my hair was this and that, I love you, do you love me? I was embarrassed. Couples were known to 'like' each other. 'Betty and Barbara like each other, don't they?' They sat on steps, arms around each other, talking baby language. It didn't cross anyone's mind to call this lesbian. When 'Tigger' in the holidays told her parents, as a joke, my mother was distressed, and said she would warn the House Mother, while my father told her not to be so silly.

One term I created, edited and mostly wrote, a House magazine. There was a social column, and verses about prominent or popular personalities. The pages circulated in proof through the dormitory and a deputation of older girls came to ask me to delete this or that. 'We don't want the House Mother to know that, do we?' I was shocked. I was appalled. The truth was all, the truth had its rights, the truth must be told . . . but I took out the bits they marked. My first experience of censorship. Tyranny had won.

What I *see* looking back is a busy hardworking girl, ready to conform, anxious to be liked, wanting a best friend, like the others. But how could one be a real friend when you could not share the best of yourself, that is to say, the self made by the farm? There was no one here I could even talk to.

What I *feel*, when I put myself back in this or that scene, is a raw loneliness, isolation, anxiety. I was a defended observation post. I felt, in short, like everybody does, until we make little places for ourself in a group, a family, a gang, where the cold air does not blow so cruelly on tender skins. It was 'Tigger' who saw me through. I made the girls laugh, and the teachers too. They were in every way different from the nuns. Nearly all were English, young, and taught for only a year or two before getting married. The farmers' wives joked that they were running marriage agencies, because their governesses and mothers' helps married inside a year, and here it was the same. These young women were brisk and matter-of-fact, and taught us more than was on the syllabus, because all talked of what they had fled from, which was unemployment, the dole, people leaving home to go anywhere, Australia, Canada, Africa, to get away from the grimy, graceless, bleak, ugly, dreadfulness of the England described by Orwell and by Patrick Hamilton, whose books I read later. But I had already heard it all from the young teachers who had escaped.

Sometimes they failed to get a husband. There was one, not

young, perhaps forty or fifty, a grey cold lean woman who always wore flat walking shoes and thick stockings. She taught history. When we heard her steps outside the classroom we all froze, including the girls who had such a bad tone. Why? She never hit us, or even threatened to. It was her personality, contemptuous, sarcastic, angry. She would stand, ruler in her hand, and direct her gaze at the bad-tone girls, first on one, then another, saying calmly they were riff-raff, they were rubbish, they thought they were clever but that was only because they did not see how silly they looked to someone with education. The cold eyes would then travel slowly, all the time in the world, from one face to another through the classroom. Then she laid her ruler down precisely on the desk and began. That year she was teaching us something called Ancient History. She taught by dictating sentences which we had to take down. My mother sent me archaeological magazines, for that was the great time of archaeology. The teacher took the magazines from me, said she would return them at the next lesson. She did so, saying, 'You may cut out the pictures and stick them in your exercise book. Then show it to the other girls. Possibly one or two of you may get some benefit.'

At the end of term exam I got 100 out of 100, because I had learned it all by heart. And forgot it all in a month. I had inherited my mother's talent for exams. She kept insisting, 'I always came first, because I had a good memory. You are just like me.' Her so frequent *You are just like me* made me white-hot with rage.

For one term this teacher taught English Literature, and I wrote an essay, or rather 'Tigger' did – about her methods of teaching. I thought she would applaud it because it was funny, but she called me in and made me stand in front of her while she slashed me with that tongue of hers. I thought I was clever, didn't I? Well, compared to really clever people I was nothing. I stood trembling. I could have murmured about Justice and the Truth – for had she not been feared by generations of girls, and had she not made sure she would be? But I was a sullen coward. Outside girls waited to hear what she had said, and I boasted about how I had said to her face this and that. This is how subjects are with tyrants.

I was still doing piano, by now a couple of hours a day, mostly scales and exercises. The accompaniment for hymns or the House Assembly was always played by a pupil. They came to me and said the girl who usually played was sick and now I must do it. Terror and panic. It was not that I could not do it: the music teacher showed me how easy it was to fit bass and treble together. Obviously it was easy – look, you are doing it, aren't you? So what is the fuss about? Yet when the time came I played the melody only. I simply could not make my hands make the match. That *I can't, I can't* . . . as if my hands were stammering like a tongue.

At the end of term concert one of the 'skits' was of me, playing the accompaniment to the hymns, but with only one hand. I was seared with mortification. Coming first in class and being clever Tigger Tayler was nothing, because of that failure. I gave up the piano, just like that, as I had given up religion, after years and years of expensive lessons. That's that.

One morning I was called out of a lesson, and there was my mother, in her smart town hat, her gloves, her good dress, on the verandah. 'Your father has diabetes,' she announced. And then, lowering her voice dramatically, 'He may die at any moment.' I stood staring, while she waited for an appropriate response. The thing was, the drama of it. Long ago I had ceased to respond to her always theatrical announcements. I could not bear them.

Since I did not say anything, she turned and led the way to the old car where sat a very thin man, holding tight to the door handle. Where was my father? I almost said it, but did not. I looked for help to him, who never failed me, but his eyes, old and faded and sick, did not seem to see me.

'He is very ill. He must have insulin three times a day,' announced my mother, and got into the back seat beside him. A neighbour was driving. The car went off under the jacarandas. I stood there. As usual I had failed in a decent response. Now I wanted to say the right thing – to him, not to her, whom I would always disappoint. 'Poor Daddy. I am so sorry.' But that was not the right note for an announcement of death.

The letters she sent me, twice a week, now had more in them than farm news. They were all about him. If he had contracted diabetes even a year ago, he would have died, just like the Latty child. The newly discovered insulin, extract of cow's pancreas, would save him. But he could not manage the farm, and what were we going to do? Of course Harry could help in the holidays, we couldn't afford a farm assistant, some of the samples taken from the reef at the edge of the Big Land had looked promising, and perhaps this time we would find a gold mine and . . . The letters came hurrying in, she must have sat there every evening, writing, filling pages of her blue Croxley pad, the big lamp at her elbow, the dogs at her feet, the cats . . . And my father? I could not read those letters, after a while. Who was writing them? I didn't recognize the writer. 'Of course you don't care about me, you never did, you don't care about your father, all you care about is yourself, you are selfish through and through, and giving up your piano lessons just like that, I've made a record of what they have cost us through the years, you know we couldn't afford them, we can't afford . . . we can't afford . . . and now I have to give your father his injection. I've reduced the dose a little and we'll see.' Pages of it, ten, twelve, twenty pages, of hurrying desperate sentences, mostly accusation.

I got measles. There was an epidemic. In those days the quarantine was six weeks. Several girls were sent off to the Isolation Hospital, a large old house in a garden. A nurse came in twice a day from Salisbury Hospital, not far off, to give us medicines and said, 'Now behave yourselves, you are big girls now.' There was a 'boy' who made our meals. Suddenly I discovered that a problem I was beginning to think insoluble was not a problem at all.

Food at the High School could not have been more different than the Convent's. Coming into the dining hall for breakfast, lunch and supper the room glimmered with whiteness, for a glass of full-cream milk stood at every place, and on each table were several platters of thinly cut white bread spread with glistening white butter. At breakfast, white mealiemeal porridge and white sugar, white bread and golden syrup. At lunch, cold meat and boiled potatoes, then 'cake' pudding with jam. At supper, macaroni cheese and white bread and butter. Once a week we had an orange, once a week a bit of lettuce. We learned a useful lesson in the ways of the world, because we knew when the food inspector was coming, for as we entered the dining room there would be an egg or a jelly. In fact, white bread and butter was the staple of our diet. Years later, when I compared notes with a fellow sufferer, we agreed this was the only time in our lives we had been constipated. And we were always hungry.. We ate nasturtium leaves, we spread mustard on the bread, and we begged food parcels from home like refugees.

In the Isolation Hospital, with food sent in from the General Hospital, the affliction disappeared. It was a very good time for me. Nothing was being asked of us. We were supposed to do lessons, but no one had time to supervise us. We were not really ill. Who were the other girls? I don't remember, only laughing a lot, sitting about on the wooden verandah gossiping and making up stories, and above all, being alone, if I wanted. Six weeks is a long time if you are thirteen. Six weeks of freedom, with only those terrible accusing letters with every mail, but I didn't read them.

When I went home for the holidays, their bedroom had become an extension from those long-ago days at the Royal Free. There was a tall table made of newly knocked-together petrol boxes, full of medicines and vitamins, and on the top a tray with his spirit lamp and the tubes to test urine for acetone and for sugar. Syringes. Capsules of insulin. Bottles of meths. Cotton wool. Teaspoons and desertspoons for measuring. Beside his bed, the table was crowded with medicine bottles. He was obsessed and so was my mother. The fact was, he was dying of starvation. In those early days of insulin they had not worked out how to use it, and he was supposed to eat a little cold meat, tomato, lettuce, dried bread. That was about it. He was so thin big skeleton hands trembled at the end of arms that were bones with a little sunburned skin clinging to them. His face was all bone, with deep anxious eyes, like a monkey's. Yet that was the right diet, the doctors said so. My mother decided to defy them. She said it was nonsense to take a medicine that was from a gland that digested, and then give it nothing to digest. Trial and error. She worked out with him how to eat the forbidden bread and potatoes and butter, and balance them with insulin, and slowly he put on weight and returned to the living. When I went back to school at the end of the holidays he was, with difficulty, taking back the control of the farm from the bossboy – no longer Old Smoke, who was too old now and had gone home to Nyasaland – and from my brother, who at eleven years old was spending all day on the farm trying to be as good as his father. That is, until he went back to school too.

At school I got pinkeye. There can be few afflictions more horrible to look at and to suffer, yet which do so little damage. A lot of us got it, and as every girl woke in the morning with swollen eyes glued together, the rest of us teased her; she tried to laugh at herself and retired to the sickroom. It is very infectious. I was afraid. My eyes were enormous inflamed lumps, pus oozed, and behind thick bandages I watched the fireworks caused by unhappy nerves, falling stars, rockets, a variety of dazzling pyrotechnics. I was so afraid, and oh how Tigger jested.

In came my mother from the farm and I was taken to all kinds of experts who said there was nothing wrong with my eyes, now the pinkeye had gone. But I said I could not see properly. I would not stay at school. I went home with her and that was the end of my life at school. A drop-out, long before the term had been invented.

9

MY FOURTEENTH WAS a make or break year, a sink or swim year, a do or die year, for I was fighting for my life against my mother. That was how I saw it. That was how it was.

The main arena was clothes, what I wore. I had a cousin in England, who had everything my mother wanted for me. She was at a good girls' school, and her mother's every letter was about money and smart friends. Her clothes arrived in beautifully packed packages for me. Layers of tissue paper held dresses as exquisite as my mother's, long since cut up for play. I remember an apple-green silk dress, with little frills and puff sleeves. My cousin was not only much younger than me, but these good little girls' dresses could never be worn in the District. Where? How? Everyone would have died laughing. Alice Larter, being appealed to about my unreasonableness, was worried – for my mother. She was tactful, tried to rescue me, invited me over – it was no good. I knew what it was my mother wanted, when she nagged and accused me, continually holding out these well-brought-up little girls' clothes at me. 'Well, try it on at least!' They were sizes too small for me. She's mad, I cried to myself. And she was, a little, at that time. We fought over food. It had begun to dawn on me that the reason why I was fat was because I ate too much. Was I fat? Not really, but I was 'fat and bouncy Tigger'. At every meal, I tried not to eat so much and she, her face tense with worry, tried to pile my plate. Suddenly, she took it into her head to start worrying about my being alone in the bush. All those years I had wandered about sometimes miles from home, and had tactfully not said how far I went, but now it was Principle, it was the Truth, and we fought. 'If they are so dangerous why is it no one has ever attacked me?' 'Yes, but there is always a first time.'

Mad, mad, all of them, I would say aloud, stamping out of the house.

There is something about adolescent girls that does the strangest things to their parents. Nancy Mitford has described how their father nagged the girls about not becoming victims of the White Slave Trade. I suppose there is some sense in well-chaperoned and guarded girls being considered at risk in wicked London? But what can account for it when a father, who has been sitting in his desk chair gazing out at the mountains for hours, summons his daughter in from where she was lying reading under a bush down the hill to say: 'When you get to England, never talk to a stranger. Particularly not a woman. They can sit next to you, stick a hypodermic into you, and the next thing, you'll find yourself in a brothel in Rio.'

Mad, mad, mad – I cried to myself, but silently, for he was too ill to quarrel with. And I raged around the bush, sick and angry and in a blaze with frustrated pity. When away from my mother I was capable of achieving a decent level of pity. These two people, these sick and half crazy people, my parents – it was the war, it was the First World War, that was what had done them in. For years I kept bright in my mind, like scenes from a film, what they would have been without that war. She, a jolly, efficient Englishwoman, probably running the Women's Institute for the whole of Britain, or some nursing service, not the kind of woman I would have much in common with, but the point was, she would have become herself, not this harried over-wrought victim. And I only had to look at pictures of my father, before the war – these idealizations refused any of the other chances and choices life comes up with, but I was sure about one thing: my father had been strong, vigorous, in command of himself, this is how he would have gone on – and now he was an invalid, with no hope of ever being well again. I wandered about the bush or sat on an antheap, angry to the point of being crazy myself, seeing my parents as they were now and what they should have been – and from here it is only a step to the thought, If we make war impossible the world will be full of whole and healthy and sane and marvellous people who . . . In my mind I lived in utopias, part from literature and part the obverse of what I actually lived in. Into these lovely and loving societies I had begun to fit black people, particularly black children. Kindly, generous, happy people, in cities where no one went to war, black, brown, white people, all together . . .

Daydreams . . . and my parents were lost in daydreams and

imaginings, both of them. My father had been panning for gold, sinking shafts and digging trenches looking for reefs, for years. But now this had become his occupation. Farming was something that went on routinely, to produce enough money to keep us going, but was no longer expected to produce a windfall. No, a lucky reef would do that. They waited anxiously for the results of sweepstakes, even bought tickets to the Irish Sweepstake.

We listened to the News from London on the radio every night, introduced by the sonorous notes of Big Ben, as portentous as the bells of the Roman Catholic cathedral. My father was raging against the stupidities of the British government and its blindness about Hitler, whom we also heard ranting and raving from Germany. I could hardly bear to listen to the radio, and it was hard to sit quiet and listen to my father. The personality of the diabetic was taking over. He was hypochondriac, querulous, self-pitying, splenetic, . . . *where was my father?*

Oh my God, the unforgiving clarity of the adolescent, sharpened by fear that this might be your fate too. 'I will not, I will not,' I kept repeating to myself, like a mantra. Meanwhile both parents observed the other, and saw something not very different from what I did. My mother the nurse, knew all about the diabetic personality; she was watching the inevitable deterioration of her handsome and brave husband. As for him, a scene repeated itself, and would, for years. He has asked for something, or she has missed an article on a tea tray or the dining table, and she is off out of her seat as if pulled by elastic, she charges off with a lowered head but with a lifted harried face to fetch teaspoon or sweater, as if escaping from fire or a reprimand, while he says, full of unease, 'For the Lord's sake, old thing, sit down, it's only a plate.' For she did wear herself out. She was never still. And now I heard how she spoke to the 'boy', the scolding, insistent, nagging voice full of dislike that so many white women used to their servants – and still do in some white households in South Africa. My father hated how she was with the servants. He remonstrated with her. 'But they're just hopeless, hopeless,' she cried, her face flushed and desperate.

'I will not, I will not,' I repeated silently, and took myself off out of the way. And away, too, from my mother's nagging that I had no future, what was I going to do now I had disgraced myself, leaving school?

When my brother came home she nagged at him, because she

could not reach him. He had become polite, cool, appeared to listen but took no notice. He went with his father down on the lands, and did everything he could, but after all, he was at school, that was his place. And his task. She told him continually that he had to go into the Navy, or into the Army, he had to 'get out of this second-rate country' – in everything she said was implicit that he had to go away from *his* country, which was the bush, the land, the landscape. I would say to him, 'Why don't you stand up to her?' But he said, 'Oh she's all right, really.' He did not seem to notice anything much. That is, anything in the realm of the emotions. He had blocked all that off. As for his parents, he was already calling them M. and D. Not Father and Mother, or Mummy and Daddy, not even Maude and Michael. No, it was Em and Dee, and so it was until they died, M. and D. M.ummy and D.addy. M.other and F.ather.

I took myself off again. I threatened to run away from home, laughing, or sullen or defiant. Rather, Tigger laughed, but I did leave. I went to stay in Umtali with the Jameses. Mrs James was the kindly matron at Rumbavu Park, and now she was in a tiny railway cottage in a lush tropical garden full of guavas and mangoes and peaches and granadillas, with her daughter, Audrey and her two sons, and a little docile husband. Again, the atmosphere of genteel poverty – what can possibly be worse? I wrote about this in *Going Home*, but the essence of my situation was that there was a 'gang' of young people, on the American model, all older than I was, dating and 'going out' and picnicking and dancing, and I was too young for them by two years.

By every mail arrived letters from my mother. They were frantic. They were disorganized. They might be ten or twenty pages long, and accused me in every paragraph of my usual crimes, selfishness and obstinacy, but now threatened me with an inevitable end in the brothels of Beira. 'That's where girls like you end up, you'll see.' Pages and pages of it. But I was living in a most conventional family where Grace was said before meals and God sighingly appealed to in every other sentence. I could not bear to read these letters, and could not have imagined that one day they would seem just another little symptom of general human lunacy.

Something else happened, which I have had to think about ever since. At a Mission in Old Umtali there was an afternoon's fête, and black people as well as white wandered about under the trees drinking tea and eating cake. I had never been with black people as an equal, in a social situation. I was delighted. I was curious. I was threatened, and did not know how to behave. I went up to two old black men standing each with a teacup in his hand and began chattering, social stuff, of the kind my mother was so good at. I chattered and they listened, looking gravely down at me. Then one said gently, 'You see, I am very old and you are very young.'

Nothing very much, you'd think. I had been given the mildest of snubs, with a smile that forgave. But that was not it. There was something about the occasion, the old men, the words, that 'got to me'. I knew they had. But what? What happened? Yet not for years did anyone say to me anything as powerful, making me think, forcing me to use words, incident, old men, as if hidden there was some kind of original excellence, which I must refer to. But nothing had been said, judged in terms of simple sense. And yet everything had. Long after, when something of the same kind happened, and then again, and again – I understood it doesn't matter what words are used, if a person waits, unconsciously, not even knowing it herself – himself – wanting to hear something, be struck by something, needing it, then words as apparently empty as 'It's a fine day' can have the same effect. But time was needed for that little incident to lodge itself in my mind as a paradigm, and meanwhile I was in a seethe of confusion, rebellion.

Back home I went, to the same situation, but worse.

My father was less on the lands: once he had been there, watching the labourers all morning, all afternoon, with half an hour after lunch for a nap. Or out in the bush divining for gold, all day. Now he slept after lunch for a couple of hours, and so did she. I tiptoed into their room and saw him lying there, on his back, as he always did, his wooden leg stiffly out, his hand gripping the bedrail as if he feared to be swept away by a wave or a wind. She lay huddled, on her face infinite distress. Beside both, on their night tables, were glasses of water with their false teeth in them. There was always trouble with these ill-fitting teeth.

Both parents were getting deaf.

'*I will not, I will not!*'

My eyes were still supposed to be bad, and although I read all the time, my father began to read to us both, in the evenings after the News. They were always books about the war, the Trenches.

He would come to some terrible passage, and then lay down the book, his face working with rage, with tears.

Meanwhile I was succumbing myself.

When I went off into the bush, I became afflicted by a mysterious exhaustion, and sank under a tree not a hundred yards from the house. If I went down with my father to the lands, I found a shady place and read. But I was reading and re-reading the same books: I was reading as an aid to daydreaming. I was slowly going under.

I dreamed of glamorous futures. I was a dancer, I was a singer, I designed clothes, I ran nightclubs in wicked cities. Mostly I was a bohemian in Paris or wandered on the shores of the Mediterranean. That was long before any youngster anywhere knew they could go any time they liked, everywhere they liked. To dream of Marseilles or Nice was . . . there is no comparison now. Timbuktu or Chimboraxu are at the end of a plane journey. The world had unused places then.

My mother did not know of my fantasies about nightclubs and clothes and amorous adventures. But if she found me playing the piano, or heard me sing a tune from the radio, if I happened to make a drawing, or if I danced by myself in the bedroom and she caught me, then she would at once announce that 'when we got off the farm' I must have proper lessons and then . . . it was implicit that I would have a great career.

We have come to *Martha Quest*, which begins about this time – and a need for explanations. Readers like to think that a story is 'true'. 'Is it autobiographical?' is the demand. Partly it is, and partly it is not, comes the author's reply, often enough in an irritated voice, because the question seems irrelevant: what she has tried to do is to take the story out of the personal into the general. 'If I had wanted to write autobiography then I would have done it, I wouldn't have written a novel.'

One reason for writing this autobiography is that more and more I realize I was part of an extraordinary time, the end of the British Empire in Africa, and the bit I was involved with was the occupation of a country that lasted exactly ninety years. People no longer know what that time was like, even those who live in Southern Africa. My own children can be surprised when I tell them this and that, are perhaps disconcerted at the roughness of those times, karosses as blankets, furniture of petrol boxes, flour sack curtains. The sometimes paternalistic, sometimes brutal relations between white and black then have changed. African friends, white friends, may be infuriated or amused at my father and Old Smoke philosophizing for hours sitting on either end of a log while they watched the 'boys' at work. Whites will refuse to believe in the brutalities of Cyril Larter or Bob Matthews, or that white adolescents were not rebuked by their parents for pretending 'as a joke' to run down a black man or child in the car along the track, or that the more stupid white men played cruel jokes on their employees.

Last year, a television company was prepared to make a series about Martha Quest up to the time she left for London. At the very beginning of the negotiations I joked that only one person would be able to write the scripts – me, for no one now knows how things were. A South African writer, a very good one, did some trial scripts, from which I learned that indeed that time had most irrevocably gone, for he got small things very wrong, and there was a South African flavour to the tale. There was another thing wrong. When I wrote *Martha Quest*, I had already had a decade or so of being asked, How is it that a girl isolated in the bush had all those clever ideas about life and race relations? My explanation, that she had been soaked for years in the best that has been said and written, was not considered adequate. I therefore put in the Cohen family, as storekeepers in Banket, but intellectuals, and politically minded. There were no Jewish shopkeepers in Banket, and probably not anywhere in the rural districts. The Cohen family came from later experiences, for indeed I had Jewish intellectual mentors. The film script gave a large place to the Cohens. And here is a contradiction. If the series was simply a story – why not? But my interest in the series was a historical one. Truth. Facts. All that. I was caught on my own cowardice. Over and over again in my life I've been sorry when I softened or changed truth for some reason, to satisfy outside pressure, or make things easier. And certainly the novel would not have been nearly as informative if I had not objectified my mental battles, embodying ideas in people. The long slow processes of coming to an understanding through reading, or what you observe or overhear, are not as satisfactory as Mrs Cohen feeling wounded because her sons don't respect the dietary laws, or as Martha Quest being 'asked the catechism' but in political form. This did happen to me ten years later when my scornful mirth cost me a friend.

There were also the Afrikaners, the Van Rensbergs. The novel begins with two women, middle-class English and a rural Afrikaner, talking, while Martha listens. The Van Rensbergs influence Martha. Mrs Van Rensberg was created from Alice Larter – but one kind and maternal woman is much like another, no matter what the culture. I knew people like the Van Rensbergs. They were at gymkhanas, at dances, keeping themselves to themselves, clans of people whose daughters were at school, both at the Convent and at the High School.

In short, when I wrote *Martha Quest* I was being a novelist and not a chronicler. But if the novel is not the literal truth, then it is true in atmosphere, feeling, more 'true' than this record, which is trying to be factual. *Martha Quest* and my African short stories are a reliable picture of the District in the old days. That is, from a white point of view. Once I met a man who had been a 'piccanin' with his father and big brothers, working on the farms in Banket, at the time I was a child. We talked. We talked. We tried hard. What he remembered most was moving, for the black farm workers, either in groups attached to a chief or a clan leader, or in families, were always on the move from farm to farm, after better conditions or because a senior relative had arrived from Nyasaland. Most of the labour in the District was from Nyasaland. It seemed to him all the farm compounds were the same – poor, ugly, badly built places. He was getting old, and was humorous where he was once bitter. But an anger burned deep in him, the historical anger, the same as my feeling about the War: '*But how could it have happened?*'

Martha Quest has a simple plot. She has a childhood in the bush, quarrels with her mother, is taught politics by the Cohen boys, reads, escapes into the big town, Salisbury, learns shorthand typing, plans all kinds of attractive futures, but is swept away into dancing and good times, and marries a suitable young civil servant while the drums are beating for the Second World War.

While my mother talked of my brilliant future, I was lying on my bed, daydreaming, or reading books I had read twenty times before, particularly the anodyne girls' books from America. I succumbed to something called Low Fever. Is there such a disease? I ran a temperature day and night of about a degree, and could hardly get out of bed, where I lay with the door propped open with a stone, looking out into the bush, a cat on my bed, and the two dogs on the floor, wagging their tails with beseeching looks every time it seemed as if I might at last get up and go out with them, as they felt was their right, for hours and hours into the bush.

A doctor came from Sinoia, not once, several times. Quite an expedition in those days, not the trip of a few minutes it is now. My mother told him I had Low Fever, and demanded he should prescribe quinine. He did. In the rainy season we took five grains of quinine night and morning, but now I was taking more and more. The tablets were large, bright pink, five grains each. The quinine had no effect at all. My ears rung so I could hardly hear. I was inside a bright ringing clarity, poisoned, crazy with quinine.

And now we began to take trips into Salisbury, often lifts, because of the price of petrol, to a faith healer. She was a woman of thirty-five or so, English, unmarried. My father said she was in love with the man who was the best-known faith healer, for there was as it were a group practice, and said, in the way he had – a countryman's approach to such matters, but in this case mingled with sorrow for her – that not being married was what afflicted her, not religion. This figure, the unmarried woman of a certain age whom everyone pities or scorns, has gone from our culture. So there is progress. Miss – I forget, let's say Burnett – had a Red Indian guide who sometimes allowed coins, which guaranteed his authenticity, to materialize during seances. 'Why not Egyptian? Why not a black witch doctor, while she's about it?' demanded my father, while my mother lowered her eyes and suffered. She, a nurse, and on the side of science, was uneasy.

The process itself was soothing. You sat in a darkened room, Miss Burnett behind you, while her large hands passed lightly downwards over shoulders and arms, and she gently expelled her breath, and, with it, the poisons that caused the sickness. The soft hissing of her breath, like air escaping from a puncture, the rhythmic hand movements, were hypnotic. But the diabetes did not get better, nor did my Low Fever. Nor my mother's ailments. A year later my mother claimed that X-rays at the hospital had revealed a brain tumour cured by faith healing, but my father said they had mixed up her X-rays with someone else's.

The family were on a 'prayer circle': at certain precise times of the day people all over Rhodesia and South Africa 'joined up' and said the same prayers, asking for health and grace. I lay on my bed and linked up, ears shrilling. By the time I was rescued, the dose was sixty grains of quinine a day, while my temperature remained unmoved.

A charity paid for me to go for six weeks to the mountains south of Umtali called the Vumba, to a guest house run by a remarkable old woman famous for miles around. Granny Fisher was eighty and had only recently given up walking down through the mountains to Umtali once a week, miles of wicked rutted paths, striding in front of a string of native bearers, who carried supplies for the guest house on their heads back up the mountains. She was a short, fat, bossy old lady who ruled all her lodgers with a sharp eye and tongue, but was kind and generous. It was an old thatched spreading house, full of rooms and verandahs. There was a vegetable garden and an orchard and she kept cows. But how did she make any money? She fed us – no one, anywhere in the world, could eat like that now. No pesticides, fertilizers, poisons. The cows, when they got sick, died if they did not recover on old-fashioned remedies. The compost made from their dung was dug back into the garden. The air was clean. We ate our way through five big meals a day. Great jugs of cream stood at intervals down the table at every meal. Everyone's favourite pudding was granadilla pulp stirred into cream: half and half. It was my job to collect the granadillas, then growing wild everywhere. 'Young people should eat plenty of cream and butter!' she ordered, leaning forward with fierce looks, holding the cream jug so we could watch the thick smooth yellow curling down over porridge, fruit, pies. We laughed and pretended to shield our plates with our hands. There might be twenty or thirty people around that long table. Friends of guests arrived for a meal, friends of friends, and she fed them, free. Or they dossed down in the corner of a verandah. This scene of unlimited abundance, giving, generosity, was repeated forty, fifty years later when my son, John Wisdom, farmed in the Vumba.

The verandahs of the guest house overlooked mountains and lower mountains, hills, lakes, rivers. Down from the house was a large pool or small lake where the cows spent most of the day under the trees near the water, or standing in water to their knees, chewing and content, swishing their tails. When they lowed it was like the conversation of gossips, never the wild raging bellows of beasts complaining about their lot. Their calves were not taken from them, for Granny Fisher did not hold with it. The calves were with their mothers in the day, but during the night were kept separate, so the main milking was done in the morning, when the cats, the dogs, the chickens, the geese, waited at the cow byre door for their share of the milk.

I spent many hours of every day in the pool. It was believed that bilharzia got in through cuts and sores, or through the urethra, up to the kidneys, thence making its way deviously through the body. Now we know it can get in anywhere through the skin. To keep children out of water is impossible, so we were allowed to bathe only if it had just rained and the waterholes or rivers were full, or if our skins were whole – but children's skin is always cut and scabbed – or if we made sure our private parts did not get into the water. But Granny Fisher, like Alice Larter, did not believe in bilharzia. 'Nonsense, it will do you good to get some exercise.'

Reeds and grasses stood tall around the pool, except where the cows made pathways. I stood with the sweet brown water washing over my feet, then slow careful steps, with the mud squishing through my toes, until I lifted my legs and let the water take my weight. I did not swim, but floated, lying on my back, limp as a drowned person, rocking on the ripples. Above me a flawless sky where kites and eagles hung. On the trees around the pool dangled hundreds of weaver bird nests. Kingfishers flashed through the reeds. Swallows skimmed the water. Then the cows stepped carefully down through the reeds and stood not far from me, and their swishing tails sent ripples over my face. Their calves stood together under a tree. Not a week had gone when I decided I was not ill, I refused ever again to be ill. From this distance I could clearly see how at home two unhappy and desperate people used real and imaginary illnesses to make life bearable. Never again! And Granny Fisher, who was instructed to keep an eye on my Low Fever said Rubbish! What's wrong with the girl?

When not afloat in the pool I wandered over the hills and high places above the house which were unlike our part of the country, the highveld, exhilarating reminders of the great world where one day I would go. Short sheep-grazed grass, little springs and wet places, winds blowing around the edges of the higher mountains, and clumps of sheep, white on green, everywhere you looked.

I wrote a poem.

Take what moonpaths he will
Over the sea-ways,

or where sheep-known outcrops, grass-flowers
shimmer in a thinner air,
let him move disdainful on the heights there,
glance down once at the populated valley,
pass to the higher ranges, if he will.

He will seek towns at night,
linger in square-lit doorways,
and when arms gleam, quick-glance eyes invite,
will saunter through the night-haunts or
take what love he can wait for.
He will change tales of travel in a lighted bar
hear many strange true stories in the night.

But there comes a time for all
men who must follow new ways:
for will old fevers chill him,
coming when the sun is low?
Shall the nights deceive him as the years go?
He will return to the small unchanging valley,
learn to talk to children, after all.

These verses are not here for their worth: they interest me. First, the writer was a fourteen-year-old girl . . . but wait, that cannot be true. Some Ancient had moved in, taken temporary possession of that many-tenanted young mind. Then, it is *he* who takes moon-paths, *he* goes adventuring. If it had been *she* who frequented night-haunts and took what love she could wait for, it would have been a very different matter, going to the heart of male–female difference. And then, there is nowhere even a hint of a recognition that 'modern poetry' was at least twenty-five years old. Lastly – and most important – when I was being a writer, that is, at my most truthful, I did not believe in the efficacy of the spell or rune I was repeating every time I remembered. *I will not*, I simply *won't*. The poem comes from a different level of knowledge. Nor is this the easy and pleasurable melancholy that is one of the habitations of poetry.

Health was one thing that long visit, nearly two months of it, gave me. The other was not on my parents' agenda for me, and was Granny Fisher's gift, for her own life had been so hard it seemed to her simply silly, wanting to shield children. As a poor girl, she was on the farms and mines of South Africa. She had known Johannesburg when it was a gold rush town, people living in streets of canvas tents or iron shacks lit by hurricane lamps. There, chastity was the least of her worries, keeping alive was more like it, with brawls, murders and muggings, and above all the drunkenness, which was what all the women I've known from those days talked about. The men got drunk every night, and the women had to drink with them or keep out of their way. She had been married. There was an elderly daughter somewhere. A picaresque history had been absorbed into 'Granny'. Good old Granny, now *she's* a character if you like.

Two of her guests were engaged. In the District this meant, or was supposed to mean, chastity until the wedding night. Appearances were kept up. These two shared a bed. He was an agricultural expert, working down in Umtali. He drove up the twenty miles of those terrible roads at dusk to spend his nights with her. I forget her name, Lesley or Jackie or Billy, some clever little 1930s name. She spent her days sitting on a verandah overlooking leagues of country and made her trousseau, satin knickers and camiknickers and petticoats. Also copies of Kestos bras, in scraps of silk and voile. But what for? She was rangy. She was flat in front. She had short sleekly cut yellow hair. She was not pretty, but, said Granny Fisher, attractive in a hard way. She had scarlet satirical lips and cold grey eyes. Whenever her fiancé arrived from Umtali she was sewing. I knew that she was a siren, but did not understand the hostility she radiated at everyone and everything, not least her loved one, a good-looking young man, but too soft. (Granny Fisher: What she needs is a man with a sjambok.) He adored her. He could not keep his eyes off her. His hands kept moving, disobedient to his will, wanting to touch her. But if he tried, he was repulsed with a shake of shining yellow hair, a laugh, or the sharp withdrawal of a long hard thigh. Did she hate him then? She seemed to. When she looked at him it was always with cold laughing insolence. She let everyone know she did not like sleeping with him.

How, if she disliked him, disliked sex, could she drag him after her 'like a poor little puppy on a leash?' ('Oh Granny, you are so cruel,' says a guest.) But it was not only her I watched, using the new eyes given me by this situation and by Granny Fisher, but, retrospectively, women from the District, for I was now able to pick out from admirable devoted hard-working housewives – the farmers' wives – this other kind of woman. Scarlet women. Hussies. Sluts. Sirens. Yet each one was irradiated by this half-secretive half-flaunted hostility to men they enslaved. I could not understand it. It was not that *they* were slaves to sex, loved sex, loved men, and that was why they kept men running after them like puppies on a leash.

I was able to turn over, as it were between my fingers, more than one couple from my now rapidly receding childhood, and restore to my mind conversations heard on the verandahs. For instance, there was Reggie, who came to my parents for advice. He had arrived in Southern Rhodesia, unable to find work in England, the youngest son of a middle-class family. He got his farm from the Land Bank. He was twenty-three or so when he first became my parents' protégé. He was half crazy with loneliness, and said he had to marry, had to, he could not stand it. A tall, very thin, stammering young man who worked so hard my father said he would get ill. He had more barns than one man could manage, but he stayed up half the night, worked all day, got thinner and crazier – and went down to the Cape for a holiday, to stop himself cracking. There he met Vera, the half-English, half-Dutch girl who had already been engaged more than once. He married her down there and brought her home to the big stone house built among boulders on a little kopje. There she refused to sleep with him. He came driving over to my parents, driving like a drunk man – no sooner had we seen the dust of his car whirling along the track than he had arrived at our house, frantic, stammering, his blue eyes reddened with sleeplessness and strain. He was so thin that my mother ordered the servant to bring out everything in the pantry to feed him. But he needed to talk. Vera would not, she would *not* have sex . . . that is a word for now. He said, 'make love'. She hated it, he said.

Vera was a large solid olive-skinned woman. Every movement was lazy, self-possessed, full of disdain. She had brown eyes, black bobbed hair, wore smart tailored dresses. She did come to visit us once, but our house was too shabby for her, so it was Reggie who came.

'She won't let me touch her,' he said, his hands clenching and opening, clenching and trembling. 'She hates it, she says, I asked her, why did you marry me then?'

We all knew why she had married him, a catch with his own farm, but he could not face that.

It went on. He got thinner. His stammer worsened. He decided to take Vera Home to England to see a psychologist. This was not yet a fashionable thing to do. He did not really have the money, but he borrowed from the bank, and they went.

Vera sat in the office of a precursor of today's confessors.

'Well now, Mrs B., your husband tells me you want to talk to me.'

'I don't want to talk to you. He wants me to talk to you.'

'But I understand you agreed to come here and see me?'

'I am seeing you, aren't I? You're sitting there aren't you?' And she lazily lit a cigarette, tilted back her head, and sat breathing out fragrant smoke.

'Come, come, Mrs B. You aren't being fair to your husband.'

'Why aren't I? I keep a good table for him. The house is clean. I don't waste money.'

'But Mrs B., marriage is supposed to include sex.'

'I don't see why it should. I just don't like it. I never have.'

'Have you never enjoyed – anything in that line?'

'No. Don't see the point of it.'

'I see. Do you think it would help if you came to see me while you are in London?'

'Help who?'

'If you like, your husband. Aren't you fond of him at all?'

'Fond?' She considered the word. 'I must have been, mustn't I? I married him.'

Reginald was told it was a waste of time and money, nothing could be done with Vera, because she did not want to change.

Vera announced she had enjoyed her holiday in England, she had always wanted to see it.

There were two beautiful children. One of them was a little girl I yearned over with anguish and with love. 'She fell the two times we slept together,' said Reggie, baring his teeth in what was meant to be a grin. 'She wanted children. She's got them. And now – that's the end of that.'

Reggie became a very rich tobacco farmer. Long after I came to England people visited me from the District. They said Reggie was a terror with the blacks, why didn't they kill him? He deserved it.

He hated them. They hated him. By then he had divorced Vera. She had gone to the Cape where she lived alone, drinking.

There is a little stained snapshot of Vera leaning against the old Chev. She is wearing a late 1920s dress, showing her knees. One hip is indolently, insolently, thrust forward. She is smoking through a very long cigarette holder. She smiles at the camera, calm, confident.

The bad woman. The bad wife.

Beside her stands the wife of the farmer across the river. She is a plump jolly young woman, smiling with enjoyment. Everyone loved her, everyone disliked Vera. But she, too, is a bad woman and bad wife, because if Vera won't make love with anyone, let alone her husband, then she makes love not only with her husband but with other men too. She, like Vera, was later divorced and went to live in the Cape. The two children of Vera, of Reggie, grew up as they should, so Evolution (or Nature, or the Life Force), so indifferent to the miseries of that marriage, must have been satisfied. If someone had said to Reggie – even my parents, whose advice he valued so much he might drive across to see them two or three times a week – 'Don't marry her. Can't you *see*?' then he would have laughed his guffawing hee-haw laugh and stuttered that he loved her.

Betty – or Franny or Jamie – could not lean against a verandah pillar, or bend to pick up the morsel of pink silk she had carelessly dropped without a lazy flaunting of hip, or thigh, and a smile that told the world she would never, ever, give anybody an ounce more of herself than she had to – and, furthermore, she was proud of it. She sat by her heap of silks, satins and coffee-coloured laces, and looked across at her lover, just arrived all dusty from Umtali, and laughed, and said, 'Why didn't you stay the night in Umtali then? I'm sure I never asked you to drive up here every night?'

Then his face, pale with hurt, his eyes blazing and pleading, while he stammered, 'But d-d-darling, but my darling . . .'

'Oh never mind!' And she smiled at him, in a way that made my heart go dizzy and soft, let alone his, as if he had been forgiven for something. Forgiven for what?

'She's a common little piece,' says Granny Fisher, 'but there's no parting a fool from his folly.'

I spent hours with her. She liked me there, the poor clumsy avid adolescent, dazzled with her sophistication. She was a secretary from England who had come to the Colony to get herself a better husband than she could hope for at Home.

Quite soon, they married. Later, they divorced. Women with this kind of lazy mercenary sexuality end up alone, though usually well-heeled.

I see Granny Fisher as one of my missed opportunities. From her I could have learned more about a South Africa that is not in histories than anyone I have met. Any woman, that is: men have a different story. I have never met a more remarkable woman. Even then I knew that the matter-of-fact, even casual way she dealt with the crowds of people who passed through her house, her life, meant few people could have had her experience. Not only did she feed and shelter: all kinds of people made their way up the Vumba to ask for her advice or her intervention with the highly placed, for she seemed to know everyone in the Colony. The well-known and even the fashionable came to her old house for holidays from the rigours of social life in Salisbury. I saw the wife of a cabinet minister and her eighteen-year-old daughter sitting at supper next to the surveyor who was mapping the Vumba mountains, and then in conversation half the night on the verandah, the three of them, kindred spirits who would otherwise not have met. It was this woman who shone, just for a moment, a light on Granny Fisher's past. She was in the grip of a need to proselytize a new diet creed: growing girls should eat mostly meat, preferably half raw. Granny Fisher listened and said, 'I was in the northern Transvaal on a mine during a drought. We had nothing to eat but salt beef. That's what I and the kaffirs ate for two months. And another time I was on a poor kind of farm near Stellenbosch and we ate pumpkin and mealie porridge for half a year. Your body will put up with anything you put into it,' she told me. Now, if I had asked her . . . but I was too fascinated by Phyl – or Pat or Tony. But perhaps I didn't know enough then to ask the right questions.

I returned home healthy, full of energy. I took with me a bra, made by myself. My mother, confronted by this antagonistic newly breasted young woman, switched into the fighting mode, and called, 'Michael, Michael,' and went on until he came, when she pulled up my dress to show what I had on.

'Lord, I thought it was something serious,' said he, going away again.

I was consumed with rage and hatred, just as I had been when I began menstruating, and she rushed through the house to announce it to my father and brother.

Out of all proportion was my anger, my disgust at her, so strong that for years I put it all out of my mind. And then something happened which made me think again. Years later I was living next to a demented old woman whom half the street was trying to keep out of a lunatic asylum. She had no physical shame, or even ordinary sensitivity. She would thrust her filthy smelly feet at you and demand you wash them, or sit picking bits of skin off them and putting them, with relish, into her mouth, or sit scratching herself all over her body with her tongue protruding and a look of voluptuous pleasure. She lifted her great flopping breasts to examine the rash under them, inviting you to have a look, or rubbed her crotch as vigorously as if towelling down a wet dog. There was no end to the disgustingness, but after all, she was demented, she could not help it. The violence of my disgust and rage was unreasonable. It was disproportionate. It made no sense . . . it *does* make sense if you imagine a very small child shut up with a gross adult. Who? Most parents employ a nursemaid or baby-sitter as casually as they buy groceries. Perhaps it was the drunken Mrs Mitchell, with whom I shared a room. Perhaps even my mother, but the smallest infringement of a child's sense of decency can seem an enormity. A visitor comes, bends with smiles over the child, and on her chin is a round shiny lump like a nursing rabbit's teat. There is a ginger hair growing out of it, and that hair seems a revelation of dirty secret practices, even cruelty. 'Mummy, why . . . ?' 'Don't be silly, it's only a mole.' 'Why has she got a *mole* on her chin?' 'Not that kind of mole, sillybilly.' But if the pimple or sweat-soaked underarm hair is part of a parent's body, the child shrinks away, silent, but staring, full of disgust.

It is not tactful for a mother to whip up her fifteen-year-old daughter's dress to expose her breasts to the father, but it is hardly a crime.

Physical insensitivity was compounded by something else. Again, it was a long time afterwards, and I became friendly with a therapist whose speciality was mothers and daughters. It is common, said she, for mothers to be so identified with a girl-child she can hardly tell the difference between her own body and the child's. One said, when reproached for hitting her daughter, 'But it's the same as hitting myself.' Another, having shouted and screamed at her little girl, said to the therapist, 'But it's between us, *she* knows why I do it.' I am not suggesting my mother was anywhere near this level of neurosis. Yet she did handle my limbs as if they were hers, or at least her property. After all, she had been a nurse, when she had made free with the bodies of her patients.

When I shrank from her, defending my body, refusing to let her touch me, I knew I was saying, 'I will *not* be infected by your illness, by your hypochondria, the diabetes, the scarred pitiful shrunken stump, by the war, the war, the *war* – the Trenches, I will *not*.'

'I am not a child any longer,' I said to her.

I had put on so much weight at Granny Fisher's I had to lose it. Now people would be muttering about anorexia, but I was equipped with a fine organ of self-preservation. I might be eating 'nothing at all' as my mother accused, but I had worked out a healthy diet. I ate tomatoes and peanut butter, and lost weight fast, and my body gave me much pleasure, while it shocked and frightened my mother. Disproportionately, but there is little in this tale of mother and daughter that has measure or even ordinary sense.

The point was, I was every bit as sensible as – surely? – she wanted. I've seen this sense of self-preservation since in girls you would think were bound to do themselves in. They brink-walk – my shorthand for that habit of behaviour where young people create situations where they *almost* come to grief – they do come to grief, mildly, and you resign yourself, she will certainly get pregnant and need an abortion or she'll have a love-child, or he will get himself arrested, but not at all, crises and alarms of all kinds continue, but the girl has secretly visited a doctor and been equipped with birth control, the boy's follies stopped just before the point of serious consequences.

In a corner of the bush near the big land, I stood with my rifle loose in my hand, and suddenly saw my legs as if for the first time, and thought, They are beautiful. Brown slim well-shaped legs. I pulled up my dress and looked at myself as far up as my panties and was filled with pride of body. There is no exultation like it, the moment when a girl knows that *this* is her body, *these* her fine smooth shapely limbs.

I was every bit as good as the models in the magazines. But my clothes . . . we had no money. None. We never bought anything, my mother made every stitch. But what she made were dresses for a little girl, and all day I saw how her sorrowful eyes condemned my new slimness, my new body.

I taught myself how to use the sewing machine, but there was no money for materials. It was then I began taking into the butcher on maildays half a dozen guineafowl, which I had shot that early morning, running from the house to the lands when it was half-dark to catch them before they sailed down from the trees to feed. My mother was beside herself. She raged and accused and stormed, but what she really said was, 'You are escaping me, you are leaving, and I am stuck here in this awful, miserable life of mine, and I shall never be able to get out.'

Dardagan's had shelves stacked with bolts of linen and cotton. Six guineafowl bought enough to make two dresses. I wore my clever new dresses, and my mother cried out that in England I would still be in the nursery, this was a horrible country if it let girls grow up at fifteen.

I had to get away. This time it was to Johannesburg, the big city down south.

Once, in Norwich, in my father's young time, he had danced and flirted with two sisters. He had been in love with both, he claimed, but even more with their mother. When he said this his regret, his irony, were tributes to vaster losses to time.

One girl had married a promising young man from the Chamber of Mines, and he was now an important executive. She had written to my father a keeping-in-touch letter, remarking that her husband did not have an affectionate temperament or, as we would say now, was not much into sex. When my father reported this, as with other sad stories from the same shore, he did not look at my mother, and spoke with a sorrow, mingled with irritation only just held in control, which protested again, at much more than this personal deprivation.

My mother wrote to (let's call them) the Griffiths, asking if her daughter 'in a difficult phase I'm afraid' might visit. A train journey 'down south', my first. For families like ours, that is, settlers from after the war, there was a concession on rail travel. A journey of two days, second class, with six in a compartment. At every station, where the train might stand panting for as long as an hour, black children brought carved animals, or an orange, or apricots or a few guavas, the superior arrogant whites bargained over tickies (threepence) or pennies, and laughed when having held the toys high out of reach of the children who were afraid they would be cheated, threw down the coins just as the train moved and laughed again and jeered as the children fought in the dust. This scene is described in Nadine Gordimer's story, 'The Train from Rhodesia'.

In Johannesburg a chauffeured car took me to the world of wealth, a large house in the richest suburb, with servants (here they were female, not as with us, men), barred windows, and an atmosphere of siege which was then new but has intensified ever since. Mr Griffiths was Scotch, had retained his accent, was forthright, downright, clever. He was nearly always at work. He came into his house like a guest, while his wife, a pretty middle-aged woman in expensive clothes, jewellery, her fluffy grey hair perfect, attended to him with an air of doing her duty, but her eyes were always on him, full of reproach. Here again was a couple, bound to make each other unhappy, opposite in every way. Meals were brief, English food, in a dining room I thought stiff and unpleasant – like in fact all the house, full of expensive things. I never left our shabby falling-apart house without thinking how much more pleasant it was than anywhere else I went. Mrs Griffiths did not work. In those days rich women did not work. She was bored, bored to silliness. She would say, 'Oh God, then let's go for a drive.' The chauffeur was Stanley, a young South African, a lean, burned, cold-eyed man from the world of real poverty. Easy to recognize this in him at once. For one, there was his bitter hatred of the blacks and the Coloureds, and, again, the anxiety always there behind the eyes. He did much more than drive the car, care for it, run errands, shop. He was like a son, but respectful, in the South African casual way, and always on the look out for something that needed to be done. When we went for a drive it was around the rich suburbs, full of similar houses. Or to the expensive shops.

It was considered good for me to work in a smart dress shop for a week, the owner being a friend of the family. It was. All day very rich women came in, sat about, looked discontentedly at suits and dresses held out for them. Not by me, for I was assistant to the chief sales lady. Most often they did not buy, but the shop was always full of women gossiping. Much later I met women who had been married in Johannesburg, but had fled to London from the place and the people. They were bored, these women, bored to sullenness and hysteria, in their little world which from outside seemed like a prison or an expensive boarding school. They knew only each other, met every day at parties and charity affairs, and consulted the same fortune tellers. Quite soon it would be Indian gurus and ecology. I thought the clothes were horrible. It was the life they were created for which was horrible. I said to Mrs Griffiths – well-brought-up children did not call their elders by their Christian names then – that I did not think this shop was doing me any good, that is, if the aim was to improve my character: I was simply getting cross and spiteful. 'Tigger' spoke. But Mrs Griffiths was not one to find much worth laughing at.

There was an exhibition in Johannesburg. Visiting it was one of the reasons Johannesburg was expected to improve me. It was certainly intimidating, with its enormous flashy pavilions, the vulgarity, and crowds of people, many white and from the country in their best clothes, cowed by this sophistication, and Africans, multitudes of Africans different from the people I knew, sharper, more self-aware, better dressed, aggressive, taking a good look at the display of goods and opportunities not meant for them. Stanley said the kaffirs were getting cheeky, and needed a good hiding: that at least was just like home.

I went to the bioscope, a more flashy version of Salisbury's. The titles of films were in lights outside, and inside, people were dressed up in cocktail dresses and jewellery.

Sometimes Stanley might be told to drive us to morning or afternoon tea at a certain very smart tea room. Again women, women, women, chattering, fiddling with their gold chains and bracelets. 'You can never have on too much gold!' This was the voice of Johannesburg.

My visit 'down south' was an entrance to a world of bored and unhappy women. Except for once, when Stanley, instructed to drive me around to see the sights, took me out of the rich suburb, through increasingly poor streets, until I was in a street like those of Salisbury when first built. Along this street, in a thousand Westerns, have lurched wagons and whooped cowboys. Along this street in little towns in the Andes, Indians lean against walls, chew coca leaves and stare at tourists. In a fringe of Los Angeles it peters out in a mountain ravine, by way of pinball machines and Mexican eateries. Here it had been built wide enough for wagons to turn in it, forty years ago, and was of one-storey buildings where tiny rickety houses were interrupted by a cinema, a Chinese laundry, a dance hall, cheap eating places and bars. Very few black faces in this street, for it was an area for poor white people. (Not 'poor whites' – a phrase used for very poor farm people, usually Afrikaners, living in 'rural slums'.) There were Indian shops, with bright flashing displays of colour outside.

Stanley parked the great rich car and said to me that if I liked I could just sit in it, he'd only be gone a minute. But I went in with him. It was like a barn or hall, full of snooker tables, each surrounded by men, unemployed, because of the Slump, and at the back were tables where men played poker. There were two tables for *vingt-et-un*. These had girls with a lot of make-up and Veronica Lake or Jean Harlow hairdos, in evening dress, moving the chips about with long rakes apparently made of gold. Like the chips, made of gold, like the gold glasses and bottles wrapped in gold paper that enticed customers to the bar. Cheap Cape wine was being sold, and Cape brandy. All the men were shabby, in over-washed and thready white shirts, cheap grey and brown slacks, but sometimes with a cowboy's scarf tucked in at their necks. All the women were dressed for glamour, in cocktail dresses or dance dresses. There were a couple of hundred men to perhaps thirty women. Once, so Granny Fisher had let drop, in a way that would have led to interesting reminiscences had anyone taken it up, she had run a gambling den on the Rand but the police had closed it down. I could not at first see Stanley, he had disappeared into a group throwing dice at the bar. His smart khaki chauffeur's uniform marked him out for me. He was there about twenty minutes. No one took any notice of the girl in her cotton dress and sandals standing by the wall, feeling a bit sick with tobacco fumes and the smell of drink. When Stanley came out he was grim, and only nodded at me to follow. In the car, 'I have nothing but bad luck, just bad luck, I'm telling you man, it gets me down.' He told me that at nights when he had finished with the Griffiths he worked as a barman. 'Just as well Mabel doesn't know what I get up to, but I've got to live haven't I?' Mabel and Stanley, but Mr Griffiths and Stanley: democracy as far as it went.

Mabel Griffiths was always making tart little remarks about her husband's meanness.

The other sister was somewhere around in South Africa. Later I met her too. A large highly coloured woman, in 'good' clothes.

These were the women my father said he could never have married because they were rude to waiters and servants. They both snubbed me, but did not know they did. They lacked the delicacy of Stanley, who did not in words ask me not to 'tell' where he had taken me, and the instinctive kindness of the heart it turned out Mr Griffiths had. He had hardly spoken a word to me while I was there. I thought he did not like me. But he sent me all the way from Johannesburg to the farm, a typewriter, a tall heavy clattering machine that weighed so much the cook laughed when he carried it to my room, pretending to stagger under it.

Mr Griffiths sent me a short dry letter. He had been a poor boy from Scotland. He had made his own way. He wanted to give me some advice. I should learn everything I could, it didn't matter what, it would turn out useful some time. And there was something else. Young people often did not notice there were opportunities all around them if they looked for them. With kind good wishes, sincerely, Allen Griffiths. He also sent me a Writers' and Artists' Handbook.

I went home pleased to be out of the Golden City. My mother was pretty demented. I am surprised now she wasn't worse. I was surprised then. It is possible to know that someone 'can't help it' but at the same time be driven crazy with rage and frustration. Money, it was all money, money in every sentence, money, day and night. My father was spending his time and the money we didn't have digging trenches and shafts, buying dynamite. He did not have energy for the farm but could stand for hours at the back of the house panning samples. This passion turned out to be disinterested: when a wandering prospector found a workable reef and there was a small gold mine not a couple of miles away, my father was delighted, and went off to share knowledge about reefs and soils and divining. The farm went on, somehow. Now I wonder why my mother, that supremely capable woman, did not take over the farm? There were women in the District who farmed. I think she did not want to undermine her husband's self-respect. Meanwhile she raged at me not only for being selfish but for wasting money. On what? – I asked. Why did I not contribute the money I earned for the guineafowl to the household? This was crazy and she knew it. The debt at Dardagan's was for over a hundred pounds. It was a deadlock. She was a kind and generous woman and would have loved spending it on me. But this war had nothing to do with money. She made me think of a bird or animal flinging itself against bars. She made me think of a badly treated small girl. I was sick with pity for her and beside myself with hate.

Money, why do you waste money, you know we don't have any money, you don't care for anything but yourself . . .

In Salisbury, while my parents were again enduring a humiliating interview with the Land Bank for an extension of the loan, I hung about a furniture store in Manica Road, giving myself courage with Allen Griffiths' letter. The furniture was all 'store' furniture, glossy and smart, and lifesize cardboard housewives, several to a window, offered decorous smiles to the pavement, because of their happiness over a table or a chair. I went in and asked to see Mr Hemensley. He was a lean but paunchy man, in shirt-sleeves because of the heat, worried because of the Slump. I told him he had no idea how to present his goods, but I would write him verses he could put in the *Herald's* Personal Column every week. They would catch people's eyes. No one had advertised in this way before. He was tickled at my cheek and asked to see samples. I had some folded up with Mr Griffiths' letter. Each jingle ended with the exhortation, Furnish Your Home the Hemensley Way. I wanted ten shillings for each one, but we settled for seven. For a couple of years, whenever I needed money, I dashed off verses and went into Hemensley's, where he gave me tea and cake, asked about my doings with the wistfulness of a man already regretting his youth, and laughed while 'Tigger' entertained him. He said my adverts had brought in customers. At the Indian stores you could buy perfectly good cottons and muslins for one or two shillings a yard. A decent pair of shoes cost ten shillings. A good handbag cost a pound. No one wore gloves, or stockings.

The Hemensley money did nothing to improve things at home. Each dress, or skirt, or bra was another nail in my mother's cross. And now we were quarrelling about the Colour Bar, the Native Question. The trouble was I had no ammunition in the way of facts or figures, nothing but a vague but strong feeling there was something terribly wrong with the System. For one thing why were all these people working for so little money on our farm? Already there were letters in the *Rhodesia Herald*, signed Commonsense or Fairplay, saying the Natives were inefficient because they were not properly fed or housed, and anyway they should be educated. Each such revolutionary letter drew fire, and the next letter page was filled with Indignant, or Thirty Years in the Country, or Pioneer's Wife, saying the Natives would not appreciate anything better than what they got, or they needed a good hiding, or that education would spoil them. But while I was in the right, I had never met anyone who agreed with me, nor yet found books that would help me. Yes, *Oliver Twist* was as much about a black child here as one in England, but it was not an argument that would appeal to Cyril Larter or Bob Matthews or Mr McAuley.

My arguments were noisy and foolish, and I knew they were.

Soon my father said he could not stand one more day of a ringside seat at the quarrels between his two females. He said I didn't have a good word to say for either of them, and why didn't I leave. Having been a parent myself with difficult teenagers I sympathize.

I went as a nursemaid near Salisbury. (In those days 'au pair' meant rich families exchanging daughters so they could learn languages and foreign ways.) I longed for new experience. This one sounded romantic enough to satisfy even the most avid teenager. (A word not yet in use.) Mrs Edmonds, a beautiful rich girl from Society in Vancouver, had fallen madly in love with a poor but well-born farmer from Rhodesia, and married him despite family opposition. She was having her second baby, and I would look after the little boy, four years old. The house was on a hill across a little valley from Rumbavu Park. There I began the business of looking after children that went on with minor interruptions for years. The child who had never been able to resist any baby or toddler within reach ('Come and cuddle me') now had as her charge a delicious, clever, and amenable little boy. I adored him. He liked me in the easy-come-easy-go way of small children who have already had a succession of nannies and minders. Mrs Edmonds was indeed beautiful, with buttermilk gently freckled skin, auburn hair, a willowy figure – the same physical type as my friend Mona from the Convent, who was thin and angular, so thickly freckled her pink skin was hardly visible, her hair dull and stringy. Mona with the drunk for a father and a broken home, always seeming to apologize for her existence, I knew could be as lovely as Mrs Edmonds if she had the money. Mrs Edmonds was another woman who taught me how competent my mother was. She wailed and winced through the tasks of life in crepe de Chine negligées, while people waited on her. Her new baby was a week or so old, a 'good' baby, and had a nurse. Having observed not a few women making a good thing out of being helpless, now I wonder about Mrs Edmonds, whose attractive husband, attentive and anxious, treated her like an invalid. They were so poor! – they never stopped saying. I had heard so much talk about poverty I did not – could not – listen. Middle-class poverty is never as simple as it seems. I am returned to my query, my quest: what had they been promised, and by whom, to make them feel they had been let down?

Soon I was doing much more than look after Marcus who needed so little looking after, but who ran around after me like a jolly, glossy little puppy. I ordered food from the shops, cooked, instructed 'the boy', assisted the nurse who was in attendance with breast-pumps and nourishing foods.

I daydreamed about glamorous futures, none dependent on being properly educated. My fantasies were of handsome lovers, ambiguous heroes familiar to the readers of romantic novels, all of them anxious to whisk me off to magic islands, shores, cities I knew from books.

I was also writing short stories, and sold two to smart magazines in South Africa. Coming on these in some drawer years later I so burned with shame that I had to tear them up on the spot. I had written to suit a market. I had succeeded. But later I could not do it, even when I badly needed money.

I was relieved when the Edmonds said they couldn't afford me. There was talk of a real English nanny, to be paid for by her family. Later they got divorced. I remember these charming people and their little boy as if they were pretty children.

I was three months with them. (Time had by now settled down into adult time – that is to say, early adult time, very different from middle-age time, or elderly time.) And now a year in Salisbury itself. This family was already in the second generation. The father had been a Northcountryman, had started one of the country's best-known firms. The well-off daughter had been married for her money by a young moneyless engineer, a refugee from the Slump in England. She was described as unattractive, wore maidenly unfashionable dresses, and her hair in a plait around her head. So much did she not conform to the current taste, she had evidently decided not to compete at all. But she often had a small smile which managed to be both knowing and naïve, and was probably thinking, 'I might not be pretty, but just look at what I've caught,' while she flirted gravely with her handsome husband. The unmarried brother lived in the house, and ran the family firm. He was a Christian, and a pattern of conventionality. It was in this house that through a long day of some royal occasion on the wireless – Princess Marina's wedding, I think – this young man stood to attention every time the Royal Anthem was played, which meant dozens of times, and yet there was no one there to supervise him but his conscience. Derision, from me, or from 'Tigger', while the husband, who agreed with me, allowed his eyes to engage with mine, but at the same time made some conforming remark designed to affirm his place in this God-fearing, patriotic household. Jasper (let's say) was paying heavily for his secure future. He saw himself as an intellectual, was the equivalent of my parents and their friends before the War. He was well-informed about the state of the world, with strong opinions, which had to be presented carefully.

Suddenly I understood, as these revelations so often occur, perhaps while changing a nappy, that I was no longer earning the epithets used for me, 'Such a clever girl, reading all those books,' and so on. I had become mired in dreamy repetitive reading, long ago – or so it seemed. The wholesome girl heroines of America's Midwest had given way to the easy popular novels written so plentifully in the 1930s, aids to fantasy, but what Jasper read was different fare.

The book that happened to shake me awake was Wells' *The Shape of Things to Come*, but it could have been anything. There was a world of ideas I knew nothing about, and I was ignorant. Probably everyone looks back and sees a graceless bewildered floundering child, sees too that older people recognized need and were kind. Jasper was unflattering, forthright, and salutary. He kept me supplied with political and sociological books he ordered from Home, certainly not left-wing, for he was not, and introduced me to the Everyman's Library. For the first time I heard views that had to be taken seriously – that is, not from a 'crank' – about the Native Question, on the lines of enlightened self-interest. When Jasper said that the Natives should be properly fed and educated and housed, because in the long run the whites would benefit, he put forward these views in a mild and judicious way, as if he had only just thought of them himself – not, as I knew from private discussions with him, because he was burning with impatience because of the inefficiency of it all. So did seditious ideas enter this household, years before they were respectable.

My charge this time was a new baby, then four months old. He was in fact my baby. He was no trouble at all. In those days middle-class babies not on the breast were put on to milk powder. The business of preparing a bottle with boiling water from a vacuum flask, cooling it with boiled water from another, feeding the baby, winding the baby, while the washed bottles sterilized themselves boiling in a big saucepan for the next feed, was simple, quick, efficient. This was a contented and friendly baby. He did not cry, loved to be cuddled, and slept all night and a good part of the day. It was not that the mother did not like her baby: she was proud of him. She was simply not maternal. Nor was she domestic. I was secretly full of my all-too-familiar corrosive derision because she could have studied anywhere, studied anything, but had not been interested. She had taken a degree in Domestic Science in a college in northern England, and used notebooks from her class to order the household's meals. Sunday: roast beef. Monday: collops with sippets of toast (mince). Tuesday: beef stew. Wednesday: brawn. Thursday: steak and kidney pie. Friday: stewed oxheart. Saturday: tripe and onions. To be a white housewife was hardly arduous. There were four servants, cook, houseboy, gardener and 'piccanin'. The grocer and butcher were telephoned every morning with the day's order, and a man soon appeared on a bicycle with the goods. Jasper, a man who liked his food, made humorous suggestions, such as, 'Perhaps we might change the order a little – like collops on a Saturday?' With calm and maternal smiles she soothed his importunities, with, 'But it's so easy, just ordering the same thing.'

He did not complain. Nor did he ever actually say anything to me, but it happened that I took over the ordering, and often the cooking, though the cook did not like this invasion of his domain. The lady of the house sat on her cretonne sofa and sewed a fine seam. She made little floral dirndl skirts which she wore with the Hungarian embroidered blouses which were then so cheap and so flattering. She made panties and petticoats out of expensive silk and satin from the high-class shops: no cheap Jap silk for her. 'Pretty,' her handsome husband would say, coming to sit by her among the scraps of pink and mauve, his head on one side. Then with a consciously whimsical smile, and the swiftest of glances at me, he put his hand on her cheek, to make her raise those wonderful eyes of hers, flower-blue with black lashes, so he could look deep into them. 'Croodles!' he murmured, 'Joodles!' Past her smooth braids and again-lowered lids he looked long at me, where I sat on the other sofa dandling his baby. Our rapport was perfect. I knew he would have liked to seduce me. I knew he would not, rather, could not. Meanwhile I tried to seduce his brother-in-law. He came into my bed most nights, unless he had a Rotary meeting or a Mason's occasion, or was giving a speech in his role as – if not a father of the city, since he was in his twenties, then a father on trial. There he lay beside me, brotherly, allowing himself tentative and inept kisses. I could not see why this was moral, while actually having sex was not. I found the fine points of his theology unconvincing. The point was he was tepidly sexed. For the year I was there this went on, while I got more frustrated and angry. Jasper saw it all, and remarked that, in his own private opinion, I was wasting my time. This meant the man's nature was in question, not his morals.

Twice that year he was happy to be taken home to the farm, where he enjoyed being seen in the light of a potential suitor – by my mother. My father, knowledgeable, said that he was a funny chap.

This story of me and my half-hearted suitor has hidden depths, surely comforting to moralists. If when I first arrived in that house I knew no one in Salisbury, by the time I left it there were plenty of alternative possibilities. You may often see a girl choosing to stay with a young man not likely to be of much use to her, at least not in that area where she seems to be most interested. And why is this? She is protecting herself, not knowing, or only half-knowing what she does.

Twice I enticed him into the bush on moonlit nights, and I admire the ingenuity of my attempted seductions, which included do-it-yourself birth control, found in some history book lent me by Jasper. Again, kissing and cuddling was in order, but he intended to present his virginity to his wife.

My other activity in that house was making dresses. I was being paid £4 a month. The trouble was, just as a couple of years before I was too young for the gang or group life of the teenagers of Umtali, so now I was too young for the grown-up scene. I did go to one grown-up ball, where everyone was kind to me. The dresses I made were smart, and inappropriate for my life as a nursemaid – or even a housekeeper. While I sank or swam, or splashed about the shallows of the grown-up sea, the lady of the house, pitying, offered me what suited her, flowers and frills, and was surprised when they were not a success.

And there was another thing I did. Babies who at first sleep eighteen hours a day like cats become wide-awake crawlers, and then outgrow the playpen – they need entertaining. I took the baby to the park every afternoon for a couple of hours, pushing the pram everywhere, because I did not want to sit down and become part of that scene, females with prams and babies. I was bored. Quite soon I would be pushing the pram around the park for three other babies, one after the other, my own, and I remember those afternoons as the apex, the Himalayas, of tedium. Time crawled almost as it had years before. While I pushed the pram, and chatted to the friendly baby, I wrote poetry in my head. I was in a trance. Over the next few years I wrote a lot of poetry, nearly all of it from that country of rich, pleasurable melancholy, which is like a drug. I think perhaps half a dozen of all those poems were any good.

By the time I left that house I was running it, cooking, and entirely responsible for the baby. I devoured Jasper's books, had clandestine conversations with him about politics and the Native Problem, and spent the first part of every night fighting the virginity of my placid suitor.

I was in a fever of erotic longing, which had succeeded the romantic fevers of my childhood.

I could truthfully say that I spent my adolescence in a sexual trance, of the kind so well described by Christina Stead in *For Love Alone*, probably the best novel written about a girl growing up. But this is what I say when in that part of my mind marked Love.

I could with equal truth say I spent my childhood, girlhood and youth in the world of books. Or, wandering about in the bush, listening, and watching what went on. Here we are at the core of the problem of memory. You remember with what you are at the time you are remembering.

It is my belief that some girls ought to be put to bed, at the age of fourteen, with a man even as much as ten years older than they are, with the understanding that this apprentice love will end. Of course I can see the objections to this. Hearts would get broken – but then they do in any case. Is there something about this idea that ignores the realities of life? Like school and homework? But this paragon of a loving mentor would insist on homework being done, and a proper social life and . . . In some parts of the world, India for instance, it was once usual for very young people, thirteen, fourteen, to be married and then shut up together for sometimes months. Presumably neither was at school, nor expected to be. Enough of theory: this is what should have happened to me. The problem is not Lust, an appetite satisfied by one means or another, but Erotic Longing which is for possibilities at last fulfilled, a transformation, an entry to – it could be argued what. This kind of longing is like homesickness. It is a kind of homesickness, perhaps for past and not future Edens. It is an illness, incapacitating.

I do think the unfulfilled dreams and desires of parents affect their children. I am sure my father's frustrations affected me. That he was blocked in his sexual nature was no secret – at least to me. Not only was there the wistful way he spoke of women who attracted him and with whom he had an instant sympathy, and the regretful way he spoke of men and women married to cold partners, but more than once he said things to me that made his situation plain. Of course I wished he had not, although I was flattered I was his confidante. But I was too young for remarks like 'That kind of thing was left clean out of your mother.' No girl, her mother's rival, will hear this without a pang of triumph, but I was sorry for her, identified with her and so I was in conflict. And when my mother confided in me, I didn't want to hear it. Illness and tiredness, it seemed, made sex too much for her. The things they actually said aloud, and what they told me but did not know they had, together with simple deduction, made up a picture of them, confirmed by a little scene I actually saw when I was seven or eight. Night. As usual my brother and I are in our parents' beds near to the room at the front of the house, because otherwise there would be a large room between the parents in their lamplit evening room and us two, in the great shadowy room under the rafters and the little glimmer of the night-light, later my bedroom. I am in my father's bed, my brother, who is asleep, in my mother's. My father and my mother come into the room together. She carefully sets the lamp down. He puts his arms around her and turns her about to face him. He is gallant and shy, like a boy – or like a rebuffed man. 'And now the children are older perhaps things can go back to how they ought to be, and we could have the door shut between our room and theirs?' He kisses her, and she laughs, but she has turned her face away and is looking over her shoulder at her two little children in their parents' beds, whom she will carry asleep to their beds next door. The door between the bedrooms was always propped open at night until I insisted on its being shut.

My father was very ill when he came to my mother's ward. He was in bed for over a year. He nearly died. He was then seriously depressed. Meanwhile the doctor my mother hoped to marry was drowned. When my parents married both were 'run down', and both in 'low spirits'. My mother at once got pregnant and it was a difficult pregnancy. She then had to cope with an 'impossible' baby. Hardly conducive to love-making, any of this. And then she had her second baby, the longed-for baby boy. I think she fell in love with him. Women do. I've seen it, several times. They may very well love their husbands, all sex and kisses and then Crash, a baby, girl or boy, and she is in love – besotted, obsessed. The husband, poor thing, who is he? He is out in the cold. I think my mother remained in love with her baby boy until he said, No, no, no and walked away from her, a lean, loping, athletic, laconic child, taught coldness by boarding school, who called her Em. Em for Mother.

Easy to say, here we have a passionate sensitive man, married to a cold and sentimental woman. He was certainly passionate and she sentimental, but I wonder. From things both of them let drop, not knowing what they said, it seems that Marie Stopes might have been a good guide to birth control, but was hardly a mine of information about sex.

It cannot have been of help to my father that he had watched her heart break over the drowned young doctor. Or that over her bed throughout their married life was an outrageously sentimental picture of two inadequately draped upwardly soaring souls in a pastel heavenly landscape with the words, 'Guess who holds thee? Not Death, but Love.'

It cannot have been a help to my mother that she had nursed him for months as a very ill, mutilated man.

In short – but enough. Her passions went into her children, his into dreams. Dreams of love. Nightmares of war.

I link this query, for it is one, with a memory. I am reading Bernard Shaw, and he says the human race is over-sexed. I must be over fourteen, for I am conscious every minute of my delicious body, that fits me like a new and longed-for dress. I am outraged. I am furious. I am threatened. Even at the time I knew my reactions were out of all proportion. I felt as if Shaw was taking away something that was my right. Now, this was the spirit of the time all right, pure extract of *Zeitgeist.* No one had promised me sex, love, as a *right*, as my due. Yet I had already learned to define myself in this way. From where had this wind blown? From whom? For all of my life until recently, when AIDS said no, sex, and good sex at that, has been a right – for everybody. But why has it?

A friend of mine, an historian, sage and savant, said to me once, 'The trouble with you people [he meant the West] is that you see everything through sex and politics. These are your imperatives. And incidentally, this makes it almost impossible for you to understand the past when people had quite different priorities.'

10

AND NOW THERE WAS a sudden change of course, like shedding religion, abandoning school, leaving home to be a nursemaid. I was going home to write a novel. These changes, or 'conversions', are not really abrupt, but the result of slow but out-of-sight accumulations of a substance, or feeling, different from the one that temporarily dominates.

I had become fond of the household, particularly the baby. I was grateful to Jasper. I had come to see my bed companion as something like an importunate dog or cat, needing to be stroked or patted. As for the lady of the house, she now seemed to me not at all the simple *jungfrau* in her peasant skirts and braids. For one thing, who had done all her work for her for over a year, while she sat on her sofa, with her little air of matronly complacency, her eyes lowered over her sewing?

Many years later in an airport lounge I saw her and her Jasper, opposite each other. She was an elderly *mädchen* with her grey braids in a coronet, her wonderful flower-blue eyes always on the immensely, grossly, horribly, fat man he had become. Once, he had patronized her. Now she was saying, with her gentle candid smile, 'It is time for your pills, darling, I put them in your pocket . . . can you reach them, shall I get them for you?'

Sometimes I watch a friend whose adolescent daughter is behaving abominably, find myself full of outrage on behalf of my friend, and dislike of the girl . . . but wait a minute, I tell myself, looking at those lowered angry lids, the cold tight mouth, have you forgotten what you were like? – and tell my poor friend, Don't you see? She finds you a threat, you're too powerful for her, she's afraid of being swallowed up. 'What, *me* a threat?' For very few indeed of us experience ourselves as powerful, rather as some frail creature who by good fortune is still afloat in a cruel sea. It was not my parents' strength that threatened me, it was their weakness.

'I will not, I won't be a victim of war.'

And I sat in the evenings listening to the News from London on the wireless and heard the Germans roaring their approval of Hitler, *Sieg Heil! Sieg Heil!*, while he ranted and raved. I was afraid. The man's voice, what he was, operated on the nerves, worked subliminally; useless saying, 'What's the matter with you, he's thousands of miles away?'

There sat my father, a man betrayed. Everything he had said for years was coming true. No need for him to say I told you so: history itself was saying it for him. He had said, Watch the Germans, they're going to get their own back for Versailles, and no one listened to him and to the other old soldiers, no one in Britain was listening now to Churchill, *he* knew what was needed . . . There was something called Stephen King Hall's *Newsletter*, which told the truth. But into that house too came the inflammatory pamphlets of the British Israelites, a sect peculiarly attractive to the British upper class, for it claimed that it was to Britain that the Lost Tribe of Israel wandered, that we were the Chosen People (The Jews? Well, they were mistaken, that's all!), and God had elected us to rule the world, which we were doing through the British Empire. Armageddon was on schedule: there would shortly be seven million dead around Jerusalem. Russia and Germany would become allies and as the representatives of Evil fight the armies of God which were – naturally – Britain and America. I am giving the merest sketch of a credo worked out with all the relentless logic possible when you have a multi-purpose text as material: you can prove anything from the Bible, or from Nostradamus. I can hear my father's nagging peevish obsessed voice – the voice of illness, of failure, see my mother sitting eyes lowered, her fingers fiddling, fiddling with the grey hair at her ear, her body stiff, as if she were restraining an almost irresistible need to leap up and out of the old shabby chair, and run anywhere out of the nightmare she was stuck in, this husband every day more ill, this rude antagonistic daughter, and her polite, armoured son.

There was not only Hitler, but Mussolini in Abyssinia, and the fighting in Spain which my parents said was the beginning of the war in Europe.

I stood leaning against a tree on the edge of the field that had unsuccessfully grown tobacco, that crop which was making some farmers rich, and looked into a red and rioting sunset and thought so hard about Spain it was as if I were there. Why wasn't I there? There was a Brigade of Britishers fighting to save democracy and if I tried, perhaps I could get myself over to England and then . . . but where was I going to get the fare? I was seventeen, and they wouldn't have me . . . The books I had read about women driving ambulances, nursing, running field hospitals in France, Russia, Serbia, my mother's reminiscences of the old Royal Free – all this fermented into that rich enjoyable melancholy which is the natural food of adolescence – and the first signs of that exaltation, a secret pride in suffering, which is war's poisonous food, the other face of the old soldier's 'I've never known any comradeship like that since!' Cadences of sorrow, elegiac laments . . .

Leaves falling
Each one I have known with my fingers,
I walk over them
Feeling the dark veins burst as I tread,
Each one I have known through the days and the nights,
My blood theirs . . .

This was called *After a War*, but *Before a War* was more like it.

But really I was writing prose, my first novel, on the mountain of a typewriter sent all the way from Johannesburg. It was a short satirical novel, mannered, stilted, making fun of the gilded youth, the young whites whose ways I had after all only just glimpsed. I would become one of them within a year. Their pretensions, their privileges were contrasted with the lives of the blacks. I did not know enough to write it. This production too was later torn up in transports of embarrassment. I then wrote another novel, very fast, in a sort of trance, and in longhand. This time the inspiration was Galsworthy, whose novels were everywhere. I was uneasy because his rhythms were setting my pace, and because I knew he was hardly the best exemplar. Everyman's Library was supplying me with much better: parcels of books arrived at the station, and were brought to me in the mail sacks. I opened the parcels with a beating heart, longing for the newer shores of literature. I was reading, particularly, D. H. Lawrence, but the intense physicality of that prose, the evocation of England or Italy, was so immediate, so violent, that I would emerge from *Aaron's Rod* or *The Rainbow* as if I were standing there, looking at the tigerish hills of Italy or in a bluebell wood in England – all this was no help to me, was too much its own world and place. He must have influenced me, but I could not read back what I had written. I tore up thousands and thousands of words and went back to practising on short stories.

Other futures were being energetically offered to me. My mother said I should be a nurse, like her. I drove her into Salisbury. This meant leaving my father alone for the day. By now I had my licence. On my sixteenth birthday I drove myself to the police camp at Banket. A very young man, about twenty, the newest British South African Police arrival, got into the seat next to me. 'Drive down there,' he ordered. I drove fifty yards down a bush track. 'Stop' he said. 'Now reverse and turn.' I did. 'Now drive back to the camp.' That was it, the driving test. The police at the camp were used to the children of farmers, who had been driving for years, arriving for a test and, often getting the certificate without even being asked to drive down a farm track for a few yards.

At the Salisbury Hospital my mother, in her smart hat, gloves, her bag neat on her lap, sat opposite the Matron behind her desk. A sullen girl watched those two antagonists, full of secret derision, longing to be accepted, while at the same time hoping she would be refused.

My mother said I was physically as strong as a horse, and clever too, and the fact I had no certificates to prove it did not matter, nor that I was too young. The Matron did not take kindly to being patronized by the Royal Free in London, and she could see that I would make an insubordinate nurse who would not accept discipline.

Next, would I like to be a vet? enquired a pink and white young man just out of England, and I jested that I had a special feeling for koodoos. He punished me by inviting me to watch a peculiarly repulsive operation – but I will spare the reader – and I decided that if this was being a vet, then no. Yet I could have made a vet, or a doctor. I could have even made a Matron, since I am capable, like my mother. Best of all, I could have been a farmer. We know what we could have been good at. And with what wistfulness I have secretly wondered how I would have done if lucky chance had made me one of that company of dazzlers, the physicists, who have been the adventurers of our time.

That year 1937 is described by me according to what memory-mode I am.

It is the year I wrote two bad apprentice novels.

It was the year of my first great love. He worked as an assistant on a nearby farm, twenty-five years old, and was unlike all the other farm assistants. He was reserved, proud, with an air of being the possessor of a secret. This came from not speaking his mind. He got newspapers from Home, listened to the BBC, and, while he could not afford to be critical of the manners and mores of the District, let drop remarks that made the farmers and their wives look suspiciously at him. He was tall, he was dark, he was handsome. His skin was Mediterranean olive. They said on the verandahs that he had a touch of the tar-brush. Everything they found disturbing about this aloof, saturnine man found expression in the whispers, 'Take a look at his eyes . . . at his hair . . . at his nails.' Needless to say this aroused a passion of protectiveness in me and I loved him even more. When I did go wandering in the bush that year, to shoot pigeons or guineafowl, I was in an amorous trance. The sunset light washing over feathery pink vlei grasses, the passionate cooing of the doves, transported me into . . . my dreams of love might have lacked the detail experience would have provided, but they were intense to the point of real illness.

Few girls have not written such letters. There are letters that come out of stock: 'I love you, please take me away from here.' 'Your letter was, I confess, something of a surprise to me.' (Hypocrite! Liar!) 'I confess I do feel something a little warmer than friendship but of course you are much too young.'

My letter was unforgivable. I said I did not care about his tinge of colour, everyone was stupid and prejudiced. His reply was cool, correct: he probably did not know the District had so finally diagnosed him as not one of them. For years I writhed with shame, but what is the use of that? If it is felt I was writhing with shame a lot then yes, I did. It is a young person's affliction, partly because of social embarrassment, partly because it seems there will never be an end to the ineptitudes of being young. Anyway, this goes on in one form or another all one's life. 'Do you mean to say . . .' a middle-aged, an elderly, an old person may demand of herself. 'Are you *really* saying you were as stupid as *that* five years ago?'

When the war started, two years later, he joined up and was soon killed in North Africa.

I had an admirer that year I made use of as ruthlessly as girls do.

He was the Watkins' older boy, my age, very fat, slow, amiable, and I hope not as painfully in love as I was. When there was a dance or a gymkhana he was my companion, and my mother was agonized that I would marry this boy who was perhaps a thousand degrees lower on the social scale than she was hoping for.

The Watkins' house had been a place of fascination for my brother and me. Built among granite kopjes, it was always very hot or very cold, and when there was a storm balls of light bounced down the verandah and vanished into the telephone with a *ting*. In those days the existence of these electrical balls had not been admitted by science. Once my brother was there during a storm. 'Just like a small football,' said he, 'they bounce like anything.' The Watkins boy was not interested in the lightning or the farm. I recognized my state in him: he only wanted to leave, get anywhere out of the confining scenes of childhood.

There were dances that year at the village hall. Suddenly there were a lot of young people, a new generation, who drove in to Banket from miles around to dance to the insidious intoxicating music of the 1930s played on a wind-up gramophone. My young man and I, unable to dance, I too awkward, he too shy, stood by the gramophone, wound it up, fed it records and watched the older young men and women steering each other about, stiff in each other's arms. The young men, who spent every minute of their ordinary lives in old khaki shirts and shorts, wore ugly suits, but the girls shimmered like ice-creams in that ugly dusty hall. The fashion was for crepe de Chine or satin dresses in white or pale blue or pink, cut on the bias, smooth over every curve, with pearls in the 'vee' in front, or, daringly, knotted at the nape to dangle over a back cut down to the waist. These were the sons and daughters of the post office, the garage, the store, the railway station, together with the farmers' children.

When the horses flashed past at gymkhanas, the Watkins boy and I stood at the rail and he stared at me, while I looked past him at my unattainable love, who was being attentive, a perfect gentleman, to his employer's wife, a gentle unhappy middle-aged woman, who was in love with him – my father said so too – but he could not stop himself sending glances at the averted face of the slim blonde rider, leaning against the rail a few paces away, waiting for the next race to start. Before she sauntered off, trailing her riding whip, she turned to give him a cool smile. This scene has remained in my mind, entitled, The Comedy of Love. Goya, I think.

Or we sat side by side when a touring concert party came from Salisbury, in the same hall, but furnished with chairs, and I put a lock on my tongue not to say what I thought of the clumsy entertainment, for he thought it wonderful. He too was killed in North Africa fighting Rommel, soon after the war began.

All that year my father suffered crises with his diabetes, when I drove him and my mother into Salisbury over the awful roads, and he disappeared into the hospital while I sat in the car under the trees and waited. For hours. By then it was not only diabetes, but its attendant complications, which they manage much better now but are still grim enough.

That was the year I taught myself speed-typing, and shorthand too.

That was the last year when I was part of the bush, its creature, more at home there than I've been since in any street or town. My last year as a farm girl able to turn her hand to anything: the technology of that time has gone, and now all farms have proper electric or gas cookers, electric light, piped water, refrigerators.

It was the year when – it cannot be said that it was an important incident, that is, compared with the momentous affairs of nations, but I think of it, sometimes, and it has an always wider application, for we are not only back in the realm of Timing, the exact clocks that rule life's process, but cruel Chance.

I was left on the farm, while neighbours drove my parents to Salisbury, because my father was ill. There was an incubator full of eggs due to hatch. Both parents kept repeating helplessly that I shouldn't be left alone, a girl of seventeen . . . and so forth, but off they went and I was full of triumph alone in my house that was like my other skin. It was cold, very, the dry bitter dusty cold of the highveld winter, when the bush becomes a ghost of itself, and in the mornings the stones numb the soles of your feet and your hands go thick and clumsy. The incubator was in a little room at the end of the house. The old window and misshapen door let in every breath of cold, and the thatch seemed to hold a pool of cold under it. Beneath the incubator was a candle-lantern, its heat dispersed through ducts among the eggs that were only five days off hatching. Suppose the candle went out? – and I could see the flame sway and surge about in the draughts: if it did, six dozen chicks would die in the shells, I would be the murderer of seventy-two little lives. I packed the sills of door and windows with rags. The dogs, who were never shut out of anything, whined and waited. The cats, their feelings hurt, miaowed and sulked. I scarcely left the room where the drama was invisibly developing. I lay on the bed and read, continually lifting my eyes to make sure the candle still burned. Suppose we ran out of candles? Or matches? Near the incubator I put a soap box lined with an old eiderdown where the chicks would be transferred as they hatched. Outside in the chicken run hens still unaware of their futures ran about after the cocks. Every three or four hours I gently damped the eggs with tepid water and imagined the embryos within, hideous, I knew. I turned the eggs as a hen might, making sure those great feet of hers did not miss a single egg, and I brooded over those eggs as if the future depended on seventy-two chicks.

Then on a blowy and bitter night I woke, but there was no candle flame, only the lamp illuminating the thatch, the white walls, the eggs. I rushed to the eggs. They were chilling, but not cold. I lit the candle. I spread the eiderdown over the top layer of eggs. Were they dead? I would not know until the moment came for hatching. Meanwhile there was a telephone call from Salisbury: my father was very ill, he might die, and I would just have to hold the fort. My mother's voice, for these so frequent announcements, was dramatic as always – but I no longer listened, could not listen, calamity should not be declared too often.

The day came for hatching. I sat with my eyes on the eggs. This was the moment of truth. Never, ever, has time gone by so slowly – that is, adult time. I sat in a little room that seemed to be full of cold air like currents of cold water and then . . . there appeared on an egg the little rough place that meant . . . I held the egg to my ear, and heard the tap tap tap of the hidden chick, and wept with excitement and relief. Out flopped that hideous chick, dried at once into adorableness, out flopped another . . . soon all over the cracking eggs lay and sprawled the wet monstrous creatures, and between the eggs trembled the pretty dried-out chicks. I ran out, found the oldest and most experienced hen, and put her into a pen where the nest box was already lined with straw and feathers, and when I brought out a couple of dozen little chicks and put them one by one into the nest she seemed not to know her own mind. Then, her brain switched gear, she clucked, and delicately trod her way among them and became their mother. So with three other hens. But when my parents came back one look at my mother's face told me the seventy-two chicks would be my private and personal drama. I used to watch those broods of chicks following their mothers over the rocky crown of the hill, while the hens kept an eye cocked for the hawks, and I thought, Suppose I had woken even ten minutes later?

A travelling salesman remarked that there were jobs for girls at the telephone exchange in Salisbury. Knowing that I would be stopped if I said anything, I got a lift from Mr McAuley, flirting with him as my contribution to the journey, 'Tigger' being good at this, and when I got to Salisbury, went into the telephone exchange, and although I had followed none of the correct procedures, got a job at once from the two gentlemen who ran the place. I was too young by some months, but they thought I could manage. Besides, they did not get enough applications from the right class of girl. I then got a room, from an advertisement in the *Herald*, in the house of a widow – but it is in *Martha Quest*. That house, old style, is now 'listed', because it occurred to the authorities they had pulled down so many of the old houses there soon would be none left to remind people of the old days. What they are putting up now, and in some of the most savage and wild parts of Zimbabwe, are houses inspired (distantly) by the thatched cottages of England, tight little houses with prim little windows and dinky dormers, without verandahs or shady places, or sheltered walkways. No writer can come up with anything as merciless as what Life Itself, that savage satirist, does every day.

11

THE HOURS AT THE TELEPHONE EXCHANGE were not long. I enrolled for evening classes in a secretarial school to improve typing and shorthand speeds, and I answered advertisements. My only qualification was a driving licence. Girl chauffeurs were then unheard of. Twice I found myself being interviewed by a surprised employer, who was intrigued in the wrong way. I tried for a job on the *Herald*, but would have been expected to do Society Notes. Then, remembering my success at Hemensley's, where I still might drop in to have a chat, I got an interview with Mr Barbour, who owned the biggest women's shop in town, was a town councillor and an important public man. He would be happy to employ me as a window dresser, but at a rate of pay which meant I would have to live in a hostel for poor girls. He recommended one subsidized by the city. 'Tigger' did not fail to inform him he was using his public position to get cheap labour for his shop. The trouble was that cheek is its own antidote. He was only too happy to argue the thing on its merits for as long as I wanted. Never had I heard self-interest so persuasively presented as a public good – and after all, I had been brought up with some of the most convoluted thinking in the world. Mr Barbour put it to me that to make a profit was to prove success, that commercial success was in the interests of this new country, that when the Town Council subsidized the rent of his girl employees, it was only paying for general prosperity.

I worked in the telephone exchange for a year. Most people, when I say I was a telephone operator, are uncomfortable: much better, they seem to think, if this unfortunate episode was forgotten. (The genteel 'au pair' redeems being a nursemaid.) It suited me very well. For one thing I understand absolutely why women, asked why they put up with boring and repetitive work, go on with it, saying, But it allows me to think my own thoughts.

It took me a day to learn the procedures: a process like learning to ride a bicycle, which at once becomes automatic. I think such an exchange would now be possible only in some remote town somewhere in Africa or South America. You wore a headset. In front of you was a flat board with plugs on wires that you pulled out and fitted into sockets on the upright panel in front of you – connecting citizens of Southern Rhodesia with Johannesburg or Cape Town, or even London. And there were the party lines, shared by farms strung out along a road. To connect one farm on a party line, let's say, near Salisbury with another near Sinoia might take half a morning, because on farms people tend to be out on the lands or checking the animals, and do not hear the telephone. In slow times I read, not too surreptitiously. That is where I read *Resurrection*, knowing that every word could be matched here, in this 'young' country, by people I probably knew. What interests me now is that it was to literature I had to go for such thoughts: Tolstoy's landowner's crisis of remorse, his view of himself, came from religion.

There were ten or so girls, and a supervisor. Matey, that's the word for life in the exchange. Recently I met a woman who said she had gone to the exchange to replace me when I got married. Asked how she found the place, she said at first terrifying, but I was patient with her. I was a quiet and thoughtful person, so she said. I was glad to hear this, for what I remember is the chatty brightness of 'Tigger' – who was certainly the person who dealt with the social life that at once swept me away into drinking and dancing.

Meanwhile, an encounter, described more or less as it was in *Martha Quest*, with the local Reds, spoken of with dislike, in lowered voices, for they were seditious, dangerous and above all kaffir-loving. Dorothy Schwartz, who later became my friend, waited outside the exchange one afternoon, to say she had heard I was interested in the Native Problem, and perhaps I should meet the Left Book Club crowd. They were a lot of spineless social democrats, but better than nothing. In the provinces, don't bother to ask how someone has heard about you. Boredom lends the brightest of wings to gossip.

My disappointment that afternoon exaggerated my reactions. But remember, these people were the last word in social daring, they were the cutting edge of thought. The women, they were the worst. They all looked like gypsies, in flailing coloured beads, and trailing linen skirts and Hungarian blouses. They complained all afternoon about the hardness of their lot, and these complaints were directed against the apologetic men, whose fault it was they were on these lawns, with children who dragged them down and prevented their real selves from developing. Men were the villains, men were criminals. Had they not chosen to get married? – I was (silently) accusing them. Had they been forced to have these children? Who had held a gun to their heads? (Two or three years later I would have said, It was the guns of war.) Were there not three black nannies in attendance? Perhaps their self-pity was triggered by this very young, attractive, unencumbered girl, who was sitting watching them so critically.

The three husbands, the three wives, were malicious about members of their group not present. All of them complained that really they ought to be fighting in the Spanish Civil War – I found this an uncomfortable mirror.

As I left I was instructed to cancel my subscription to the *Observer*, much too reactionary, and at once order the *New Statesman*, of which I had never heard. They reminded me of the mentors, every one absolutely sure of their credentials, who had ordered me to eat only vegetables or meat, or eschew milk products, or eat nothing that wasn't steamed.

Their unlikeability postponed my involvement with the Left for four years. But my main reason for disliking them so much is not in *Martha Quest*, for it had nothing to do with her. What I could not forgive was how these women demeaned and diminished their children, talking in front of them, as nuisances, as burdens – unwanted. Yes, they were tiny children, but I could remember myself, as a tiny child, hearing this talk.

Plans to equip myself to become an efficient secretary did not last a month. Young men soon were knocking on my door on 2nd Street. There were too many young men for the girls in the town, or, to put it another way, too few girls for the men – that is, of the class of people who used the Sports Club and were the golden youth of the town. It is not unusual for attractive girls to misunderstand this situation. If there had been too many girls for the men? I would have done all right, but no more. I was eighteen. Dark-haired. Dark-eyed. A good body, already established in that rhythm it would follow for decades: slim, then plump – strict dieting, then slim and then plump. I was full of health and raw vitality. I had chosen my parents well, according to Bernard Shaw's prescription. I do not see how physically I could have had a better start, and I abused the gift, as if health were inexhaustible. I was sixteen, on the farm, when I lit my first cigarette, after having for years despised my parents with their stained yellow fingers, with the bits of tobacco dropping from their hand-rolled cigarettes, their look of avid need for the smoke that poured from their mouths. So much for 'I will *not*. I simply will *not*.' Far from feeling sick, that first delicious breath of smoke told me that this was what I had been born for, and I smoked with pleasure till I gave it up a quarter of a century later. As soon as I arrived on the Sports Club verandah, I drank. Everybody did. Everywhere. In every country. It was smart. It was clever. It thumbed the nose at authority. But the patterns of drinking in Southern Africa might have been especially designed to do the maximum amount of damage. The men went out from their offices to drink in the hotel or bars, at lunchtime. Often they didn't eat. Serious drinking began for everybody with the famous sundowners, at six o'clock, and went on, with nothing at all to eat except perhaps a peanut or a potato crisp, until dinner two or three hours later. If we went to the bioscope, we might drink and not eat. We danced often, and drank all night. We drank a lot of beer, from the Castle Brewery, but also a horrible mixture of Cape brandy and ginger beer. The men drank whisky when they could afford it, and the women drank gin. Gin and lime, gin and lemon, gin and tonic, Pimm's Cup. We drank a lot of liqueurs, probably because they are sweet and blood sugar was low. Most nights we went to bed at least tight. I often had hangovers, not the kind when you can't move, but when you feel bad and listless. The young men were always 'going on the tack'. Being drunk was a joke. Bevies of solicitous girls put to bed incapacitated youths, sometimes one after another. Maternal: boys will be boys. Not long ago in a country town in Ireland I watched a wedding party reaching its noisy culmination. The men drank themselves stupid, playing the role of gay dogs, while the women sat bored and patient around the walls, each holding a glass of sherry, until the moment when they had to support the men home and put them to bed.

All that is in *Martha Quest*, the manners and mores of the time, and it is 'true', well, more or less – the atmosphere yes, taste and texture and flavour, yes, but sometimes several people have been put together to make one, and of course the story has been tidied up. Every novel is a story, but a life isn't one, more of a sprawl of incidents.

What a lot I did pack into that year. I did not only connect telephone calls and dance and make dresses and go to the bioscope, I read. How I did read. Lawrence still, for he had an intoxicating quality that is not there at all now, at least for me. Thoreau and Whitman. Virginia Woolf had joined Olive Schreiner. I felt I had two elder sisters – a role I am told I now play for some young people, not only women. If the people I was surrounded by did not understand me, then Virginia and Olive would. What would Virginia Woolf have made of Olive Schreiner, I wonder? Or Olive of Virginia? It is a thought one may entertain oneself with. I read books I was too young for, Carlyle and Ruskin and Renan, for instance. But the Russians, they entered my life like a thunderclap, Tolstoy and Dostoevsky and Chekhov and Turgenev and Bunin and the rest, and here they still are. Proust. Thomas Mann. Stendhal – an ever-accelerating roller coaster of exploration, as the parcels arrived from London. Mrs – but I've forgotten her name – brought in yet another parcel and said, 'I saw the light under your door last night. You can't burn the candle at both ends.' But what she wanted was to talk, and so another thing I did that year was to sit in the back verandah and talk with – a widow? A deserted wife? She had spent her life on farms and mines and now she was alone and she wanted to make me cups of tea and tell me my dance dress was lovely and she had just such another when she was a girl. She mothered me. When my mother came flying in, all anguish and agitations because of her fantasies about what I might be doing, she could find me talking to my landlady. Whom she then cross-examined about my doings. 'She doesn't get in at night till twelve or so?' 'Yes, well, but young things must have their fling.' 'You are drinking too much!' she accused. Nonsense, I said, meaning, But it is what everybody does. And certainly there is something contradictory about the drinking I did in those years. The fact was, I was a bad drinker, and that probably saved me. The idea that one might learn how to drink had not occurred to me, along with other things that I could usefully have learned. Like, for instance, that attitudes to girls vary precisely with amount and quality of the supply of girls. Or – any of the other things I had to learn painfully for myself.

In 1938, 1939, my idea of myself, my possibilities, had little connection with reality. I was in a crescendo of emotion, public and private – as if they can be separated. The conditioning of my childhood for war was being reinforced by newspapers from Home, the BBC, the local radio – and how people talked. In the voice of every man (and shortly this would be women too) who has fought in any war, or been near one, there is always that tone of regret for intense experience. We are sensation junkies, predisposed to excitement, and if that means danger and death, we are ready for it. Every generation has been talked into war by the nostalgic voices from the one before. All that year I was dreaming of rushing off the moment war was declared, to be a nurse – a soldier – a parachute jumper into enemy territory – a spy for my country – an ambulance driver. And what stopped me leaving Salisbury there and then, so as to be in the right place, London, at the right time? For one thing, money. I had none. I could not ask my parents for my money. Not only pride forbade. I sometimes wonder why it is that our lot – my peers – would rather have died than ask our parents for support, and left home the minute we could, only to be succeeded by one generation and then another whose one idea is to prolong dependency as long as possible. This is not a criticism of either. Both ways a price is paid. If you cut ties with home very young, emotional ties are cut too. If you live at home, that doesn't come cheap. But the interesting thing is, why does the one generation's imperative – so taken for granted there is no need even to spell it out – change into its opposite for their children?

Not having money was only part of it. My experience was the farm, the Vumba, this little colonial town, and – briefly – Johannesburg. I was as raw and inexperienced as a black girl now of the same age in Zimbabwe, whose lack of money and opportunities makes Britain, Europe, seem as far as the stars. But in theory I could have left. Instead I read, I danced, I flirted, and dreamed of heroic adventures. I would explore the Gobi, I would live by myself in a hut in the Kalahari.

The other intoxication was my body. Is there any pride fiercer than a young woman's? Now I read, am told, that girls are all dissatisfied with thighs, waists, breasts, legs, something, everything. I had not been subjected to years of advertising, of beauty magazines, of fashion. It never occurred to me to be ashamed of what I had, even when in a plump phase. I used to stand among people, knowing my body was strong and fine, under my dress, and secretly exult, or look at a naked arm, or my hair in the mirror, and thrill with pleasure. This hidden strength sustained me as much as anything through months which were like shooting the rapids.

And now, a small sociological, as much as a literary, note. In *Martha Quest*, I describe Martha lying in the bath, looking at her nakedness, while a thunderstorm crashes and bangs outside, and her landlady waits to make her a cup of tea and scold her for – something or other. When I wrote it, for a long time I hesitated about describing her joy in her pubic hair, young and glossy, and growing in three perfect little swirls. But I knew there would be a fuss and if this was a question of principle, then it wasn't my principle. Later, in the 1970s, I wrote a story called *One off the Short List*, and in it a woman is described as having golden fringes of underarm hair. An American publisher, and then magazines, would not print the story because of that hair. Yet in America you might describe any killings, tortures, rapes, horrors of war, cruelties. Not underarm hair in a story about seduction and sex. But I insisted, for by then it was a matter of principle.

And, the strongest intoxication of them all, dance music. When I came into Salisbury from the farm, I was at once possessed by music. As were we all, dancing to the heady, seductive tunes of the 1920s and 1930s. Has anyone ever studied – but I mean *really* studied – the probable effects on whole generations of young people perpetually driven by the rhythms of drugging music? And – but now we enter the realms of what they call 'the mystical' – it is surely not without relevance that the whole world was dancing to the same tunes, often at the very same time.

I've got you under my skin,
I've got you deep in the heart of me
So deep in my heart
You are really a part . . .

Well, sometimes I wonder.

A scene. I am wearing a black velvet evening dress I had made that afternoon. It was cotton velvet: within a year I would finger it once, reject it. The dress was a classic shape for that time, the back bare to the waist with a halter neck, low in front, fitted to the thighs and flaring gently. A man much older than the boys of the Sports Club is sitting on the arm of a chair, examining me with a smile which I am too young to know holds all the regrets of an ageing lover of women. The dance music is throbbing from the dance room and I am restless, already half dancing, wanting to abandon myself to it.

Heaven, I'm in heaven
And my heart beats so that I can hardly speak.
And I seem to find the happiness I seek,
When we're out together dancing cheek to cheek . . .

He says: 'Who's taking you to the dance?' I say, So-and-so. 'That dress is wasted on a little boy like him,' he says, smiling, his mouth bitter. He turns me about, with the authority of male sexuality, and then becomes, in a breath, a different person. 'Are you wearing a brassière?' 'No.' 'Panties?' 'Of course,' I say indignantly. 'Well,' he pronounces, 'you have a perfect figure. But it is a pity your left breast is a third of an inch lower than the right.' 'I daresay I'll manage to live with it.' 'I daresay you will.'

This little reminiscence could be seen as the equivalent of the pictures of their youthful selves old women put prominently where visitors can see them. What they are saying is, Don't imagine for one moment that I am this old hag you see here, in this chair, not a bit of it, *that* is what I am really like.

Years and years later I was sharing a room for some reason with a pretty girl of twenty who was more than usually dedicated to her body. Deliberately she allowed the towel wrapped around her to fall, displaying a beautiful back. She half-turned, smiling, at an off-stage invisible recorder of her perfections, so I could see breasts that might or might not have a third-of-an-inch fault. She smiled at me, cool, triumphant, and went out. Pain was slicing through me for what I had lost. And, too, because I knew I had been every bit as arrogant and cruel as that girl.

A young woman sensitized by music, and every molecule simpering in abased response to the drums of war, a young woman in love with her own body – she did not have a chance of escaping her fate, which was the same as all young women in that time. If, then, I could have seen myself with as cold an eye as I use now . . . but no, it would not have saved me. No point in saying, I will *not* – when the Fates are playing war music, dance music . . .

Nature (Gaia? The Life Force?) was preparing us to replenish the population, due to be decimated. But it wasn't, not in Britain and America at least, so perhaps Nature (The Great Regulator? The Great Mother?) was responding to the last war, World War One, the War to end War, with its millions of dead, just as generals are always well equipped to fight the last war, but not this one.

A question: in Russia, in Germany, in Japan, where in fact there would be millions of dead, was Nature (or the *Zeitgeist*) more actively softening up the girls, by fair means or foul, so minds and wombs would be ready to co-operate?

A plot, one of the ten basic plots: young woman, or man, disadvantaged, comes to the big city. After vicissitudes he finds a benefactor, she a husband. I found a husband, Frank Wisdom, a civil servant. I was not in love with him, though such were the intoxications of the time it was easy to think so. He was not in love with me. In fact he was engaged to a girl in Britain, where he had gone on leave a year before. I could excuse myself by saying he was ten years older than me, and did not mention the girl until it was too late. But that is not the point, which is, my calm ruthlessness as I took her place. This is a basic female ruthlessness, female unregenerate, and it comes from a much older time than Christianity or any other softener of savage moralities. *It is my right.* When I've seen this creature emerge in myself, or in other women, I have felt awe.

If I was part of the general delirium of excitement, I was also quietly miserable. The undertow, a feeling of being dragged or propelled, of not being myself, of long ago having lost control, that was as strong an emotion as I have felt. Emotion? – no, it was a lack of emotion. Perhaps the same as the numbness, a kind of chloroform, that overtakes someone being eaten by a lion.

My parents were understandably surprised when I arrived on the farm with Frank, having announced for years that I had no intention of getting married or having children to tie me down, at least not for a long time, perhaps never. But they were relieved, because Frank was as near to their prescription for a suitable husband as could be found in the Colony: the doctors, lawyers, or military men my mother wanted for me were all in England. My father took it for granted I was pregnant. I was but didn't know it. 'It can't happen to me.' Young women believe it can't happen to them. There is an absolute barrier between the idea of oneself standing there inside one's young strong body (every cell of it is silently dedicated to the business of getting pregnant and then being pregnant), a body that is *yours*, so you think, and really, but *really* knowing how easily you can get pregnant. Similarly one may sit by a dying person day and night, but no matter how hard you try, the actual consciousness, the knowledge of death, as this dying man or woman is experiencing it – this friend not a yard from you – is out of reach, you cannot enter it.

If my father had said to me, using the knowledgeable male authority that had saved me more than once, 'You are making a mistake. You are going to be sorry. And you are much too young and uncooked,' then I would have been secretly relieved.

But in fact we were quite well suited, at least at that time. For one thing, we both had to conceal our seditious thoughts about the Native Problem. We both subscribed to the *New Statesman*, regarded by the white citizens as something not far off the *Communist Manifesto*. We were both rational, and un-religious, perhaps I should say anti-religious. Quite hard to find now that particular tone or timbre of thought, for it was a long way from 'Science, not Religion', more a question of personal integrity. We wore our atheism, or agnosticism (we might debate the exact degree of either) like religious medals. Because we were in a minority we felt close, intimate; believed ourselves to be made of the same stuff and substance because we could exchange sarcastic comments over a cartoon in a newspaper or ironical looks at a 'reactionary' remark. And we were apparently alike in character, for we shared a way of behaving, of presenting ourselves – a style, if you like – which was no-nonsense, practical, impatient with difficulties. This was essentially the confidence of young people who have recently understood they can cope with the machineries of adult life, which most of us secretly doubt, when we begin.

There was a graceless wedding, which I hated. I remember exactly how I felt: it is not a question of the inventive memory. In the wedding photographs I look a jolly young matron. It was 'Tigger' who was getting married.

And off we went on a honeymoon to Beira. The Sports Club rugger team was playing Portuguese East. A young married couple accompanied us, Joyce and Bill Blair: they had monitored our courtship, if that is the word. I thought of them as worldly, sophisticated. She was from Singapore, all glamour and sleek clothes. We drove down to Beira, fast, drunk, dangerous – the road from Umtali was then only a bush track. We saw elephants, and stopped to cheer them. Luckily they were indifferent. Beira had streets of sand edged with flame trees, and one-storeyed houses and shops, most of them Indian shops.

We were entertained by Portuguese friends of the Blairs for a lunch that began at one and went on until five or six in the afternoon – food I did not know existed. At sunset we were swimming, drunk, in a warm muddy sea and then went into the hotel, a vast wooden structure that stood on piles in the sea, and went on drinking among clouds of mosquitoes and midges. The hotel, the whole town, was full of Rhodesian rugger players and their admirers singing rugger songs and making jokes about dagoes. They climbed up lampposts, threw down a statue or two, behaved like louts. Which they were. But were expected to be, even approved of.

There was one intimation of more civilized ways. At the dance for the rugger players I sat near a young Portuguese woman, and admired, making conversation, her evening bag, an affair of gold and red sequins. She at once presented it to me. I was upset, because I knew they were poor. But there could be no argument. It was explained to me that there were societies where admiration was inevitably followed by a gift, and where one must be careful about what one admired. Portugal had been colonized by the Moors, been taught by them the chivalrous ways of the Arab civilization. I kept that bag for years, like a talisman, and when I came on it in its place at the bottom of a drawer, remembered that there were places in the world where reigned the grace of the heart.

We drove back all night through the forests of Portuguese East, the four of us, to Umtali and then up to Salisbury just in time for the men to get to their offices, hungover, unwashed, hungry, stinking of beer.

I had understood that I had married one of the boys, the lads, but now it seemed as if the whole town was celebrating not only our marriage, but others, for every day yet another couple appeared bashfully on the Sports Club verandah, surprised by love, while everyone in sight began whooping and shouting and calling for drinks.

We were in a small flat owned by Frank's friends, a middle-aged couple who owned property in the more raffish parts of the town, and a bar or pub which they ran as if they had never left England. They were both short, stout, with fair stubbly hair, high-coloured cheeks, little blue eyes. They kept a watchful, shrewd, non-judging eye on me, while giving me tips on how to be a good wife. Rather, how to adjust myself to Frank.

When we bought a table made of local wood, she stood by me while I fed it linseed oil and polished it. 'You aren't going to get any good out of that lovely bit of wood unless you feed it with elbow grease, my girl.' And when Frank bought army boots, since like every young man in the country he thought only of how to get into the army, months before there was any question of a call-up, she fitted her bulk, sighing, into a chair, to watch me knead and soften those boots between my hands. 'You know, my lovey-dove, you mustn't mind if all that work you're putting in doesn't get what it deserves.' Thus did she warn me, while she jested with Frank. 'What are you going to do with those feet of yours on route marches, Frankie boy? In the army you have to use your feet. You can fall out of a rugger match, but you can't get out of the army.' And then he, 'Oh come on, give me a break, don't be like that, these boots'll see me through anything.'

Making thick leather as supple as suede, polishing a table top till you could see your face in it – this devotion to married life, you'd think, was total, but all the time I dreamed of escape, not from Frank, who I liked well enough, but from a life which I was every day scrutinizing more critically. It was not too late for me to reach England, and then fight. Women did, didn't they? I could shoot, couldn't I? I was tough, wasn't I? Fitter than Frank, certainly, who suffered from years of drinking too much.

His story was a common enough one then. His parents came from Australia to Southern Rhodesia, to try their luck, and were farmers and miners. They had three children. Frank, his brother George, his sister Mary, had a childhood of ups and downs. When Frank was fifteen the family were in a straitened time, and he left school and joined the Civil Service, as was then possible, provided you undertook to pass the necessary exams. This he had done, studying at night. He had lived in furnished rooms, watched pennies. In his first office he found Dolly Van der Byl, who was older than him by several years, and she befriended him, showed the poor country boy the ropes, told him he should eat better, not drink so much. He always said how much he owed her, how kind she was.

While I dreamed about getting out of the country, cooked for

Frank, and danced and sundowned, it was borne in on me that I was pregnant: a fact that had been apparent to some of the grown-ups for weeks. When they suggested I might be, I only laughed. My doctor said no, he never did abortions and healthy young women should have their babies young. This was Nature's plan. Now I think there is a good deal to be said for this view.

It was taken for granted by both of us that there should be an abortion. The boys and girls of the Sports Club were unanimous that it was irresponsible to have children, the world was too dangerous, too precarious. Women wanting abortions went south to Johannesburg. But Frank did not know anyone in Johannesburg except a student in the medical school with whom he had played rugger. I took the train to Johannesburg, six people in the compartment, second class, on a concession for the wives of civil servants. I found a cheap hotel, and took a taxi to the medical school. It was the midday break and what seemed like hundreds of students, all male, stared at me. Trying to match the sophistications of the big city I wore a smart dress, a black shiny straw hat, carried a new black handbag in which there was so little money I did not know how I was going to eat. I asked a passing youth if I could speak to so-and-so, watching how sniggers and smirks spread through the crowd. At last a young man came unwillingly towards me, announcing that he was very busy. I said Frank Wisdom had sent me, I was Frank's wife, and Frank had said that he, Frank's friend – did he remember the match in Salisbury last year? – could tell me where I could get an abortion. He replied that he didn't understand why Frank thought . . . but perhaps the forlorn desperation of this nineteen-year-old matron clutching her handbag touched him, and he said nicely, no longer sniggering, that he would find out and leave a message at the hotel.

In a dingy room, full of furniture that could not be found now, so thickly varnished it looked as if it had been made out of new toffee, I sat at the window, and waited for the telephone call. I was listening, too, to the Ossewa Brandwag (a Nazi organization) rioting at the end of the street against the possibility of the South African government supporting Britain and America in the forthcoming war. A message was left for me at the desk that I should go to such and such an address. Next morning I was in an even dingier building, waiting my turn with several women. At last I went into an office where a Coloured woman sat behind a desk, examining me with sharp hostile little eyes. She certainly did not like what she saw.

'What do you want?'

'I was told you did abortions.'

At once she began screeching at me, scolding, banging the desk, How dare you, who told you these wicked lies, she was an honest doctor, she would never . . . etc. Only later did it occur to me that the door into an office where a nurse sat was open. Or perhaps she thought I was a spy for the government. I found myself on the pavement, weeping, while I could still hear her shrieking abuse inside. I do not remember how I found the address of a real doctor who did abortions, but there I was in a sleazy room in a wretched building, in the same part of the town where I had watched Stanley gamble away his chauffeur's wages. Dance music pulsed from every part of the building.

Do you want to be better off than you are
Carry moonbeams home in a jar . . .

I used this place in a short story, *Road to the Big City*.

The doctor was young – well, youngish – with a look I knew well from the veterans of the Sports Club, as if he were being eroded from within. He was pleasant. He was drunk. The company were his friends, everyone was animated, singing, dancing, having a good time. A woman took me into the kitchen and said that I should not let this man do an abortion. He was a friend of hers, he was a good type (modern equivalent: a good person) but he had been struck off the Medical Register for operating drunk. If I valued my womb I would thank him and say I had changed my mind. This I did. He was rueful, wry, and generous, he must have known I had been warned about him, and by his friend. There I was back in my hotel room, watching from the window groups of young men and their girls hanging about the cinemas, the dance rooms, the gambling places.

And now there was a telephone call from Mabel Griffiths, saying her husband had told her to say no one wanted to put any pressure on me, but they suggested I should visit a certain doctor – theirs. I could have every confidence in whatever he said.

I found myself in a bright, clean, serious consulting room, opposite a serious man who examined me, told me that I clearly did not realize it, but the baby was already four and a half months old. He showed me a statue on his desk, of a slim diving girl. 'It is that size,' said he, pushing the statue gently towards me. While I registered that I was being manipulated, and resenting it with that part of me that so readily flared into rancour, I knew it was all over. And I was relieved, the conflict was done with. 'I wouldn't operate on my wife or sister – not on anyone,' he said, 'not so late.'

I thanked him. The Griffiths paid the bill. I did know how narrow an escape I had had.

There is a saying, 'Every woman has an abortion story.' This is mine, and why when debates rage about abortion I don't know where to take a stand. I think that my son John might never have lived if I had not been – thank God – so incompetent. Now it seems to me obvious I knew all the time I was pregnant, was in alliance with Nature against myself. I think of the women I know who have changed their minds about having an abortion and been grateful ever since. I think of poor women who have a baby every year and there's no help for them, and they get old and sick, and their babies die or their children go hungry. I think of that dirty office with the dishonest woman in her greasy white coat and know that desperate girls have to trust themselves to someone like her.

So I returned pregnant, and pleased about it, and Frank was pleased and the Sports Club whooped and shouted and drank the baby's health, and I went on dancing, but in the daytime I sat on a sofa communing with the foetus who shared with me my long, slow, fatalistic thoughts about the war, the ineptitude of our rulers, and fear of Hitler whom we listened to ranting and raving on the wireless, while the German mobs howled their oneness with him. In the Sports Club, in the hotels, silent crowds listened to the Nazi rallies broadcast by the BBC, and slowly became fused in a determination very far from the wild whooping dances and singing that were already seeming anachronistic. It is a strange business, sitting for hours, hypnotized by contempt for one's government – for the purposes of this time the British Government was ours too – apparently paralysed by Hitler, able only to watch an invincible enemy grow stronger. Winston Churchill was still being treated like a wrong-headed maverick. On 25 August the Anglo-Polish Treaty of mutual assistance was signed in London, but Hitler called our bluff and invaded Poland on the 1st September.

On that day I was on a farm just outside Salisbury, for Sunday lunch, with another just-married couple. The man had been Frank's friend for years. The wife, like me, was being absorbed into a group of old male friends. A good deal of joking went on about sex. While we joked we sat listening to the German armies invading Poland. I felt the low, helpless brooding anger, but, too, the exaltation in accepting disaster I had been listening to all my life. Slowly, the secret pleasure in pain would weaken and die, while anger, rage, sheer incredulity, strengthened. My emotions at the end of the war were not those I began it with, nor was I feeling left out while real life was going on somewhere else.

Frank's friends were his age, already substantial and established men. I thought of them as old. One was Tommy Wolton, who had recently married Ivy. She was pregnant like me and became my special friend or, as we used to say, my half-section.

We spent days together, curtains drawn, listening as it were to the growth of our foetuses. She was a nurse. We both owned instructional handbooks of that gently hectoring kind considered suitable for young mothers, and we knew to the day what our infants were up to, putting out fins or fingers, losing a tail, acquiring layers of fur and shedding them, growing little nails. Ivy was a thin nervy pale-blue-eyed woman with fair hair that was soft and pretty when she was happy, but dank and lank when she was not. She is in *A Proper Marriage*, but if that book is to be considered as a personal testimony, then she is not there enough. She was my first real woman friend, only because we were having the same experiences at the same time. I could not share with her anything I believed, nor anything I read. 'There she goes again,' she probably thought, if I incautiously ventured on a literary or political idea. Uninterested in the human condition, she might contribute her mite. 'The natives are all right if they don't get out of hand.'

There is a companionship of women who are having their first babies that is like no other. They share a journey through revelations, yet the stages of what is happening to them is written in a book they have lying open on the table, for what they are doing every woman has done. Timidity, or a sense of proportion, stops them from claiming to be extraordinary, but that is what they feel they are, and only one other person can understand this. They ground each other in the commonplace, while what is going on in them threatens to dissolve them in its enormousness.

There was another bond then: an alliance against doctors. In those days, you did not dare say your baby had 'quickened' long before the official three and a half months, and that when still in the womb the creature responded to your thoughts and your moods. No use saying that your infant knew your voice from birth, and was aware of what went on around him, listened intently, tried to focus still blurry eyes on familiar faces. Some people coming too close clearly had an abrasive presence, others a soothing one, for the baby reacted with tears and apprehension, or with evident pleasure. To such claims the doctors would say patronizingly you were imagining it, women did imagine things, you mustn't let your fancies run away with you. Now science has justified these old wives' tales. Have doctors stopped patronizing young women? I doubt it. Has any doctor said to a woman whom he has told is an old wife, with the implication that she is hysterical, 'I am sorry, we were wrong, you were right all the time?'

Husbands in those days went along with doctors. That meant women kept quiet about what they thought. A woman friend was essential for balance if not for survival. Ivy and I sat for long morning hours comparing sensations, insisting that our babies responded energetically when we danced, or made love with their fathers, or permitted ourselves apprehensive thoughts about the war. Were we troubled that what we felt and thought accorded so little with official scripture? Not really. 'Oh well, have it your way,' we thought, or something like it, as we pursued our private researches. The companionship was interrupted when we both had to move. Why did we? But everyone moved, all the time. And now the war had actually started, the Rhodesian young men had been told they would soon be called up for training. Everyone knew they would be sent 'up North' to the war in the desert. In Britain plans were being cooked which would send millions of men to Australia, South Africa, Canada, Kenya, Southern Rhodesia, to RAF camps, to produce pilots, bombers, navigators. Even more than tourism, war moves masses of humanity about and around the globe.

Our little flat was needed for some war purpose. Frank found a place – only temporary he kept assuring me – some twelve miles outside Salisbury, a little shack of a house that had been left to fall down when a better house was built. There I was alone all day and half the night, since Frank was after all working in his office when not trying to pull strings to get himself into the army – he was too old – or drinking with the other men. It was very hot, the rainy season, and the bush had grown back around the house, sending outriders of saplings and seedlings into the rooms, pushing up bricks, and announcing that soon the house would fall, would be overtaken by trees. I was by now pretty large. I 'carried big' as we used to say. Besides, we were told then to eat for two. I was uncomfortable, and could not get cool. I filled a galvanized bath with water – there was no running water – and sat in it sometimes for hours. The water was tepid. I was agreeably cool as long as I stayed in the water. I sat and communed with the creature just behind the wall of my great stomach: energetic, much more so than Ivy's baby. I listened to the news from Europe on the wireless, with my hand on my stomach, assuring it the war wasn't going to hurt it, and thinking about the mothers and children running before armies in Europe.

My state of mind was in every way different from that of girls having babies now. It never occurred to me that anything could be wrong with this baby. Ivy was possessed with thoughts of possible calamity: being a nurse she knew what could happen. She thought it was hubris, when I assumed this baby and any other I might have would be hale and whole. It turned out that I was right, and for both of us. I had refused once and for all to be ill – and, after all, that was only six years before – and this set of mind governed my expectations for the baby, making it an impossibility that it could be born defective or, once born, die. I was full of a calm and confident elation. I sat in that tepid water, while doves cooed, called, crooned all around the house and in the branches of the tree over the house, and I smoked, or got out to cut myself a sandwich and got back in again. I read the books stacked up beside the bath. I listened to the bush with the practised ears of my childhood. I listened to the war quietly boiling up in Europe on the radio. Since Frank would soon be off to the wars – so I thought – I would be alone, with the baby, and then . . . but I liked being alone. I fantasized a romantic affair with one of the English already to be seen in Salisbury, uniformed or in civvies, spying out the land – that is to say, its suitability for various types of war effort. I would at last have time to write my first novel. I would write some more short stories, but this time, real ones.

This blissful time of being alone for hours, long dreamy hours interrupted by the arrival of Frank and his boon companions, all boisterously and optimistically drunk, ended when back I went into

Salisbury into a furnished room I have forgotten. The thing was, one furnished room then was very like another, all with chintzy or flowery curtains, and the toffee furniture. No point in finding anything better when Frank would soon be off. Back into the exuberant conviviality of the Sports Club, the verandahs, the parties, the talk of war.

My half-section Ivy and I resumed our mornings together, but it was not the same. Her Tommy was going to be called up, though she had told him, 'You can't join up, I can't manage by myself.' 'But perhaps I'll have to,' said he, with a light in his eye. 'And I'm not going to have you sleeping with all those women.' 'What women!' 'Oh, I know you, you old dog!' Giggle, giggle, while he looked rakish and was flattered. She went to the authorities, men with whom she had spent the last ten years flirting, dancing, drinking, but they were now transformed into Majors and Colonels with power over her Tommy. She always burst into tears as she went into an office. 'I can't manage without my Tommy,' she announced, her blue eyes red with these and many earlier tears. It was being remembered that Tommy was no spring chicken (like Frank), and they promised to remove his name from the call-up list. Meanwhile she became clinging and helpless, a sight that convinced not one of her female friends, but we were wrong. She was even thinner than usual, the lump of her pregnancy small and prominent, her hair was lank and straggly, and she smoked all day and all night. Laments on the theme of 'Who would be a woman' now took the place of our friendly silences, and being with her was no longer a comfort.

I was in a mood of triumphant accomplishment, and looking forward to the birth. I did not believe it would be as painful as they said, because I was so healthy and at ease with myself.

My gynaecological history would be appropriate for that fabled peasant woman who has never had anything wrong with her. I had my first period when I was fourteen. My periods lasted two or three days and were never excessive. They were sometimes mildly painful. As for pre-menstrual tension, no one had heard of it. I gave birth three times, normally, was never torn, stitched, forcepped, caesared. I have never suffered from thrush or herpes. My periods ended in my early forties, as is common for women who smoke. The dreaded menopause did not happen: my periods ended and that was it. I do not see how I could have been more fortunate.

Women with this kind of history – quite a lot of us – are sometimes made to feel guilty, as if womb troubles are the proper fate of females.

I am saying this for the benefit of young women, because all the propaganda at this time is for misfortune, their lives as females presented as an obstacle course with falls all the way, and a débâcle at the menopause. There is something not far off a secret society of women who have had an easy time at the menopause, and without the aid of drugs, but they hardly dare say so, for their sisters will accuse them of lying, or suggest they are letting down the side.

Here is the point, and if I am labouring it, then this is because I believe it to be important: When I – my generation – looked forward to our lives as females, we were not full of fear and foreboding. We felt confident, we felt in control. We were not bombarded with bleak information from television, radio, newspapers, women's magazines. We now know – we, the human kind – that we respond to what is expected of us. This information is usually presented in the context of children in the classroom, but it has a wide, indeed general, application. If girls are told, from very young, that they can expect bad times of every sort from pre-menstrual tension to menopausal miseries, is it not possible that they are attracting bad times? Whereas we, who had never heard of – let's say – pre-menstrual tension, might say no more than, Damn, I'm a bit irritable, I must be due for the curse. If you have spent years in secret dread of cancer of the breast and womb, are you more disposed towards getting it? This is a query, not a statement.

The surge of energy that announces imminent birth carried me up – and up – and into the Lady Chancellor Nursing Home where any child one was likely to meet was likely to have been born – white, of course. It was a large house, in North Avenue, that had a couple of rooms on either side of the entrance, rooms down the sides, an inner verandah around a court, and a long room where the babies were kept, well away from their mothers.

I was greeted by a very young nurse who announced there were too many babies being born that night – it was the fault of the war – and I must be a good girl and look after myself. The time was eight or nine. I wandered about, ignored, listening to the screams of women in childbirth, and stood over the cots of the newborn in the nursery, wishing I could cuddle one of them. At one point

I was told to take a bath, at another, shaved: they did this then.

The place was run by a large woman, Miss something, untrained, who always wore a nurse's uniform and was present at most births, assisting. She had a matey relationship with the doctors, after all not more than a dozen or so, who were in and out of the Home every day.

She came to me and with condescension told me that she was glad I wasn't making a fuss like some of the other girls. I did not get into the labour ward until early morning and there I was deposited on a high bed and left. I was in great pain. This is described in *A Proper Marriage* – well, more or less, it will do.

Sometimes women say, It is not true you forget labour pains. But I think you remember you did have bad pain, but not the pain itself: you forget the intensity of the pains in between each pain. Real remembering is – if even for a flash, even a moment – being back in the experience itself. You remember pain with pain, love with love, one's real best self with one's best self.

What interests me now is the fact of pain, the strength of it. I was not yet twenty. I was healthy. And if expectations govern physical experience then the birth should have been as easy as the two later births.

Perhaps it was that I was so alone and uncomforted. The only person in that long first labour who gave me support was the black cleaning woman, who was washing the floor. Again and again, in reminiscences, novels, autobiographies, one reads of how white people have been given ordinary, decent human warmth by black people when they needed it.

Where was my husband? He was whooping it up with the boys, as then was obligatory. The idea that husbands should support their wives – I cannot imagine what Matron would have said. 'You don't want *them* – they're just a nuisance.'

The babies were a nuisance, and so, too, were we, the mothers.

When the baby was born, that is, my son John, they held him for me to see, a long lean baby fighting in the arms of the nurse. 'You've got a real rugger player here,' I heard, as I was wheeled out of the labour ward and he was taken off somewhere. I lay sore and forlorn, longing to hold the baby. When I asked timidly to see him, I was told, 'You'll have more than enough of him soon, what are you in such a hurry for?' Later I was told I was not to worry, they were giving him sips of sugar water, and I would see him next morning. I tried to assert myself, Tigger weakly making a joke of it, and they brought him that night, that is, nearly twelve hours after the birth, and only for five minutes. The Matron stood over us, and no sooner had the baby's lips made contact with the nipple, she took him away again. 'That's enough for the first time.'

The spirit of Dr Truby King pervaded the Lady Chancellor. Right from the start, a regime of four-hourly feeds was standard, unless the baby was under a certain weight: mine was over seven pounds. A baby wanting food before the right time was left to cry. 'He's got to learn who's boss.' 'He has to learn he can't have his own way.' When the babies had finished their feeds – they were never left with their mothers longer than half an hour – they were taken off on trolleys or balanced here and there on the arms of the nurses. Silence reigned for a short time when, if you were lucky, you slept a little. Soon the babies began screaming. They might cry for an hour or two, while the mothers lay in their beds longing for them, but they could not go to them, or have them brought. In those days women were supposed to stay in bed for a week. Well, for my mother it had been six weeks. I lay helpless in that bed, breasts stinging and full of milk, listening to the babies' frantic crying down the verandah, and I was full of rage and frustration.

Mary McCarthy, in *The Group*, describes a similar regime. The Lady Chancellor Nursing Home was not unique. But if a committee ever plans how to make sure that mothers do not 'bond' with their babies – or, as we used to say, love them – then they could not do better than study the Lady Chancellor. I am told this is how they do things in Japan.

We often see on television this picture: there is a trolley or shelf or table, and on it may be ten or more babies, identically wrapped, arms and legs bound, and over these hapless infants stands a warder nurse. That anarch, the new baby, full of explosive and wonderful possibilities, is being taught its place in the world, taught what's what. I do not expect things to change much. Something very deep and nasty is fed by this business of keeping new mothers and their babes apart, making sure babies cry for every feed, and that the women are restless and uncomfortable. '*You've got to let them know who's boss.*'

I was in a room with two other women. One was having her third. She was large, shapeless, slack-fleshed, and I lay watching her, secretly full of horror. I lay there inside my now loosened body, willing it to return to its proper shape. I hated my great bursting breasts. (Which did not prevent me from being proud of all that milk.) A girl defines herself against her mother first, then against the world, by her bright tight shapely body, her silky little breasts, and, of course, the triangle of glossy pubic hair in its pretty whorls – she is invulnerable to criticism inside that new body of hers. And then – she is lying in bed, a sack of bruised flesh, like a snail pulled out of its shell.

At the bedside of this domestic and complacent woman – so I saw her – the black ward maid who had befriended me came to stand whenever she was unobserved by matron and nurses. She admired this woman, and said, 'With us a woman is not a real woman until she has had her third baby.' 'So now I am a real woman,' said the mother, amused and pleased. (Similarly, a Shona woman told me, 'With us, we don't think a man and a woman are married just because they have been through the wedding ceremony: it can take years for a couple to be really married.')

I was silently swearing I would never have another baby, I would never again be ugly and fat. But that woman over there, perhaps she had made the same vow, and there she was, so pleased with herself, her body like a milk jelly. I was infinitely forlorn, anxious, threatened. Frank bounced in and out, sometimes with his cronies, and after all, they were my friends too now. They were all so pleased, with me, with the baby. My mother rushed in, and at once said that four-hourly feeling was ridiculous for such new babies. This made me support the nurses: I could not afford to agree with my mother who radiated accusation and sorrow, though she did not know she did.

When I took John home it was the first time I had spent longer than half an hour with him. He did not conform to the rules in the book. For one, he lifted up his head from the start, and at the breast nursed with pleasure and energy, his legs working like pistons, and his eyes, supposed to be unable to focus, were alert and observant. He never nestled. He was always striving to haul himself up, get his head high enough to see over the edge of the pram. It was clear he soon would have to be in the big cot, with bars. He seemed to be hungry. I got fatter every day, and was miserable. The milk was only adequate. Between feeds I agonized about the amount of milk being manufactured by my breasts, for a feed drained them. The baby was satisfied for a couple of hours, and then yelled. But the rules said it must be four-hourly feeds. I sat anguished, irritated, anxious, 'letting the baby cry' according to prescription, until the clock struck the second I could lift him up and feed him. Now I know he was a baby who should have been fed when he wanted to be, not least because then I would have made enough milk. I began to defy the Mentors by lifting the baby in mid-afternoon and trying to calm him, by cuddles and conversation, hoping the contact would fill my breasts which by then were empty. I remember standing on the verandah, holding an infant who seemed to want to stand in my arms, his fist in his mouth, an embodied howl for food, and I wept, and wept, demanding of him, What shall I do? What shall I do with you? – for he was so strong I was worn out only just holding him. Yet I was a strong young woman, and this was a baby.

In short, he was a hyperactive baby, and then a hyperactive child, but the word was not then in use. I am glad. Probably he would have been sedated, filled with chemicals.

Meanwhile the proud parents were as full of theories as they usually are with a first baby. Frank had read some book about not coddling babies, who if taught early to stand cold, would later be impervious to chills and colds. The baby was put out in a vest, a thin jacket and a nappy, on the verandah, to harden off. He did not seem to mind this, though the air was chilling with the approach of winter. His crying continued to time itself according to how long ago it was since he was fed. First babies need stamina: I recently watched a baby of nine months being fed grilled cheese on toast on the grounds that babies must find their diet boring.

I was frantic with worry, with dislike of my fat body, and because of continual visits from my mother, who said that John was being ill-treated and was not impressed by his putting on weight, the formula that Mentors found adequate for reassurance. I said I was going to put the child on the bottle. She said I was irresponsible. My friend Ivy, exhausted with the miseries of watching the clock, had put her infant on the bottle and all was well.

Never has there been a regime more guaranteed to make young women anxious, feel inadequate, inefficient, lacking, nor one more likely to make women lose milk and – it goes without saying – all pleasure in the process of having a baby. But that, I am sure, is the point.

Sometimes I went over to Ivy's flat to share the business of the morning bath. I trusted her, and did not trust the Mentors. We stood side by side at the long table where she washed and changed her infant. Her baby lay in front of her, mine before me. When babies are born they may be lean, wrinkled, scrawny, dark red, perhaps covered with hair; are, simply, unfinished. We knew what was wrong with babies: they are born too soon. The proof is that at about two months they become perfect, achieve what clearly they were destined for all along. Her little girl was plump, pretty, dimpled, blew bubbles, waved soft arms about. John was long, lean, with intelligent eyes and his arms and legs were never still. He had to be held down with a hand or he would roll off any surface. He punched the air, he kept lifting up his head to see the little girl next to him.

'Hey, look at that,' says Ivy, 'he's after her already.' And she instinctively puts her hand to shield her daughter.

'Sex mad, you are,' I say. 'Ivy, at three months he's hardly likely to rape her.'

'Oh I don't know – Look at that! No, I'm going to keep an eye on her, knowing what I got up to – I'm giving her a warning now. And I'm going to keep an eye on *him*, I'm telling you. John, are you listening? Hey, John!'

The enormous prick and testicles of the newborn male had gone back into proportion, he was in scale. All the same . . . that sweet puffing dimpled little girl, that vigorous reaching boy . . . we laughed. We stood laughing and could not stop.

Recovered, I said, 'But you could have had a boy and I a girl.'

'Never! Im-poss-ible!' She was expressing that feeling so strong in us both that we could have given birth to none other than just *this* baby.

'Look at that creature! Look at that *thing* . . .' She points, with dramatic scorn, to my boy's private parts. Then she turns rapturously to her own babe, and admires the mount of Venus, so plump and perfect. 'Just like a little post box,' she coos. 'Oh, I could eat it all up with ice-cream. Oh, I could post letters into it. Oh what a sweet little cunt, how can you possibly defend *that* . . .' And she points, all dramatic scorn.

We snatched up our infants and danced around the room with them, singing, 'I'm in heaven, I'm in heaven, and my heart beats so that I can hardly speak. And I seem to find the happiness I seek, when we're out together dancing cheek to cheek.' Or, 'Night and day I think of you . . .'

Ivy, after all a nurse, announced to her husband that if the health visitor dared to come again, she would probably kill her. I said the same to Frank. He said I should do what I thought best. The fact was, our husbands were tolerating us, and I don't blame them. We were obsessed with the developing babes, their feeds, their stools, their putting on weight, their sleeping or not sleeping. Women emerge from their time of being submerged in trivia amazed at themselves.

The men spent their days and nights with other men, all longing for the moment when they could put on their uniforms, in bars, in the hotels, on the verandah of the Sports Club.

'Just imagine,' says Ivy, looking at her lank and lifeless hair, at her scraggy body, and then at me, overweight, bursting out of my dress, 'and we were the toasts of the Sports Club, who'd believe it? Oh well, to hell with it all, that's what I say.'

I put the baby on the bottle, and my mother said I was selfish, and thought of no one but myself. She was more than I could bear, and I became even more cold, polite, patient. I agreed with everything she said, and this drove her crazy with rejection, being shut out. She kept saying that daughters needed their mothers at this time, and I agreed with her and waited for her to leave.

What was going on in her life was certainly more than she could bear. My father was now almost permanently ill. He was an invalid: the word suggests a steady, low condition, but with him it was all crises, traumas. He had been in coma, or nearly in one, he had been taking far too much insulin, or too little. His liver . . . bowels . . . stomach . . . his good leg was so thin now he could hardly walk. There she was, alone with him on the farm, and he could no longer safely drive. She was always appealing to neighbours for lifts in to town, and she hated being dependent. Why did she not learn to drive? She did later, in town. Her letters to me were, in the current idiom, cries for help. To me they were threatening. She was a threat: half an hour with her reduced me to exhaustion. After a visit from her I would go to bed and sleep.

Meanwhile my brother was in Dartmouth, England, learning to be an officer in the Navy. She had achieved her great ambition – a son in the Navy. She had pulled strings, written letters to England, haunted offices, exhorted and pleaded. And there he was. Harry and Dick Colborne went to Dartmouth together. Later he said that in peace time he would never have made it. 'Those English Navy johnnies, they were just buttering up the colonies, you see. Had to have one or two of us, for show.' He found that his standard of education was nowhere near the level of the others. To keep up meant working every minute, day and night, and he scraped through the exams, but only just.

It took weeks to get letters from England. He wrote, lively cheerful letters to 'M.' and 'D.'. I have a batch of them. He wrote verses, too. Nothing of his interior life – what he thought, felt, or secretly suffered – was in those letters.

I wrote letters to the farm. 'Dear Mummy and Daddy. Yes, I really am fine, and John is doing well. He put on four ounces last week. Frank is off to the wars. Much love.'

The very second the baby was on the bottle I began to diet, and lost pounds every week. I would not have begun a diet before I stopped breast feeding, I would not by one day have advanced putting the baby on the bottle because I hated being fat. I would not cheat, absolutely not, but once my body was my own again – then . . . And I was my own neat shape again, in the smooth-fitting dresses we all wore, with my smooth shining hair, and I was ready for the Sports Club verandah, but the Army had spoken, and all the men were down near Umtali in a camp, to be turned into soldiers.

We wives at once followed them. Our men did not want us.

I was in a cheap hotel, in an ugly room, and it was winter. In that year, 1940, it drizzled and was misty for weeks. I could not get nappies dry. The baby got some infection and his stools were for the first time wrong, large white curds of undigested milk in yellow slime. He yelled or grizzled. I called the doctor, a young man, from whose exasperated voice I understood I was being hysterical. 'All you girls, what are you doing here? They aren't going to let your husbands out of the camp, don't you know that?'

I walked up and down, up and down, the streets of Umtali where every house pulsed out dance music, those streets where in another life I had been so miserable because I was too young to be part of the teenage groups of boys and girls. I pushed the pram for hours every day under the flame trees and the jacarandas, and I was dreaming of how a soldier who had escaped from the camp, would see me, come shyly over, strike up a conversation and – no, I would not be asking after the welfare of my husband. These daydreams were as sharp as scenes in a film: and they were the fantasies of a girl, not a young woman. We would embrace under the flame trees, kisses full of anguish because of the partings of war, the loss and the pain of war.

At a much better hotel, Brown's, just around the corner, was another wife. She was not my favourite, nor I hers. A particular chum of Frank's had gone home to England on holiday and returned with a bride. She dazzled us colonials, always ready, indeed, anxious, to be dazzled. A rich girl, middle-class, she had 'good' clothes of the kind we admired as a style because it so exactly expressed the women who wore them. She was affable. She was cool. Now I see she had probably not understood what kind of life she was marrying.

In the colonies, class snobberies were always at a remove, as if a bit of machinery that works perfectly has been taken from its matrix and now revolves uselessly while sitting on a work bench, to be inspected. 'Look at that lovely bit of action!' When Ivy and I talked about Mary, it was with pity. What was the point of being snobbish here, with us lot? – was our feeling. But as a performance we admired it.

Her husband was another of the Sports Club boys – rather, men, he was getting on a bit. He was hard-drinking, and in fact an alcoholic, good-natured, amiable, rather foolish. What he had brought back with him as his life's comforter and companion was this sharp-tongued, upright, no-nonsense redhead who might say, 'Now that's enough, dear, come home, you've drunk enough for one night.' In short, here was the archetypal marriage, we have all seen it a hundred times, between the good-fellow, the boon companion, the tippler, the toper, the man's man, and the efficient moralistic woman who despises weakness because she has no weakness in her. It is as if these delinquent men feel that the reproaches of their own consciences are not enough: they need to make sure of a constant tongue-scourge as well.

How else to explain that other common marriage, between the scholar, or sage, or egghead, and the prostitute, or barmaid, at any rate, a frivolous and sexy woman? Both can be sure they will have a companion day and night who is thinking: 'You're sex mad!' 'God help me, but you're a dry old stick.' 'Can't you ever be serious about anything?' 'Do you always have to be *thinking*? Can't you give it a rest sometimes?'

This good-fellow, when with his new wife, always had a humorous and hangdog look. She, with him, was like a queen brought down from her rightful place, as D. H. Lawrence would say. Mary, her name was.

I went to tea with Mary at her hotel. Its excellences would not startle the habitués of the world's great hotels, but I did enjoy the good cakes and the wood fire. Understandably, Mary did not come to tea with me at mine. She had a nice, well-behaved baby dressed in clothes from England. John at these tea parties did not behave badly; he was simply himself. Mary would say tartly, 'He's very energetic, isn't he?' watching him strive and wrestle in my arms, already putting his full weight on his feet. It was evident to me that he was impatient with the condition of being a baby: interesting that I could discuss this idea in all its ramifications with Ivy, but not with the 'intellectual' Mary. For so she was described. 'Are you sure he's only four months?' Mary enquired. Later, when he pulled himself to his feet at nine months and began running at a year, people would say, 'Are you *sure* he's only a year?' This is one of the things that are 'impossible', but which I know to be true: for years I held in my mind a list of things that were impossible – because people said they were – but true.

I felt proud of John, but abashed. I did not know why other people's babies lay calmly in their prams and allowed themselves to be dandled and cuddled. Mary made me feel incompetent. But for the sake of the great wood fire I would have put up with a lot more than that. In my room was no fire: there is a convention, in hot climates, that they are never cold. In Salisbury nappies dried under the hot sun in an hour on the washing line, but here they were going mouldy, and I bought more and more new ones that at least were dry.

Ivy arrived in the same hotel as me, there to find her Tommy was not going to be let out of camp. At the most, it would be for a couple of hours. She was not herself. Losing her husband to the war had undone her, as she had known it would. A careless, angry, hard woman sat staring, a cigarette loose on her lower lip, while smoke poured up over her face. She did not hear when I spoke to her. She was thin. Her hair was lank and dank. The pretty blonde had achieved magnificence. That staring face, the empty eyes – she could have modelled for Despair. She was removed from daily pettiness, she was in allegorical and antique regions.

I made futile attempts at, 'But Ivy, half the women in the world must be losing their men to the war.' 'But you got on all right before you met Tommy, didn't you?'

She stared at me from her far country and did perhaps notice that this idiot was prattling. The thing was, she was depressed, the real depression, but I had never heard of it. When much later I knew people who suffered, I belatedly understood what had been wrong with Ivy. It was not that I expected her to 'pull her socks up' – Mary's prescription – rather that I could not believe she was as ill as she looked. Why should she be?

She sat for hours in a chair while her baby slept – at least the baby did sleep – and when she had to, tended to her. She herself did not sleep. I would go in during the night and find her where she had been all day, the cigarette cold between cold lips, staring at nothing at all. Soon, she did not hear when her baby grizzled or whimpered. I took to sitting in her room with her, with one arm holding my own baby who strived and struggled as usual, while pushing her baby's pram back and forth with the other. If at feeding times she did not move I might feed both babies one after the other. I suppose she did know I was there.

This went on for days, while her Tommy sometimes sent messages from camp, but did not come. Then the war spat him out again. Ivy was like that frond of seaweed limp on a rock, and then the wave lifts it. She laughed, she cried, she giggled, she washed her hair, made up her face, cuddled the baby, and when Mary said, 'I see you've pulled yourself together,' said, 'Oh go on, have a heart.' And so we three wives went back to Salisbury, our husbands again in civvies, to begin a proper married life. War loves twenty-year-olds. Frank was thirty and had bad feet, Tommy had something or other, Mary's husband was unfit. The town seemed full of bitter men, knowing that life was passing them by, because the army would not have them. Frank was miserable. This was when he really knew his youth was over: only the 'old men' would be left in the town. The men who had been rejected spent time together, drinking: they needed to be understood. It was not that I was unkind, no, I was affectionate. 'Oh, poor Frank, I am so sorry.' But I was too young to know what he was feeling.

In *A Proper Marriage* the husband actually gets up North and is invalided out of the army. Quite soon men who had successfully concealed stomach ulcers and other disabilities from the army doctors came back, angry and disappointed. Their return coincided with the news of our first casualties in North Africa. When their loved ones said things like, But you might have been killed or wounded, they were not consoled.

The verandah of the Sports Club was scattered with bitter men. I listened, I listened, after all had I not been taught how by that other war? I pushed the pram back and forth with one hand, held a cigarette in the other, and listened. I was feeling that pleasure, almost an exaltation, which is how a writer may recognize that her life is matching her natural disposition – her talents. I had written very little then. But I was listening, selecting – *recognizing*.

A novelist's creatures can never go outside the range of behaviour their natures – the novelist's – permit. Obvious, you say? What happens when a writer's hand – or mind – simply refuses to write the next sentence, because Tony or Susie is about to behave *out of character*, is a far from simple matter. It involves not only the whole business of identification of writer with a character, but the many possible selves of the writer. Which may be, after all, limited. You would not find, let's say, George Meredith using Crippen as a model.

I once thought of writing a book called *My Alternative Lives*, using the conventions of space fiction, some of whose ideas are the same as those on the frontiers of physics. But the plot here would be that the lives of the doctor, the animal doctor, the farmer, the explorer would run concurrently with my life, set in other parallel universes or 'realities', continually influencing mine. As in those cases of multiple personalities, where only slowly do the personalities inside a woman or man become aware of each other, the heroine of this book – me for argument's sake – would slowly come to know that multiples of herself are living these other lives. A nice idea for a book, but time is running out.

Meanwhile, I was not – as I so easily could have been – one of the young women left behind with the baby in a furnished room or returned to her parents' house. I was the wife of one of the City Fathers. A joke Frank did not enjoy. Not my joke – I was not so unkind, but jokes flourished, a bitter crop, on the Sports Club verandah. 'Hey, Frankie,' shouts some new girl, waving her hockey stick as she strides down the verandah, or calls across the room from the arms of a man in Air Force uniform where they are dancing to 'We'll hang out the washing on the Siegfried Line'. 'Hey, Frankie, how do you like being a City Dad?'

As for me, I was suffering one of those reversals no one had prepared me for. Eighteen months ago I – and all the other girls – had been competed for by every man in sight and now I had become invisible. I was treated as respectfully as if I had been fifty, in spite of my again slim body and my girl's face. Who was the star? It was John, Frank Wisdom's son, who sat up unaided from five months, striving against the straps that held him in his pram. 'Look at this kid, man, he can't wait to get on to the rugger field with us.'

What else were we talking about on the verandah? It was the phoney war, and we fed off rumours. Hitler would sweep down over Africa from Cairo to Cape, and make us all his slaves. (The kaffirs, we were told, were saying it wouldn't make much difference to them.) It did not seem improbable: he had swept over all of Europe without much difficulty. Would the black population rise at the first sign of Hitler's armies and join with them in slitting our throats? This was said, not with remorse about any behaviour of ours that had made this possible, but with indignation, and more than a whiff of the ingratitude of the servant classes. Sometimes when I am asked by young people what the old Colour Bar was like, I suggest they take a look at *Punch* up to the Second World War: the jokes about comic housemaids and the absurdities of the working classes are the same. By now we knew the RAF was going to use this country and South Africa to train pilots. We did not know this would mean hundreds of thousands of the English, all males, arriving to live in camps near this city and Bulawayo. Soon, our men, departed up North, would be supplanted by a different male population. Our lives continued unchanged except that we talked day and night about the news from Europe, and the radio was never switched off. The BBC News stopped any conversation, dancing stopped, each radio set had groups around it. We were not short of anything, though we soon would be: no more imported goods. Men were buying up stocks of whisky, and some young women thoughtfully laid in supplies of good lipstick.

I was in another small hotel. Frank was trying to find in an already tight-crammed town, another of the little houses, which it had been so easy to rent even a year ago. The hotel was full of women, all older than me by years. They wanted to be kind to this girl and her demanding and difficult baby. But I sat in my hotel room with John and made fresh pretty dresses. With an extraordinary, not to say obsessive, care for detail. Now I ask myself what I thought I was doing, piping inner seams and whipping raw edges no one would ever see, when my usual way of going about things was a slapdash but successful improvisation. Sometimes you may see someone doing up, then redoing, a house or flat. It is perfect, it is pristine, but suddenly you hear, 'But it's not right, I'm going to restructure the kitchen.' This can be repeated, every two years, walls already perfect repainted, new kitchens ripped out and redone. They are restructuring themselves, painting the walls of their psyches . . . (Or as they put it on the Sports Club verandah, syke. 'Hey, kid, how's your syke today?') Similarly, an anxious young woman turns a dress inside-out and carefully inspects every seam, whips every raw edge, pipes waist seams and armholes as if they were on the outside and not the inside. 'That ought to make it safe,' something inside her, a long way behind that bright defensive smile is muttering. 'Yes, that's in order – I hope.' Just as so long ago she dressed and undressed her teddy, ordering perfectly folded clothes in a little case.

The women had no idea how they terrified me: why should they? They were all decent, friendly, kindly women. I watched them sitting all through the mornings and the afternoons gossiping, women's talk, husbands, children, money, money, money, who would be a woman, the munts getting cheeky, men are just children . . . I had watched the women of the District talking, talking, and had sworn Never, I won't. I simply refuse! Twenty years later this kind of talk – criticism of men, dissatisfaction with the lot of women – became prescribed behaviour in the women's movement, was called Raising Consciousness, and the activity itself was Rap Groups.

From this hotel we moved to a couple of rooms in a friend's house. From here to not a whole house, but half of one. All of these were in streets and avenues not more than five minutes' drive from the centre of town. From this time comes a flash of strong real inside memory. I am sitting alone on the bed. It is evening. I am listening for the baby asleep in the cot just outside the window. Whirring insects circle the unshaded light bulb. Regarded dispassionately they are pretty, delicate, light green, slender insects.

More and more come fluttering in from the dark to the light bulb. The room is full of them. A cat comes in from somewhere, pounces, an insect begins a tiny scream, which goes on and on while the cat plays with it and stops only when the cat crunches it up. Another squeaky cry, as the cat leaps. The girl is sitting on the bed, fingers in her ears. She is hysterical. She is wild with fear. While her intelligence tells her these are harmless insects she could easily begin screaming herself. The insect's squeak plays on her backbone, as, not more than a few weeks ago, the baby's crying did. She creeps out of the insect-filled room to the dark verandah and sits on the cold cement near the cot, watching the insects flow in over her head into the room. She is weeping, hopelessly, futilely. She holds the baby's sleeping fist and weeps.

And then the young husband returns. 'Don't be *silly*, they're just insects.' 'I know, but I can't bear it.' 'All I know is it's jolly well not like you. What's got into you?'

12

WE MOVED AGAIN. We kept moving. Nothing to it! We owned our clothes, our bedding, an odd chair and the famous table, and books, loads of books. A small van transported us to yet another of the small houses. They were all furnished the same. If many farms still used improvised furniture, karosses, flour sack curtains, petrol-box shelves, then in the towns 'store' furniture was the rule. A kaross already paid conscious tribute to the country. There were chairs of the kind whose backs are adjusted on notches, woven grass chairs. Also paintings of jacarandas, sunsets, kopjes, lions, natives, elephants and innumerable buck lifting their heads to stare at the viewer. But it all didn't matter. I was not going to stay in this life, I said to myself, desperate, trapped, but behaving beautifully, doing everything I should, though I was exhausted because of the child. From the moment he woke with a shout of greeting to this wonderful new day, until night and a reluctant falling to sleep, he was never ever still. Even now, seeing some amenable little baby cooing in its nest, I remember John at the same age and marvel. Literally, I could hardly hold him. It was not his nature to cuddle. He did not really like being dandled. One felt he put up with it because it was expected of him. He liked to lie on the floor, his legs like a bicyclist's, or being held by Frank, though even he found those striving limbs hard to control, or standing in my arms, treading down on my thigh. Every mealtime was an ordeal because he wanted to hold the spoon, and was angry because he could not, tried to grasp the bottle, and yelled when it slipped. Other women's babies slept in the mornings and in the afternoons, but this one did not. When Frank went off to his office at half past seven, I sometimes was already walking John around the streets, since movement calmed him. And by ten or so was ready for the morning tea parties I despised. Young women met in each other's houses, taking their babies and small children. It was expected that I should be friends with the wives of Frank's colleagues. There was a group of ten or so women. I have described the morning tea ritual in *A Proper Marriage*, but if I were describing that society now, I'd give it more emphasis, because of the role it played in making sure new babies were born. One of the group has a new baby, and there she is with the little thing, its head helpless on her shoulder. Suddenly your own toddler looks enormous, even gross. You remember the sweet intimacy with a new baby. You might have said, 'I'm *not* going to have another baby yet – or perhaps not ever,' but suddenly, holding the infant, you are 'broody'. 'Oh, have a heart, I'm getting broody' – and you hastily hand the dangerous creature back to its mother who seems the most enviable person in the world, although she is misshapen from the birth and breast feeding. But it is too late. The hormones have received a jolt and you're off. Soon you will be announcing at a morning tea, 'I'm pregnant!' 'You're not! But you said . . . Oh, I'm jealous. When?'

It doesn't matter whether you like these women or not, or they like you. 'We have nothing in common!' Don't make me laugh, you have the biological basis in common, you are young women together and that's enough. These days we all know about periods coming into sync at the same time with women who continually meet, and that's just the start. Now we know that within a few moments of any group being together, their brain waves are in sync (*The Dance of Life*). Oh, indeed, we should be careful of the company we keep, but young breeding women will spend time with others. If the birth rate is a matter for concern in a country, then see to it that young breeding women meet every day for a couple of hours. I was bored, I was rebellious, I hated the morning tea parties. I craved them, and hated myself for craving them. I would go home and tell Frank I would rather die than ever go to another. But next day I went. For one thing, John, sociable from the start, liked them, was interested, had to watch what went on. 'John, just look at John, he's going to be crawling any second now.'

We had a 'boy', a servant. Everyone did. By eight in the morning the two or three rooms were cleaned, and he was gossiping with his friends at the back. He cooked lunch. Frank brought people home from the office for the lunch hour and we ate and, more importantly, drank. After lunch I pushed John for ever, for ever, around the park and the streets. In the late afternoons we took John with us to the Club or friends. There was the little boy, all alert intelligence, watching what went on, and always trying to pull himself up, or clamber up and over anyone who held him, 'Hey, Tigger man, look at this, you must be worn out.' 'Oh no, it's all right,' I'd say modestly, though worn out. If we did not go out, but – rarely – stayed in, then people dropped around and the servant made a meal. It was the custom for them to hang around waiting until their employers came back from sundowners, in case a meal was needed. This meant he would bring the morning tea at six, have nothing to do most of the morning and afternoon, but might be up till nine or ten at night. There was no such thing as legal working hours. Frank and I paid the servants everywhere we were much more than the custom, risking the anger of neighbours. 'You'll spoil them. You mustn't let them get out of hand.' The same words as used for small babies, in fact. 'You've got to let them know who's boss.'

All the men Frank brought home were at least ten years, often more, older than me. I was Frank's pretty new clever wife and he was proud of me. I liked being admired, for myself and for my exuberant baby. The man I remember best was a small, sandy, thin, ironic Scot, a stockbroker, Sonny Jameson, whose comments on this provincial little town came from a viewpoint far removed from the stereotypes of Preserving White Civilization or Uplifting the Natives, which still informed the *Rhodesia Herald*, and most people one met. He read. From me he borrowed my Everyman books. He brought me the Romans. If I wanted to understand Southern Rhodesia, said he, then read the Romans: the attitudes of our administrators were no different from those of a Roman proconsul in a colony in North Africa or the East. Heady stuff for a civil servant's wife. He did not make this kind of comment in company.

'*When in the provinces, you learn to keep your mouth shut.*'

The other memorable thing was that he drank, so they said, a bottle of Scotch every day. Certainly he scarcely ate. For years I assumed he must have died long ago, then heard he was alive and flourishing. This tale is not likely to please nutritionists.

Stendhal – the Stendhal of *The Red and the Black* – was my friend and ally. He is the author for anyone feeling trapped in the sticks.

'*In the provinces . . .*' so he may begin a deadly dose of hatred for mediocrity, and I would mentally add to his list, full of the relish of contempt.

'In the provinces all languages other than English are guttural.' I heard this again in Harare, in 1992: *'German is so guttural.'*

'In the provinces any girl with some life in her is sex-mad.'

'In the provinces any woman with a mind of her own is opinionated.'

'In the provinces any food not English is greasy.' This one has certainly gone with the wind.

'In the provinces women are automatically offered sweet wine, or sweet sherry. The dear little things have a sweet tooth.'

When John was nine months old, soon to get to his feet, we decided to have another baby. Yet with half of myself I knew I was not going to stay in this life. I had nothing in my mind as serious as a plan or a programme. No, I merely dreamed of a life with similarly free spirits in Paris or London. I did not belong here. Yet any observer might be deceived because I was apparently doing so well. Who was? Tigger was, bright, slapdash, amusing, a competent and attractive young woman. 'Clever Tigger Wisdom' might also make comments that caused uneasy laughter, or 'Have a heart, man, that's not fair!' But she was living this life as if born to it. Was it I who decided first to have this second baby? Probably. But it was the *Zeitgeist*. All around us young couples were saying, 'Let's have another, we'll get it all over with while we're still young.' Three or four years before it was 'I'm not going to bring a child into *this* world, no fear!' Yet while Frank and I discussed the problems of the second baby, we were still talking of how we would tuck both babes under our arms and go wandering in the South of France, or live in Paris.

I got pregnant the week I left off the Dutch cap. This method of birth control is considered now as too unaesthetic, but it works. The point is, one has to get into a routine with it. Easy in marriage but not so easy in a life of affairs and adventure. At once I had morning sickness and indigestion, which I knew would soon pass. And there was John, who got to his feet without the intermediate stage of crawling, and he was running about everywhere, and then he took off into the vlei nearby – long since built over – and although I was a pretty good runner, I soon lost sight of him. I went from house to house in a panic, begging people to despatch their servants to look for him. An hour or so later a group of black men came tossing John about in their arms, admiring this tough little boy who was already fighting to be put down so he could run off again. I did not know what to do with him. Straps and bonds hurt his feelings. If I fitted on him the harness that would enable me to attach a leash to him, he gave me a look that said, 'You're supposed to be my friend, how *can* you do this to me?' Outrage, incredulity, accusation, shouts of indignation, then tears. I would try to hold him, to comfort him. Standing in my arms, rigid with emotion, he sobbed, giving me looks of puzzled reproach. So it had to be the park, where he ran about unconfined among the flower beds, yelling with delight. Then I caught him, since I was afraid he would run away out of the park, and made him sit in his pram, where he at once stood up, with his back to me so he could see where he was going.

I pushed him around and about for hours, for hours. So it seemed. There is no boredom like that of an intelligent young woman who spends all day with a very small child. While I pushed, I wrote poems in my head.

RAIN

Rain-clouds rest on the trees of the higher town,
Here rain sweeps off the rusting tin,
Beats against the patched shutters,
Batters the leaning roofs.

Storm water scours the gutters,
Flooding away banana skins,
Straw, sweepings, filth and dirty rags,
Gurgling through the broken bottles,
Creeping beneath the crazy floors.

Already walls show patches of damp.

Thin faces of children
Peer through cracks
To their playground the street.
Soon, when the street lamp shines down
Gold, crimson and blue
Will wash across the tarmac.

But now, through the grey rain
And grey steam that drifts up from the street

A small black child runs shivering
Clutching his rags and a milk bottle
To the better house among the trees
Where the hen-voiced impatient shrew
His white mistress is waiting.

These conventional verses were the start of bad trouble with Frank. He said indignantly that I was unfair. He showed the verses to his sister Mary, and she said I was unfair too. But his indignation was theatrical. His face working with accusation, he stared at me with hot and aggrieved eyes. It was then that began an atmosphere of falseness, of unrealness, that was at first only intermittent. When a man feels his woman's work or interests outside marriage threatening, then often this is expressed indirectly. Frank always agreed that I should get a job, when it was possible, that I was going to write, when I could. But I was growing away from him, and very fast, and he felt it, although I could not have been more amiable, amenable, ready to please. Women's instinct to please confuses men, but it confuses women too. I really did not know why suddenly he was sulky, and asked me *Why are you so unfair?* 'We're not as bad as that!' he grumbled. Yet Frank thought that we – particularly nagging white housewives – were as bad as that, and was consistently critical of white 'superiority'. And later in his life he suffered for it. From now until the end of the marriage there were times – unpredictable and dangerous, when he was sulky, sometimes for days, angry, self-pitying, full of reproaches, and always about things we both knew were off the point. Meanwhile I was bright, false, 'reasonable' – hypocritical.

His angers were not only due to me, but perhaps even more because he was not up North in the desert with his friends. The moment he got home from the office he wanted to be up at the Sports Club. He was drinking a lot. Nothing new about that. Now when I try and recreate the patterns of our days, I can't believe how much we all drank – but as a matter of course, that is the point. In the 1920s – that is, after the First World War – drinking too much became not only permissible, but smart, clever – fashionable. There it all is in the novels, memoirs, histories of the time. It was not only in the colonies that everyone drank too much. Southern Rhodesia was nothing if not a drinking culture. Now we are all obsessed with food, eating it, reading about it, eating this and not that, giving it all up for days at a time. Then, we drank, gave up drinking, drank beer and not spirits, drank spirits and not beer, decided not to drink until six in the evening and sundowner time. Boon companions might have to be despatched to be dried out, but everyone knew they would be back on the verandahs drinking soft drinks, having 'given up' for ever, but within a few months they were hooked again. I was beginning to find the Sports Club almost unbearable, but Frank wanted me to be with him, wanted his son there too. I was tired. I was utterly exhausted. Never in my life have I been so tired.

And yet being tired was not on my agenda, why should I be tired?

And when my mother rushed in from the farm, to say I was being irresponsible to have another baby so soon, I defended myself, said why shouldn't a strong young woman have two babies one after the other, all the black women did, didn't they?' 'Oh my *dear* . . .' And she went off to complain to my father, but he was too ill these days to listen.

Again I was making the little dresses and rompers, filling drawers with the nappies John was already out of. He was not one to put up with having a wet bottom. Without much effort on my part, he became 'toilet-trained'. Or 'clean'.

The months dragged past. Here we are, in 1941, the Phoney War is over, the War is boiling up all over Europe, the Germans invade Russia, and everyone says that that's the end of Russia, because their tanks are all made of cardboard. Nothing goes well for us, the Allies. Hitler, it seems, cannot be stopped.

Just before the birth of my daughter, there was the Atlantic Charter, a bit of political show business that has never been equalled for cynicism. Roosevelt met Churchill in the middle of the Atlantic, during the worst time in the war, the Germans overrunning Russia and the Eastern Mediterranean, Rommel still succeeding in North Africa. The Atlantic Charter should be studied by anyone interested in how far rulers can go in contempt for the ruled. Anything that anyone has ever thought of in the way of benefits for mankind is there. Peace. The right to work. Free movement around the world. Freedom from hunger and fear. Democratic rights. In the Atlantic Charter nothing less than paradise was promised us all. Its immediate parent was the American Declaration of Independence. 'We hold these truths to be self-evident: that all men are created equal; that they are endowed by their Creator with certain inalienable rights; that among these are life, liberty and the pursuit of happiness.' I believe the cynicism of the Atlantic Charter was one of the causes of Churchill's defeat after the war. The Air Force found endless ways to deride it; these people had escaped from the grimy and dreadful poverty of Britain in the 1930s, only because of the war, to be sent out into this exile in Southern Rhodesia – they did not find the Atlantic Charter funny. I did not find it funny; my emotion was the grim *What can you expect?* Sonny Jameson was witty about it. Frank, who found it easy to admire authority, defended it. Not many things now in the way of political venality can shock me, but even now the Atlantic Charter never fails.

If you are wondering, why such contempt for these promisers of earthly Paradise, when in only a couple of years you, as a Communist, will be promising the same thing? – the reply is, we – the Reds – believed in our visions. Churchill and Roosevelt could not have believed in the Atlantic Charter. They were cynical, we were stupid.

My second *accouchement* was not what I expected. I make this point because of the claim that it is an attitude of mind that determines the course of labour. I approached my first lying-in (as it used to be called, accurately enough, when you might be in bed for weeks) without a care in the world, not expecting pain, or difficulty, because of my arrogant young health. But the pain was very bad; afterwards the baby exhausted me, probably just because of the exuberant health he inherited from me. And so this second time I was ready for a painful childbirth and another fighting baby. Once again the Lady Chancellor Nursing Home, the stupid bossy matron, the cheerful nurses making sure that mothers and infants met as little as possible. I was in a room on the other side of the entrance, the twin of the one I had before: small-town living offers continuities unsuspected by the dwellers in great cities. I went in, as before, in the evening, because of a certain pain I recognized, to be distinguished from all the other twinges, pangs, sensations, pressures of the end of pregnancy, and because of the unmistakable surge of energy Nature in her thoughtful way equips you with. I strode about the room alone, having been bathed and – of course – shaved. As usual the Home was overstretched. 'Just be a good girl,' the nurses cried, popping smiling heads around the door.

I wanted to be alone. I walked, I walked all night, around and around, went to look at the babies who at first were still asleep, but then avoided them when they began screaming a couple of hours before feeding time. I stared out of the window at the stars. I wondered how Frank was coping with John. Then, at ten in the morning, sharp pains, in came the doctor and nurses, and the baby was born within half an hour. I was still waiting for labour to begin. There had been very little pain before the chloroform. A little girl was shown me, smaller than her brother, and at once evidently made of different stuff, a pretty little thing ready to be held and cuddled. But, 'You'll have enough of her soon.' 'Please nurse, don't take her.' 'Oh well, just for a minute then.' The tiny lips fastened on the nipple, the miracle again, life knowing exactly what it must do. The nurse stands over you, frowning. 'You haven't got any milk yet, you know. It'll be in tomorrow.' And the baby is carried triumphantly away, and I am left, ready to cry my eyes out, in my bed. There was a new turn of the screw. The matron forbade brothers and sisters to come and visit the new baby because of infection. John was brought by his father to stand outside the window on the gravel, where I held up the new baby and waved to him. I was miserable. He was miserable. I cannot think of anything more guaranteed to produce jealousy of the new baby, or foment anxiety in a mother. That was the worst thing about this second birth.

In the evenings Frank came to visit with the other fathers. Exactly to the minute, when visiting hour was over, Matron would arrive in the doorway. 'Now then, Daddies,' she would cry, flirtatious but severe, 'that's enough. I'm going to ring the bell. Now let your poor wives have a rest.'

And the bell clanged through the building, while the babies yelled.

It might be asked, If the place was so awful, why did you go back? Indeed this is a good question. Well, I didn't know until later just how awful it was. And 'everyone' used it. There wasn't really anywhere else to go. I don't remember home births. I am speaking, of course, of white women. As for female passivity, and a good deal of my behaviour was that, I think that men hardly lack passivity when it comes to doctors.

I sat in a rather dark little room, nursing Jean while John tugged at me, trying to make me jettison her and choose him. I remember thinking, Then he does love me a bit, after all – and he howled and flung himself about, and I put down the baby and tried to comfort him. This went on – and on. I was so tired. Now I wonder how I did it. I swear young mothers are equipped with some sort of juice or hormone that enables them to bear it.

We moved again. The point was, houseowners could charge double prices to the army and the RAF – who were just beginning to flood the town. Frank said he was tired of wasting money on rented places, better put it into a mortgage. In short, we were about to join, as we saw it then, the privileged middle class. What did I think I was doing when I married a future City Father? I had not seen him like that. I felt as if handcuffs were on my wrists and chains around my ankles. But I smiled and chattered. Tigger was always friendly and affectionate and competent.

This house belonged to old friends of Frank. Everything we hired, rented, bought, was because of Frank's friends: he had, after all, lived in this city more than fifteen years. We bought the house furnished, with much better things in it than we had. This was a step up from the toffee furniture, but the rooms were hardly ever likely to be my taste, pale pink and pale green, with watery chintzes and dim rugs. Everything seemed very slightly chipped, or cracked, or faded. My feeling about these rooms went into some dream rooms in *Memoirs of a Survivor*. The woman Frank bought the place from was Mrs Tennent. She is in *A Proper Marriage*, but altered a good deal. But if the facts are altered to fit the story, then her role in my life was not. She treated Frank like a son and thought I would make a good daughter-in-law with proper coaching.

The house itself was larger, one of the better houses in the Avenues. All the houses first built in Salisbury were bungalows, of brick, with corrugated-iron roofs and many verandahs. Ours had plenty of space. Behind the front verandah, where there were chairs and tables, was a living room, an exercise in gimcrack good taste, a dining room, and at the back a large bedroom and two small ones. There, too, was a large enclosed verandah, like another room, where stood refrigerator, ironing table, prams, pushchairs, garden implements, chairs in case you fancied a rest there. Off that were the kitchen and the pantry and the bathroom. Three servants, cookboy, houseboy and piccanin, the last a ten-year-old who cleaned shoes, ran errands, did a bit of gardening, and might be asked to keep an eye on the baby. Again we paid them much more than the pitiful customary wage, and again white housewives arrived reproachfully or ferociously to say we were spoiling the natives. Behind the house, all along the sanitary lane, where the carts came at night to take away the contents of the lavatory buckets, were the brick rooms where the servants lived. There were two for our house. Everyone knew that many more people lived in these rooms than were supposed to. The police sometimes swept through the town to flush out wives, girlfriends or even children visiting from the country. There was no way of evading the police because every black person carried a 'situpa' which had details of identity and employment on it. Of all the laws imposed by the whites this was the most hated. Our servants might come to us to ask for a letter saying that a wife or a mother had permission to stay: when I say 'us' I mean the progressives, for our household was revolutionary in comparison with most. The people who came to it tended to believe the 'natives did not get a fair deal.' There should be a minimum wage, it was in the interests of the whites to improve conditions, pass laws were unfair. These topics would be discussed over meals, during sundowners, with the matter-of-factness that goes with holding a recognized political position: six years before it was seditious.

I never went to the native quarters, feeling it was not my business what went on. Once, asked to call the fumigators to get rid of lice, I took a look: each little room had two iron bedsteads, with mattresses and blankets and coarse sheets. The beds almost filled the room. Clothes hung on pegs on the walls. Our cook did the cooking for the servants on our stove: in most houses this was done on a fire at the back of the garden.

The cookboy cooked – nothing else. The houseboy cleaned and did the washing. Most of the day the men sat on boxes outside the rooms talking and smoking, with friends from the other houses. These jobs were coveted. Every day men came to the back door begging for work. A more inefficient system could not be imagined, but at least it did enable men who wanted to be in town to be there legally, and it kept them fed and housed. We bought their mealiemeal. We bought them vegetables, beans and peanuts. They got meat twice a week. 'Boys' meat' from the butcher was stewing steak, brisket, heart, sinewy bits from the legs. A good part of our supplies was taken, with everyone's agreement. Every day stale bread, stale cake, left-over puddings, left-over anything was put out on the kitchen table. 'You can have that, Indaba.' 'Thank you,

Nkosikas.' I hardly remember him, that man who resisted all my attempts at friendliness, doubtless thinking they were too shallow to be worth bothering about.

The vegetables I bought for the 'boys' now included sweet potatoes, cabbage, tomatoes, spinach, carrots; only ten years before my mother tried to get 'the compound' to use vegetables from our garden, in the interests of their health, but failed.

Our servants no longer wore the rags still common on the farms. The cook wore good khaki trousers, shirt, shoes, and had a jersey. So did the houseboy. The piccanin had to be bought shorts and shirt and jersey, for he arrived in rags. His name was 'Matches'.

And what did we eat? Yes, it was still the amazing diet of the British abroad. Every morning abundant porridge, eggs, bacon, fruit, toast, marmalade. By now I had shed breakfast and was continually being warned that I would suffer the consequences. Lunch was roast meat, hot or cold, or shepherd's pies or macaroni, and potatoes and greens and salads, English style, no dressings or herbs. Pudding. Cheese. I made puddings and cakes, though the cook hated it when I did. He was proud of his skills, and enjoyed saying I was putting in too much baking powder or vanilla, but was pleased when I taught him something new. The men came home for tea. Cakes, scones, sandwiches, the full English tea. When we went up to the Club for sundowners, we might bring back six, seven, eight people for dinner. 'There'll be seven for dinner,' I would say to the cook, who simply made more of whatever there was. 'Yes, missus.' 'Yes, Nkosikas.'

And what did I contribute to all this? I did what I had to do well enough. We were awakened with the early morning cup of tea at six. Having bathed and dressed, we had breakfast, at seven. Then the cook brought the order books. You telephoned the shops, or wrote what you needed in the exercise books sent back with yesterday's order. The cook would stand before me, waiting. 'We need matches . . . oranges . . . flour . . . sugar . . .' 'Thank you, I've got that.' 'And the boys' meat was no good yesterday.' 'I'll tell them.' The houseboy would arrive to say, We need polish, we need soap, we need candles.

And now I should have, was expected to, rejoin the morning tea parties. But I was no longer in cramped rooms, I had plenty to do. I looked after the garden, and sewed all morning. I made all the little outfits for John and the baby, Frank's shirts and pyjamas, all my clothes and underclothes, aprons and shirts for the servants.

When I did to go a women's party I found myself unable to keep quiet about what I thought. It was known I had all these dangerous ideas . . . but what ideas? They were more a welter of emotion. Sometimes when I am interviewed, I am asked, How was it, brought up as you were, that you understood the society you lived in? The reply is, I didn't. They have a picture, it seems, of a child, aged perhaps ten, who exclaimed, This is a profoundly unjust society, and how is it that a small white minority of a hundred thousand enslaves a black majority of half a million? But I still did not know how to describe what I was living in, only knew what I felt, and that is a very different thing. I lived mostly in a state of incredulity. *How is it possible these things can be?* I would collapse into howls of disbelieving laughter over the *Rhodesia Herald* at the breakfast table. 'What's the joke?' enquires the young husband, who is off to take his part in the running of the country. 'Look at that!' I demand, thrusting the paper at him. He stands, gulping down his last mouthfuls of tea, reading. 'Hmmmm . . . not too clever, I agree.' Frank had a slight stammer, which probably began as an attempt to give himself gravitas when a very young man working with men much older. The letter page was particularly rich in comedy. 'I want to protest at the custom of using kaffir women as nannies. Do we want our children to grow up like kaffirs? – Indignant Mother.' 'With all these foreigners and socialists from Britain coming into our country, there is a danger the natives will have all kinds of unsuitable ideas put into their heads. What steps are the authorities taking to prevent this? – Democrat.' 'Frank, just look at this!' He laughs. 'Well, I've-I've-I've got to be off. See you lunchtime.' I sit at the breakfast table drinking tea, with one hand pushing Jean's pram back and forth, back and forth, reading the *Rhodesia Herald. It simply isn't possible.* But incredulity is followed by anger, which is succeeded by But what can you expect? hypnotizing you into inaction. And what action could I possibly take? And now I was secretly contrasting the young women of the tea parties with my mother – not the woman of now, the defeated and unhappy wife of a dying man, but with what she had been. At no stage in her life would she have been happy to sit around all morning talking babies, men, sewing, knitting, food, servants and 'who-would-be-a-woman?' Between women like my mother, Granny Fisher, and my mother-in-law and these civil servants' wives was a gulf where you could lose all faith in yourself. Nothing was asked of us – that was the point. None of them worked, or expected to. They would have two or three, possibly four, children, run the house, make everyone's clothes, and in middle age take to charity. So I saw it, but in fact Fate had a different plan, for it was these women who in the War of Liberation thirty years later took up guns to defend their way of life.

And what were the babies doing? When John was in the garden he was happy playing with the piccanin. Shrieks of laughter as he escaped into the street and had to be chased back. But after all, he could not always be in the garden. In the house, he was no longer a good-humoured little boy. He was suspicious, angry, jealous. When I nursed the new baby he would rush at us both, fists raised, or howl with rage and squat glaring in a corner of the room, betrayed. When the baby was on the bottle things were not much better. I could not leave him alone with her. He was not the same child. Meanwhile Jean was amenable and amiable, slept when she should, exactly like the little boy I had looked after – well, it was not so long before – in a house not far from here.

I was not feeling well. Probably I was anaemic. The doctor said that a woman with a new baby and a toddler – an interesting word for John – was bound to feel tired. All this time our energetic social life continued. It never occurred to me not to drink, or to give up smoking. It was my right. Besides, I didn't get drunk did I? A little tight sometimes. When Frank arrived back from the office at lunchtime and at once got out the beers, the gin bottles, the tonics, if I said no, he would cry, 'But we only live once.' The food I ate could not have been more healthy, but apart from that, I was doing nothing sensible. I only wanted to sleep. I fainted several times, something I've not done before or since. I was miserable and confused, being torn apart between these two babes.

We decided I would take John down to the Cape for a month, leaving Jean with a friend. I did not feel guilty about this then and do not feel guilty now. Small babies need to be dandled, cuddled, held, comforted, and it does not have to be by mother. Any loving woman will do. In the next house lived a woman who all her life had longed for a daughter. She was now over forty and would not have another child. She could not keep away from our house where the delectable little girl babe cooed and smiled and waved her little limbs. She was not merely willing to look after the baby for a month: when we finally decided, she wept. She could not stop thanking us. She said she thought I was mad to forego even one hour of this perfect baby girl, but she wasn't going to complain. Now I would agree with her.

This trip was postponed because the ship my brother was on in the Pacific, the *Repulse*, sank when the Japanese torpedoed both the *Repulse* and the *Prince of Wales*. First came the news of the sinking. I thought he was dead. Why? It was because death by war was so much on my hidden and secret agenda. I clung to Frank and said that we must at once have another baby. Can there be a more basic, not to say primitive, reaction? I was thumbing my nose at the news of death, death, thousands of dead every time we listened to the News. Frank said I was hysterical; he was right. Certainly I was far from the jolly competence that was my mode. What had got into me? – he kept asking. The sooner I got a break the better, he said. I was weeping, and angry, because I could not say how deeply and irrationally I blamed my mother, and on a level far from the logics of: 'You worked, fought, suffered, pulled strings, to get your son into the Navy because you most passionately suffered as a girl, when your brother failed the exams to get into the Navy and you could have succeeded so easily – your son is your buried and thwarted self, and therefore he had to go into the Navy. In a war, ships get sunk. *What did you expect?*' This was not the point at all. I was staring, eye to eye, at my enemy, a deep and terrible dread, expressed, perhaps like this: If a woman has worked for years, and against such obstacles, to get her heart's desire, in this case, her son into the Navy, and then at last succeeds, *of course* his ship will be sunk. What else? If Harry had gone with all the other Rhodesians up to North Africa Nemesis would not have noticed him.

It was round about this time I began having the dream that came back for years. I was in a dusty eroded landscape. I stood by a gulf or ravine, where the layers of Time's upheavals lay one under another. This sight has become familiar to us now on the screen, but then it came from standing by prospectors' trenches, which exposed the earth's layers. Right at the bottom of the pit was a shape like a big lizard – no wait, it was a lizard, an ancient dragon, preserved there for millennia. But it wasn't dead, for its dust-glazed eye stared out at nothing, then slowly swivelled up, like a chameleon's, at me. Or, in other dreams, the eye stared out ahead, and after centuries, blinked once.

No bird's golden eye, that, not the eye of a raptor, a quick swoop and a kill and then up and away. Recently I saw a documentary made by Japanese and Chinese film-makers of the part of the Chinese desert on the old Silk Road where the sands move about, covering ancient cities and exposing them as the winds blow, revealing the frail mummy of a young woman, still beautiful, with her shreds of wind-torn silk clinging to her, and then burying her again. That is what the old lizard's eye sees while it blinks, once.

Then came the news my brother had been taken up from the sea, with some others, though most men had drowned. Later Harry told me he had been standing by the ladder, below decks, watching the men climb past him into the air, when someone said, 'Aren't you going up, Tayler?' Interesting that when later he wrote an account of the sinking to his parents, he did not mention this. It has stayed with me. Life depends on as chancy a thing as someone saying, 'Aren't you going up, Tayler?' – so he could reach the deck as the ship turned over, walk down into the sea and swim away.

The trip from Salisbury to Cape Town by train took five days. I believe it does not take much less now. I was in a coupé with John. To be shut in a space the size of a horsebox with a hyperactive child is something I do not recommend. The train crawled south through the middle of the continent, the Karroo, the mountains of the Cape, while I sang, chanted nursery rhymes, recited all the poetry I knew, made up stories. Meanwhile John swung, hung, clambered around like a little monkey. It was very hot. The window had to be up in case John threw himself out of it. Dust filmed seats, walls, our faces. My dress had films of dust, John's little pants and shirts were brown. But he was himself again, not angry, he had his mother back again, and so he was good-humoured, friendly, and curious about everything. When we stopped at the stations, again there were the black children selling wooden animals, the women earning a few pence for a mango, oranges, apricots. Time did pass. Every hour's end was a small victory. In the afternoons I locked John tight in my arms, so that he had no alternative but to sleep a little – we were stuck together by sweat and dust. At night he fell into sleep exhausted, and I with him. The five days passed, they did pass, and I was in a cheap hotel on Seapoint, festooned all along the front with coloured lights to show it was for good times and holidays, but now Cape Town was full of sailors and of troops on their way to this arena of the war or that. Also, refugees.

In the hotel was a mix of people only possible in wartime. Some had escaped the fall of Singapore. At my table was a woman who, when Singapore fell, was put into a rowing boat with two new babies, twins, reached a ship which brought hundreds of people to Cape Town, all expecting every moment to be torpedoed. She did not know whether her husband was alive. British authorities in Cape Town gave her some money. In the weeks I was in the hotel guests collected money for the refugees who had nothing. The women with babies gave her baby clothes, female guests donated a wardrobe. This very English woman, born to wear the 'good' clothes of her kind, was in the breezy and irreverent dresses we all wore, but her blue eyes, her gently freckled face, smiled equally at calamity and her reflection in the mirror, while she might remark that it was odd of 'us' to arrange that all the guns in Singapore should face out to sea, leaving the landward side, where the Japanese easily arrived, undefended. Similarly, the British, 'us', had announced that the *Prince of Wales* and the *Repulse* were unsinkable. Like the *Titanic.* Yet the warships were sunk within minutes by torpedoes and the *Titanic* within minutes by an iceberg. These events have not stopped us from believing in the pronouncements of military experts. For those weeks this Englishwoman and I were friends, learning each other's ways. Then the war blew us apart for ever. After the war she went back to England with her twins, and rejoined her husband, who was alive and safe.

While in Cape Town, Frank had said, I might as well go to a birth control clinic and get the latest advice, which was bound to be better than in poor provincial Salisbury. Untrue. At the clinic a handsome young man probed my secret parts with his gloved forefinger and breathed heavily. I enquired mildly, 'Is everything all right?' and the moment was done with. Nothing is more extraordinary than the system of switches women are equipped with. This was not an amatory situation, but alone with him on a beach at midnight I would have dissolved like wet sugar. As I write it occurs to me that this could be described as sexual harassment. His perhaps five seconds of mild misconduct would have got him struck off the medical register while I would be expected to exult at his ruined life.

In the hotel was a young woman from Windhoek who had come

especially to Cape Town because of this very clinic: she was twenty-one and already had three small children. Her husband was a railway worker and badly paid. They simply must not have any more children. She was like Ivy, with her loose dry hair, her thin anxious body, defensive, humorous, apologetic. In fact this girl and her husband efficiently coped with three tiny children, the youngest a baby, no joke in this cheap hotel. They adored each other. His large, too-thin hand kept reaching out to touch her hair, or her shoulder. She could not stop herself smiling with love when she looked at him. These two, and their babies, and I, with John, were often in their room. John was enchanted with these new friends – but not the baby, a threat. She took out her new Dutch cap from its film of silky powder and said, 'But look at it, just look, I can't use that thing.' 'But sweetheart, we've got to.' 'Oh heck, sweetheart, you mean I've got to.' 'But when I use a french letter you just get pregnant.' '*When* you use a french letter, you mean.' And they fell into each other's arms, laughing. My doctor's admonition, which I had seen as a mere moment of whimsy, made sense at last. 'It is no good my prescribing a Dutch cap if you simply leave it in the drawer.'

She was pregnant before she went home to Windhoek.

In the evenings one of the mothers volunteered to keep an eye on all the sleeping infants, while the rest went down to drink at a café on the seafront. But now I was drinking hardly at all. Drink did not seem to be an imperative, as it was in Rhodesia. Besides, this was a wine culture. And, besides, I was too interested. My friend from Singapore sat with the couple from Windhoek. The Englishwoman and the girl from the wind-dried, sun-parched dusty little town which heard of the great world from the melancholy shrieks of the trains, asked each other well-intentioned but inept questions. They laughed from helplessness, and in the end simply gave up. You may say, 'We live in a railway cottage near the lines and most of the time we are worried sick about money and everything in our house is filmed with dust, and the flies just drive me *mad*,' but it is hard for a woman from a country cushioned in damp grass to understand. 'You met your old man in school?' 'I met my husband in summer school, yes.' 'Do you go to school in the summer in England?' 'No, no, a music summer school. Mozart and Handel.' 'Catch me going to school when I didn't have to.' Nor did the Englishwoman understand the three tiny children, spaced about ten months apart. When I told her about the South African girl's inability to handle a Dutch cap, she said, 'But that's simply *silly*.' And then, brought to doubt by a culture she could not have imagined before the war, 'Well, it is – isn't it?'

One night when his wife was on guard with the babies, I was with the South African girl's husband on the beach. He at once began to make love to me. I was shocked. He was in love with his wife, wasn't he? 'Oh hell, man, you get fed up with the same thing all the time,' said he. I said, 'But I'm not in love with you.' He said, 'You don't say! Come on!' He was aggrieved and could not even look at me after that except with genuine hurt. I had no idea how quintessentially a male–female confrontation this was.

I was dreaming in Cape Town of a man, but only one who represented the bohemian life I knew this city nurtured. Painters, poets, artists of all kinds lived in a Latin quarter awash with wine and free love. But I could not aspire to these delights in Seapoint with a small child. If I did get an evening away from the hotel and the complicated systems of swapped baby-sitting, then where would I go? Besides, I was so tired. John might be friendly again, but he was running and clambering all day, with me after him.

There was a morning I remember when contemplating the implications of 'Aren't you going up, Tayler?' I went down to the beach with John, and a big sea was running, waves that towered to the zenith, then crashed, reared up and crashed, while a sharp wind flung sand and cold spray against my legs. John was jumping up and down, shouting with pleasure at the noise and the swing of the sea. He pulled himself free, and like a naughty puppy left the reins of the harness in my hands, and was running along the beach just beyond where the waves came down bang, bang, bang, then retreating and taking with them scoops of sand as if the beach was being dragged out to sea, but the whole mess, sand, water, spray, was flung back again . . . If a wave caught John that would be that, no one could swim in that sea, he would be swept out. I was running after him shouting, but the noise of the water absorbed my voice. There he ran, as fast as a bit of blown foam, and every time a wave crashed I thought I would not see him again, and I ran and I ran, but I had not been able to run as fast as he could since before he was a year old. And then, there was a man at the other end of the beach, and he saw, and went to stand where John would run into his arms. He caught him, lifted him up, and brought him back to me. He had to hold me up, as well as the little boy, I was ill with shock. He was a soldier on a day's leave from his ship. He handed me the child and said, 'Phew, that was a near thing.' His smile said that here could be the romantic adventure I craved, but a child of two is not conducive to romance.

The real event of that trip to the coast was an encounter as tonic as the words of the old man in Umtali. At our table one lunchtime, eating – critically – the curried mutton and pumpkin that was our chief nourishment, was a cousin of the mother of the twins. She was a bony, pale woman, in her thirties, wearing a cream linen suit, and pearls that insisted, modestly, that they were real. She dabbed pale pink lips often and thoroughly while she frowned at the mess on her plate, and offered me a job in her office. She had been caught in South Africa by the war, her husband was in the Army on the way – she believed – to the East, India she thought, but careless talk cost lives. She was putting her talents to use by fomenting good race relations here: she was a member of a part of the Christian church devoted to this task. Hearing I was from Southern Rhodesia she proceeded to demolish that country. It would be satisfying to report words as gnomic as, 'You see, you are very young and I am very old,' but no. 'What *can* one say about a people who have stolen all the black people's land and then talked about uplifting and civilizing them? How can one describe a country where 100,000 white people use 1 million blacks as servants and cheap labour, refuse them education and training, all the time in the name of Christianity?' But what really made them such a pain was, they were so pleased with themselves. 'Why on earth are they so conceited?'

She went on like this while I tried to hold John in his chair long enough to get some food into him, while her cousin fed spoonfuls of baby food into her twins, one, two, one, two, and said, 'Sounds a bit like Singapore to me. Pride goes before a fall.'

Surely this view of my mother country could not come as a surprise to me? Not a surprise as much as an exhilarating truth long withheld. No one I had known was capable of saying anything so simple and so obvious. This was truly a revelation, and above all, I suffered because of her contempt. My parents might have been talking about 'this second-rate little country' for years, but that did not sear me as much as she did, or a man, part of the government of South Africa, dumped for some wartime reason in this rackety hotel-boarding-house: 'Ah, so you're from our *clever* little neighbour up north?' (The Nationalists did not come to power until 1949, the 'feel' of South Africa was still British, and Southern Rhodesia had not been forgiven for refusing to become a province of South Africa in 1924, choosing self-government instead.)

My brother arrived in Cape Town for a couple of days, on his way past to join the *Aurora*, which would spend the rest of the war fighting in the Mediterranean. He sat on the hotel verandah, admiring John and me, too, while I admired the handsome naval officer. We had scarcely met for years, did not know each other. As for being sunk in the *Repulse*, he did not want to talk about it, and not for years did I hear what it had meant to him. We sat there, flirting a little, as a brother and sister may do who seldom meet. And besides, it was for the benefit of the young women in the hotel, who were pushing their infants back and forth past us far more often than was usual.

My marriage had ended, but I did not know it.

During the five days' crawl back up through the magnificences of Southern Africa I entertained John and thought of how, once home, I would . . . well, what, exactly? The Englishwoman had said that when her organization had established an office in Salisbury she would expect me to work for them. But the Church was not for me. Yet only in the missions and churches was there education for the natives.

I returned to a little girl who had spent every moment of her waking hours in the arms of her adoring foster mother, and was going to find me deficient in comparison. I told Frank I wanted to 'do something', and he agreed. That a woman's place was in the home was no part of the 'progressive' thinking he was committed to. We engaged a black girl as nanny. A simple thing, you would think. But there were already three males in our two little rooms. It would have been normal for that time if she shared space with them, but we felt uncomfortable about it. We suggested she should sleep in the room with John – Jean was in our room – when we went out, which was nearly every night. This formula enabled her to tell the police or anyone else interested that she was living in the kia – the brick rooms – at the back, allowed her to stay there in fact if she wanted, but gave her the choice of a bed in the house. Such a thing was then really unheard-of. Scandal. Shock. Horror. Neighbours let it be known they had heard 'what was going on'.

The woman next door, having enquired if it was true that a black kaffir woman really did sleep in the house, said coldly 'You're ever such a boheem dear, aren't you? She's going to take advantage.' My mother was shocked, distressed, and went to Frank's office to protest. As always he was calm and tactful, but he did not understand why my mother reduced me to frenzies of exasperation, why I simply went to bed when she had gone, or why I wept with helpless rage when she rushed into the house to be rude to the cook, insult the houseboy, and tell the piccanin that he mustn't dare touch the baby.

Years later a therapist wanted me to agree that it was a sign of immaturity that I never said what I thought to my mother, there was never a flaming, shouting, screaming, stand-up quarrel. But she would have collapsed, gone to bits. I cannot remember my parents ever saying a loud or nasty word to each other. It was not their style. I was crippled with pity for her, immobilized by self-division, behaved always with an implacable politeness – much worse than screaming and shouting. And what would I have shouted? Only one thing, *For God's sake leave me alone!*

I might have intended to 'live my own life', a formula I now find infantile, but this had to be postponed, for our house was full of people. In the dining room the table was stretched, and meals seemed almost continuous, and I cooked endlessly, unrebuked by the cook who needed help. I joked to myself – since no one else would have seen the point – that this was like a Russian family, in Yasnaya Polyana perhaps, the young married couple and their children and their serfs, the in-laws with their children, the sister-in-law in for the day from the country with her husband and children, and Frank's friends from the office or the Club.

'I say, the meat bill's a bit high, isn't it?'

'Well, so's the liquor bill.'

I had suggested to Frank we drink Cape wine but he was uncomfortable. Wine was then 'toffee-nosed' – showing off.

Frank's sister, Mary, stays in my mind like the romantic heroine of a story as simple as a ballad. She was slim, graceful, with smooth amber hair parted over a low broad forehead and worn in a knot. She had enormous grey eyes, a wonderful smile, and dimples. This lovely girl had been glimpsed by a visiting Lord, and he had instantly fallen in love with her. She had fallen passionately and finally in love with him. She had every reason to believe she would marry him and live in England. But he changed his mind. Her heart was broken: hearts do break. At once, she married on the rebound a man who was a beast to her beauty – but most men would seem that – a rough, coarse-fibred, well-intentioned farm assistant, later farmer. He would watch her, this angel who was so far above him, with a look half adoration, half rage. Every molecule of her body was sensitive, gentle, delicate, refined, discriminating. He thought it amusing to put a firecracker down the cook's shirt or shoot a bird or an animal for fun, not to eat, because she hated him doing it, or leave out money in a room where the servant would shortly clean and then watch through a crack how the man suffered agonies of indecision about whether to steal the coins and, if he did, burst in threatening the police, then roar with laughter, 'Got you then, you skellum.' Mary never showed how she suffered, was forever patient and humorous. We believed she had chosen this hair shirt of a husband as a punishment for allowing herself to be so injudiciously in love. When she visited, she would look at my books, turn a page and sigh, 'Life's sad enough, why read about it as well?' When I wrote apprentice pieces and parts of novels, she complained I had a morbid view of life, before returning to the desperate poverty of her lonely farm, the tiny house, her boor of a husband, and her two little children whom she tried to bring up to love Beauty and gracious living.

Frank's mother would arrive, stay a day or two and rush off. She was 'Mater' or 'Wizzy'. She had been left poor, was given small amounts of money by both sons, and spent her time on visits all around the country, where she played bridge. She was a short, fat, likeable woman who did not 'interfere' with us because she was not interested. She had had more than enough, after a tough, precarious life of the early days, sometimes with money, sometimes without. For me now she is a lost opportunity, like Granny Fisher. I was afraid of letting her come too close: one mother was enough.

About then Dolly Van der Byl moved in. 'Why don't we let her a room, you can't get rooms now with the war on, she's a nice kid, it isn't fair,' says Frank.

When Dolly moved in to an empty room, she sometimes shared meals with us, but she was not often in. She had been one of the Sports Club girls for years, knew everyone in the town, played all kinds of sports, was easygoing, friendly, helpful, baby-sat for friends and for us, worked for the Red Cross. In the early mornings, we all three drank final cups of tea in the back verandah, while Dolly stood at the ironing board pressing that day's dress, and Frank and she exchanged news from the offices – they no longer worked in the same department. If the children were there she was amiable, an aunt figure, always finding, with exclamations of surprise, little presents for them in her handbag. She often joked that she didn't mind she was getting too old to have children – joked she wouldn't mind taking on a man who had kids already. Frank and she often bicycled off together to their respective offices. She biked everywhere, all hours of the day and night, all over the town.

Later people suggested I would have been in the right of it, had I been jealous of Dolly, but not even at my most dishonest was I tempted to use this justification. For one thing, I was never jealous of Frank. A man whose wife is never jealous has good reason to feel aggrieved, but only in a certain kind of marriage. This bouncy, comradely marriage did not have room for jealousy. Or rather, it had been that kind of marriage. Now Frank was making scenes and sulking, and perhaps would have been pleased if I had made scenes too. This change in our emotional fortunes confused me. Suddenly he was accusing me of flirting. Yet I was behaving as I always had, and I told him he was being unfair. Not for the first time a young woman wondered bitterly, Then why did you marry me? It was my personality, what I was, that he was criticizing. I was always open, straightforward, honest to the point of tactlessness, not to say abrasiveness. This was what came naturally. It was my style, instant intimacy with everyone, the modern talent. Anything less than perfect frankness with everyone is an insult to honesty, to friendship. But why call it flirting? I continued to be *reasonable*. Under the pressure of his constant criticism I told myself that I could retaliate by saying that his matey familiarity with the girls and women of the Club was because they had been kissing, and more, on the back seats of cars for years. But I felt ridiculous even thinking this. But why did not Frank feel ridiculous making irrelevant accusations? Irrelevance: more and more I was being submerged in it. As for Dolly, I did feel chilled and excluded when she and Frank sat talking for hours, but they had known each other for getting on for twenty years.

The other woman who was often in our house was Dora, Frank's brother's wife. George had been a Rhodes scholar, a First World

War pilot, then a Colonial Office administrator in Nigeria. He was a more sophisticated version of Frank. Dora and he did not get on. 'You see,' Dora might drawl, 'we don't *really* get on.' She was a large, dark, smiling, handsome woman, full of defensive female ways. She had been beautiful: there were photographs of handsome uniformed George and his lace-swathed bride. When I think of the time when people could not divorce unsuitable mates, I remember Dora. She managed her unsatisfactory marriage like this. Clearly it was impossible for white children to live in Nigeria's terrible climate, and so she stayed in England. When George came on leave she was often staying with relatives or friends in another country or even continent. They hardly ever were together. Her criticisms of him were deprecating murmurs. He hotly criticized her, while she smiled guiltily at him, at us, remarking that if she and the children actually did live with him she didn't think he would like it very much. 'George does like his girlfriends . . . I don't think he *really* likes children very much . . . he doesn't *really* like me very much.' And she bit her lips with a little humorous grimace as if saying it was a pity he didn't like dancing or tennis. She drove him wild with irritation. He thought she was stupid.

I used secretly to contemplate lovely Mary and her farmhand husband, sighing Dora and her no-nonsense husband, and wonder as always how it was that people bound to make each other miserable so often ended in the same bed, or at least bedroom. I did not yet include my marriage with Frank in this category. In comparison we were well-matched.

George drank a lot but did not show it. 'He holds his liquor well.' There could not be a greater compliment. While he was in Salisbury we all drank much more than usual. We danced, too, it seemed most evenings, Show Balls, Sports Club Balls, and dances at the hotels. 'You only live once,' both brothers were fond of saying.

Round about now John, Jean and I all got whooping cough, very badly. They went out to the farm where my mother nursed them, and I stayed in our spare room so Dora's children could not catch it. When I drove out to fetch the recovered infants I found my father spectral and emaciated with his several illnesses. He watched the two pretty little children playing with the cats and dogs and said, 'Yes, that's what you were like too, such lovely little things you were and look what you turn into. It's not worth it.'

'Oh come on,' says my mother, who surely must have had many moments of agreeing with him. 'You're exaggerating.'

'Oh no, I'm not,' says my father, swinging out his wooden leg in front of him in his way that always looked as if he wanted to hit someone with it, but was to ease the discomfort of his shrunken stump inside the hot itching stump socks. 'What's exaggerated about it? I bet if you asked most people, they'd say it wasn't worth it, having kids, all the hard work and worrying about them and then they turn into second-raters.'

They were trying to sell the farm. Quite recently, visiting the old place again, no longer seeing it with the fantastical eyes of childhood, I realized it had always been too small. They could never have done well on it. The question is, how was it they didn't see this?

13

AND NOW I MET in the street Dorothy Schwartz, that dissident member of the group of 'progressives' who long ago (but it was four years) had thought me a suitable recruit. There she stood under the jacaranda trees, books under her arm, unchanged, a tiny dark girl, the wise maiden, listening calmly while I poured out to her in the 'humorous' style of Tigger, how I hated my life, 'white civilization', women's tea parties. I did not say I hated my husband, only that he was reactionary. A woman may say with a clear conscience that her husband is a dreadful old reactionary, but not that he is mean. Dorothy said she thought I should come to a meeting. Before I could explode into dislike of what I had seen of her friends, she informed me she didn't mean those stupid social democrats, but there was a group of real revolutionaries and they thought it was time they met me.

This kind of flattery is hard to withstand.

It never does to underestimate how everyone, in no matter how apparently obscure a situation, is being observed by individuals and groups of all kinds, who watch potentiality and performance. If this sounds self-important and even paranoid, I can only say I have seen it over and over again. The watchers may be malign, not always benevolent.

I began to meet Dorothy, always with others, who are now all a blur – mostly the Air Force, and refugees from Europe – in Meikles Lounge over beers, in the Grand Hotel, in cheap tea rooms in poor parts of the town. What I was now part of had to be interpreted for me. The old Left Book Club crowd had been 'left behind by history', and exposed as reactionaries. The lies told by the governments and the newspapers about the Soviet Union had been exposed too, by the magnificence of the Russian defence of Stalingrad. The new situation demanded an objective assessment of our resources, and, above all, of possible cadres.

I was intoxicated. They made me the secretary of some organization, I forget what. Frank was uneasy because as a civil servant he could not afford contact with sedition and because my new life did not include him.

I had become a Communist. This simple statement will immediately be understood by anyone over a certain age, or with certain experience. But an old Red may hear from some young person, who is all willingness to understand, 'Sure, you were progressive, but why Communism? Why *organized* radicalism?' At such a query I and people like me can only marvel. Seldom can the whirligigs of time have brought about, so fast and so completely, such a re-shaping of ideas. Above all, behind it is the assumption that people became Communists after studying half a dozen party programmes, side by side on a table in front of them: 'Shall I join the Labour Party?' 'No, I think Huggins' United Party . . .' 'On the other hand . . .' People became Communists because of cynicism about their own governments – that, first. Because they had fallen in love with a Communist – as Gottfried Lessing did. Because they were taken to a Party rally and were swept away by mass emotion. Because they had been taken to a Party meeting and found the atmosphere of conspiracy appealing. Because of the idealism of the Party. Because they had a taste for heroics or suffering. In my case it was because for the first time in my life I was meeting a group of people (not an isolated individual here and there) who read everything, and who did not think it remarkable to read, and among whom thoughts about the Native Problem I had scarcely dared to say aloud turned out to be mere commonplaces. I became a Communist because of the spirit of the times, because of the *Zeitgeist*.

In Turgenev's *Fathers and Sons* there is a scene where the hero Bazarov (who is the same psychological type who would later become a Communist or even a terrorist) takes a student friend to visit the past, in the shape of two little old people, survivors of that cataclysm, the French Revolution and its enlightened ideas. There they are, twittering away like little birds, a wonder to the new youth. I have no doubt that before long I, and similar relics of Communism, will be visited by young people the age of my grandchildren and great-grandchildren, and I will watch them as they go off, exchanging a smile of tolerant disbelief. 'Amazing, they had all those absurd ideas . . .'

And what was the mother of two small children doing? She was being competent. Until the moment she walked out of that house, she ran it, supervised the servants, went dancing and drinking, took the babies for walks up and down the avenues under the trees, at either end of the pram, John, needless to say, protesting. She continued to make clothes for the children and herself and Frank. She cooked. She might spend whole mornings with Dora and Mary and amiable family chat, but all the time her mind was on this other world which she belonged to *and was her right*.

A scene: a young woman wearing green linen shorts, a checked gingham shirt, is on a bicycle pedalling with long brown smooth legs she is as conscious of as if a lover were stroking them. She is exultant with her own attractions, with what she is learning, for her brain is exploding with new information and ideas. On the front handlebars is hooked a little canvas chair, and in it sits baby Jean, perhaps fifteen months old. When John is in that chair he is as elated as his mother, loving movement, excitement, challenge. But baby Jean, although she wants to be like her big brother, is a thoughtful and sensitive creature, and is not happy speeding along under the trees. I know she is apprehensive but tell her it is all right, while her lips quiver.

I am going for some reason to Nathan Zelter's house, probably to pick up a book or pamphlets. I arrive, put out a foot to the verandah edge, and sit smiling at him. He surveys me with the ironic appreciation of a man attracted to a girl who is out of reach. He makes some remarks intended to be dispassionate, but they come out sarcastic. I am furious. How dare he? is what I am thinking. Years later, at a party, in the 1960s, I watch a young woman who wears miniskirts so short her crotch is always visible in neat tight white knickers. But tonight she is sitting on the floor in a long 'ethnic' dress. It slips up and exposes an inch or so of ankle, which a man sitting near her openly admires. She flashes him a contemptuous look and pointedly pulls the skirt down over her feet.

Young women dressed and made up to look like sexual organs – whether they are aware of it or not – may mutter, 'Filthy beasts' when men respond. Similarly, my bright new lipstick, my newly slim body, smooth legs, were by some amazing division in my mind, my own business, my own property, nothing to do with this impertinent man, 'comrade' or not.

And back I cycled through the avenues with the baby clutching tight to her front bar of her little seat, her little legs stuck out in front of her, frowning with determination under the flaps of the sun bonnet.

Now comes a puzzle. I had spent a good part of my childhood adoring some infant or small child. I had been waiting for the same with John, who for some reason had to fight away from cuddles, with anyone, not only me. Jean would have happily spent every moment on a knee or in loving arms. But I had switched off. This is not to say she was not dandled and loved, but from me – not enough. ('Mummy, come and cuddle me.') This person in me, the lover of babies and tiny children, would revive later. I was protecting myself, because I knew I was going to leave. Yet I did not know it, could not say, I am going to commit the unforgivable and leave two small children.

For some weeks there was an absurd situation where everyone discussed my leaving, but I did not leave. Frank, distraught, said that I should leave long enough to come to my senses and then return. Mary was severe, and unforgiving. She admired me, had thought this was a perfect marriage. Dora, this embodiment of womanly virtues, supported me. Not publicly, when she might sigh and say, 'Oh dear, oh dear, how very very sad,' but when we were alone. She said, 'Good for you, I wish I had the courage.' Dora was in the house now. She was another who had longed for a baby girl, had two boys, and now could love Jean. She and the neighbour who had looked after Jean competed gently for Jean's favours. Meanwhile I was with the comrades every evening and some part of every day. They listened to my conversations with myself about leaving, and at last said I should either leave or not, they were getting bored with it. At least I did not have to explain to them that it was the way of life I had to leave: as far as they were concerned 'White civilization' belonged in 'the dustbin of history'. (Salutary to read this dead lexicon now and remember what power it once had.) They were less than four decades out.

I not only discussed my leaving with everyone in the house and with the comrades, but with the babies too. It was they who really understood me. As if I were their age, or they mine. What I had always had with John was a kind of friendship. We had always 'got on' even when his furious inexhaustible energies reduced me to tears. With Jean, a gentle soul, there was a tenderness bruised by unadmitted guilt. I explained to them that they would understand later why I had left. I was going to change this ugly world, they would live in a beautiful and perfect world where there would be no race hatred, injustice, and so forth. (Rather like the Atlantic Charter.) Much more, and more important: I carried, like a defective gene, a kind of doom or fatality, which would trap them as it had me, if I stayed. Leaving, I would break some ancient chain of repetition. One day they would thank me for it.

I was absolutely sincere. There isn't much to be said for sincerity, in itself.

This feeling of doom, of fatality, is a theme – perhaps the main one – in *Martha Quest*. It was what had made me, and from my earliest childhood, repeat and repeat, 'I will *not*, I simply will not.' And yet I had been swept along on some surface, or public, wave ever since I had left the farm (and my typewriter) taking my fate in my hands, or so I thought, to become one of the town's marriageable girls, then a wife, then a mother.

Looking back now I would say that perhaps one quarter of me had been involved since then, and the best part of me was in cold storage. So I felt. But behind that phrase, *looking back*, what complex processes lie. 'Oh, but that's how I saw things then,' an older woman may say to another. 'I was raw then. I was uncooked . . . ungelled . . . immature . . . I simply wasn't born.' And be understood at once. Well, I was uncooked.

Decades later I met an elderly woman who had her first baby at the same time I had mine. We sometimes spent mornings together. 'You were not maternal,' she said to me, in 1982. I looked back at all that breezy competence and could only agree. But between that time, in 1942, and 1946, when I had my third child, what happened that brought the buried or switched-off three-quarters into use again? I have no idea.

A scene: on a blanket spread on the lawn under the cedrillatoona tree sit the two children and myself. I am drunk with sincerity, and conflict, elated with ideals and with poetry.

The two little things sit watching, interested, while I recite the sonorities of Hölderlin.

Fill with yellow pears
And hang with roses
The land by the lake

You holy swans
And, drink with kisses,
Dip your heads into the sobering water
Alas, where shall I take
When winter comes . . .

Or

The World is charged with the grandeur of God

Or

Glory be to God for dappled things . . .

'But I thought you were an atheist,' I hear someone protest. 'Well, what of it? Poetry is something else.'

The two infants are entranced. Jean tentatively waves her arms in time. John bangs a building brick hard on an old tin.

Old Adam, the carrion crow
The old crow of Cairo
He sat in the shower and let it flow
Under his tail and over his crest
And through every feather
Leaked the wet weather
And the bough swung under his nest . . .

And John rolls on the blanket shouting with the words. *'Old Adam, the carrion crow . . .'*

When Rousseau abandoned his children, he thought – so he said – that he was saving them from a ruinous and debilitating upbringing by corrupt people: in the Foundling Hospital they would turn out robust and honest and useful citizens. Perhaps it is not possible to abandon one's children without moral and mental contortions. But I was not exactly abandoning mine to an early death. Our house was full of concerned and loving people, and the children would be admirably looked after – much better than by me, not because I did not perform this task exactly like every other woman around me, but because of this secret doom that was inside me – and which had brought my parents to their pitiful condition.

I did not feel guilty. Long afterwards – about ten years – a psychotherapist informed me, with the air they have of producing revelations out of a hat, that I carried a load of guilt. *No! You don't say!* By then I was regarding myself as something of an expert on guilt, both evident and devious. I know all about the ravages of guilt, how it feels, how it undermines and saps. I energetically fight back. Guilt is like that iceberg, but I'd say with ninety-nine hundredths hidden. Decades later I was giving a lecture on barriers to perception, that is, what prevents us from seeing straight, and I spoke about ten different attitudes, one of them being guilt. At question time, in this audience of a couple of hundred, people kept jumping up to ask about guilt, guilt, guilt, nothing else, only guilt. I suggested I had talked about other things, but no, it was guilt, guilt. I meet people who say, 'Your lecture on guilt . . .'

In such a guilt-ridden culture it is not easy to distinguish one's own from some burden it seems we all carry.

It took me a long time to see that it was guilt which was painting an attractive picture of myself as I might have been had I not walked out of that marriage. This picture changed with my changing understanding of myself and, too, with the changes in Southern Rhodesia. But the basic plot went like this. Instead of flinging myself heart and soul – well, not exactly, in fact I had something like nine-tenths of a heart and soul in reserve – I would have tasted Communism, until I could see the phenomenon critically, which would take a month or so, either not leaving the family at all or only temporarily. I would 'get on' with Frank, just as I had always done because of the ability of young women to adapt and please. While having increasingly less in common with him I would have understood him and been a good mother to the children, whose inner natures and capacities would be developed to their fullest. Yes, Frank and I were mismatched, but not worse than many. He was naturally a conservative, I naturally critical, but what of it? I did not like his attitudes to money, but in moments of even the greatest exasperation I would see this was natural for someone who had had to look after himself since he was fifteen, on very low pay. After all, there are marriages where they fight about money every day. It was 'the system' I hated. But I would keep to myself the thoughts I had about it. The terrible provincialism and narrowness of the life? I would make a virtue of necessity. There I would be, the wise and tolerant centre of this family and . . . all this would have required improbable feats of self-restraint, which I might perhaps have achieved decades later, when thoroughly subdued by circumstance. Well, perhaps.

The fact is, I would not have survived. A nervous breakdown would have been the least of it. In the four years I was married to Frank I drank more than before or since. I would have become an alcoholic, I am pretty sure. I would have had to live at odds with myself, riven, hating what I was part of, for years.

In 1956, returning to Southern Rhodesia after seven years – another little eternity, for in that time I had established myself in London, not an easy thing – I found that all the people still left of the old Communist, or socialist, or 'progressive', groups had changed, sometimes into their opposites. Peer pressure had converted them. It is not easy – no, it is impossible – to remain sane and ordinary while living among people who think differently. All over Southern Rhodesia were scattered people whose attitude towards race would be commonplace in a couple of decades, but now they were misfits, eccentrics, traitors, kaffir-lovers. People holding unpopular opinions for too long become shrill, fanatic, paranoid. If there are enough to form a group, then this group begins to show these characteristics. The old friends I met in 1956 had taken to drink, become bitter, were convinced the secret police watched their every move, or were even more reactionary and racist than the ordinary run of whites. I knew I was looking at what I would have become had I stayed in that marriage. Far from being that all-wise, all-strong compassionate centre of the family, an example to the children and a friend to my husband, I would have been a liability.

And now, sex. Without it you cannot write about the break-up of a marriage, not these days.

When I said I was leaving Frank because I wanted to live differently, no one believed me.

'In the provinces, if a young woman joins an organization, she is looking for a man.'

'In the provinces, a young woman leaves her husband only because she has found another man.'

In fact I was having a love affair. Rather, an affair. I was not in love with him nor he with me, but it was the spirit of the times. I cannot think of two people less well matched but that was hardly the point. My mother and all the elders reproached me for 'leaving my husband for a sergeant in the RAF.' This alienated me finally from them, for the idea was insulting to my best self. For days my mother, Mrs Tennant, Frank's mother, Mary – not Dora – arrived to confront me with this sergeant. To which I would reply with impassioned (and sincere) rhetoric about the Revolution and a new world in the process of being born. A scene for a comedy. The ingredients of my life for a couple of years were a recipe for farce, but it took me years to see it.

Everyone assumed my sex life with Frank was a failure, though I had not said so. All the old women – as I still thought of them – took me aside and with lowered voices assured me that sex was not important. I was shocked by them. What hypocrisy. Above all, I felt even more submerged in unreality. It seemed that at some point I had taken a step out of a realm of good sense into one where everything was false and no one told the truth.

The fact is, our sex life was, as they say, satisfactory. It is a question of expectation – of, in fact, information. It is my belief that the sex life of ninety-nine per cent of the world's people consists of a robust to-and-fro well described by the English word 'bonking', and most people are happy with that. For one thing, complicated sex needs privacy, and not everyone has it. For another, if you don't know what you are missing you don't miss it. Our marriage manuals were sentimental, if much better than Marie Stopes' *Married Love*. For instance, it was permissible to kiss one's partner's body, provided this was done with reverence. When it was suggested everything was permissible if done with love, one might speculate about this 'everything', but even the most vivid sexual imagination benefits from information. Oral sex? – What's that? Sado-masochism? – What can you possibly mean? For the sake of social historians I should record that the clitoris was far from the big deal it is at the moment of writing. (Things are bound to change again.) It is not that the manuals did not draw one's attention to it. When I read Balzac's 'A man has married too soon if he is not able to provide his wife with two different satisfactions on successive nights', I brooded about it, but when I masturbated in my adolescence it was the vagina and its amazing possibilities I learned about. The clitoris was only part of the whole ensemble. A clitoral orgasm by itself was a secondary and inferior pleasure. If I had been told that clitoral and vaginal orgasms would within a few decades become ideological enemies, or that people would say vaginal orgasms did not exist, I'd have thought it a joke.

As for subtle and refined sex it was years before I discovered it. I am sure many people never do. One may go in for rogering with any old Tom or Dick, but the more rarefied shores of sex may be explored only with someone with whom you share quite rare consonances of taste, nature and fantasy. Mind you, it is possible that refined sex may be born out of restriction. I was with an Indian woman friend, and she said she and her husband slept all through the hot months on a roof with the rest of her clan, children too. Seeing I would like to ask the obvious question, she smiled and said, 'There are ways.'

For a while before I left Frank I hated him. This was because I was treating him badly. I understand why torturers have to hate their victims. I am not saying he behaved well, he did not, but that isn't the point. I also went in for logic, thus, 'If your position in the Civil Service is being threatened by my politics, I don't understand why you want me to stay.' And, 'If I am so irresponsible, surely you'd be better off without me?' He became more sentimental and tearful, I more cold and matter-of-fact. It seemed we were under a curse that made everything we said or did untrue, histrionic. We did not know each other. We did not know ourselves. It was essential I leave before we both got ill. It was Frank who drove me with my possessions – clothes and books – to yet another furnished room in yet another house in the avenues. My landlady was again a lone and lonely woman, this one obsessed with theft, murder, rape. Shadows of menacing black men woke her most nights, and she shrieked and screamed at her 'boy' like a madwoman. She was probably mad. If large numbers of people are mad in the same way, it is not recognized as madness. She wanted me to be a friend, sit talking over cups of tea. I was too busy.

Frank was also not a little crazy. He put a detective on to watch my every move, even while I was telling him exactly what I did. He knew about the sergeant, who I met, where I went. I taunted him with, 'Why waste your money when I'm telling you everything?'

We have reached the end of *A Proper Marriage*. Now begins *A Ripple from the Storm*, the third in the sequence *Children of Violence*, and of all my books it is the most directly autobiographical. If you are interested in the mechanisms of a Communist or left-wing group, there it all is. It gets its title because a famous Russian writer,

Ilya Ehrenburg, a friend of Stalin, had written *The Storm*, one of the 'big' novels about the Great Patriotic War that Russians went in for then. I thought it a quite witty little title, but nobody noticed. For one thing, by the mid-fifties, Ehrenburg was being criticized, and *The Storm* had been forgotten or, at least, the 'big' Soviet novels about the Great Patriotic War had become a bit of a joke.

Incidentally, Ehrenburg had disappointed us 'progressives' badly. We continued to insist there were good as well as bad Germans, as an article of faith for a better world, and he had stood for exactly this view, but then he changed it, because of pressure from Stalin, and took the same position as the British Lord Vansittart, who said – no, shouted and raved – that there were only bad Germans. We were more than disappointed. It was a blow to our view of Soviet Communism. We excused it, as always, with 'But what can you expect? They are fighting for survival.'

Over and over again I've had this experience. Someone says they read *A Ripple from the Storm* when they were still Reds or near-Reds, and found themselves furious because I was reactionary – a renegade – fouling my own nest. And so forth. Then, rereading it later, thought, Yes, that was exactly it. And then, later again, found themselves laughing. I think it is a pretty funny book. I think a good deal of the first three volumes of *The Children of Violence* is funny.

A Ripple from the Storm gives the taste, flavour, texture and smell of the time. If you are going to say, Why write about a dubious Communist group in the middle of the war in Southern Rhodesia? – well, little Communist groups flourished briefly everywhere, and they had their consequences. Regarded from any commonsensical point of view the phenomenon was mad – crazy – some kind of lunacy I don't think we yet begin to understand. When young people look at you with ironical or amazed eyes and even say, 'But I've never been able to understand . . .' what they are not understanding is that it is impossible to distance oneself from the strong currents of one's time. Their children and grandchildren will look at them and say primly, 'But I simply cannot understand . . .' Recently on television, a young woman spoke of sex in exactly the same way my mother would have done. 'Promiscuity,' said she, 'is not really very nice.' This daughter of a 1960s mother who enjoyed free sex *as a right*, proclaimed, moreover, as a right for all women, everywhere, always, seemed to believe it was she, herself, an individual speaking, and not the fear of AIDS – the spirit of the times.

Sex, for a time, was the least of my concerns: my sergeant had been posted back to England. There have been times in my life I have been obsessed with sex, but I think it is a question, at least for women, of expectation. When as it were 'earthed' in a man and a satisfactory sex-life, I have been full of the morality and monogamy that goes with appeased appetites. Much later, in a time without a man, circumstances making sex out of the question, since all my psychic energy was going into someone who was ill, I switched off. Rather, was switched off. Sex flourished in my dream-life, but if I had been one of those who can say, 'I never dream,' then I would have said, in honesty, that I had become sexless.

I was too busy. My new friends were refugees from Europe, by definition political, and men from the RAF, from that strand in British life that now seems to have frayed away: they were the product of night classes, working men's colleges and provincial literary groups. They were of course a minority, but thousands of men, perhaps hundreds of thousands, were in Southern Rhodesia during the war. If only a few came to lectures and meetings, most I think were socialist, or at least would talk sentimentally about 'Uncle Joe', mostly to annoy the officers. When they came to the meetings, rank was dropped. In the whole country, Bulawayo and Umtali too, there might have been thirty locals, Rhodesians, some of them recent immigrants from just before the war, who had found the country stifling.

What we all had in common was this: we were people whose unused talents were being given room to develop. Most people live their lives out with, I would say, nine-tenths of them stifled, dormant. This seems to me the world's great tragedy, the worst thing: unused talent. Take a group or assortment of young people, in their twenties, give them headroom – and it is an explosion. We were all Reds to some extent, but a lot of what I learned was only peripherally political.

A young man who worked on a provincial English newspaper gave a series of classes on the Press. Twenty or so people were invited to take any item from the *Herald* and rewrite it in the style and from the bias of the *Observer*, the *Guardian* (the Communist paper from Cape Town), the *Manchester Guardian*, the *New*

Statesman, the *Daily Herald* (the now defunct Labour paper from Britain). Or take any item of news and see how facts and figures change from day to day. Or how it is given different emphasis in different papers. Or how a lead story is ridden to death perhaps for weeks, and disappears literally overnight, after which it is virtually impossible to get any information about it at all. Obviously this kind of instruction is not likely to be approved of by any authoritarian government and certainly not a Communist one. Another illumination was when I was told to take down in shorthand the speech of a certain Labour politician, famous for his oratory. He kept audiences spellbound. But when I read back my shorthand he had said nothing, but nothing at all – no sentence was finished, no idea brought to a conclusion. Literally, a tissue of nonsense.

Meanwhile, I was working in a lawyer's office, as a junior typist, on twelve pounds a month. I could just live on it, but I didn't care about that, just as I did not notice what I lived in. We all lived in rooms or boarding houses, no one had money, the refugees got what jobs they could. We were all high-minded, despised food, clothes, money. At least, we were committed to despising food and clothes. I often met Dora for tea in some tea room and heard what went on at home. This was so painful, I hated it, but yet I had to, I had to know. Frank was in the process of divorcing me, and was persuading Dolly Van der Byl to marry him. I did not contest the divorce, it never occurred to me, though I could have done: he had been unfaithful before I was. Women on the left in my generation thought it despicable to use divorce settlements to get as much as they could out of men.

The hours in the lawyer's office, eight to four, were what I paid to Caesar. From four until two or three in the morning we were at every kind of meeting, study group, seminar, or engaged in starting yet another organization, for it seemed that almost at once there were at least a dozen. Medical Aid for Russia, as it sounds, was to get money and pay for medical and other supplies for our gallant ally. This phrase was like a Homeric adjective: Russia was always 'our gallant ally'. The 'line' was that it should be absolutely unpolitical, by which we meant, not a word about the shining advantages of Communism. Cases of photographs and pamphlets arrived from the Embassy in Johannesburg by way of dangerous voyages from Murmansk. They were idealized photographs and stories. We put on exhibitions, but hid or destroyed most of the material, for it was too embarrassing. 'The Great Patriotic War' with its concomitant vicious anti-German rhetoric on the lowest possible level, simply would not go down in Rhodesia, however it might have pleased Lord Vansittart. We had crowds in for these exhibitions: the Russians were driving the Germans back across Europe, and nothing the newspapers had said for years could explain how they did it. A succession of speakers came from Johannesburg for Medical Aid, all lawyers and Communists, some of whom would later bravely defend people accused by the Nationalists of 'Communism', but what this meant was fighting apartheid and standing up for the blacks. Medical Aid meetings were 'respectable', were chaired by the mayor or Members of Parliament, and drew audiences of several hundred. The Friends of the Soviet Union were more political, not directly Communist but devoted to telling 'the truth' about *The Socialist Sixth of the World* – the name of a book then popular. Our information was taken from material sent to us. Nearly all of it was untrue. These meetings were less popular. Any one of us might address them. The Left Club had the biggest audiences, with meetings once a week, and might be on any subject at all: 'The Situation in Peru'; 'Conditions in China' – then brewing up for a Communist Revolution; 'Modern Music'. While continually reminding ourselves we had to 'keep control', the effervescence of the time meant that a habitual attender would be asked to coffee after a meeting, and then to give a lecture. This was described as 'developing cadres'. But most Left Club people, while insisting they wished 'you Communists' well, said they were not really interested in politics.

If what you remember and how much is a test of what you were really interested in, then I wasn't really interested in politics either. Do I remember one word of those dozens, if not hundreds, of lectures? 'The Second Front – *Now!*' 'The Battle of Stalingrad'. 'Sewage Systems in Great Cities'. 'Rural Conditions in South Africa'. 'Independence for India – *Now!*' 'The Mexican Revolution'. 'Fascism and Mussolini'. 'The Palestine Problem'. 'The Free French'. 'Picasso'. 'Shostakovich'.

What I do remember is a scene in Meikles Hotel Lounge, sundowner time, and around me are drinking and smoking people from three different strata of my life. The farmers and their wives, in for the tobacco auctions and shopping. The civil servants and their wives, and the RAF crammed around their tables. On the little platform with its palm trees the orchestra is playing – what? It seems now that the same tunes were played over, over, over again, from hotels, the clubs, the radio.

I'm dancing with tears in my eyes,
Because the girl in my arms isn't you . . .

Opposite me is a young man from the RAF camp, a pilot in training. He has telephoned me because I said something at a meeting. We are squashed into a corner, leaning forward to see each other through the smoke and because we can hardly hear through the din. This is a conversation of importance.

What I had said was, flippantly no doubt, because it was my style, that it wasn't necessary to have been in the First World War to have been done in by it.

He was a quite ordinary young man, likeable, nothing special about him, only his dark stubborn eyes, which he kept fixed on my face. I had told myself – and him – that I had an hour to spare, but we were both there three hours later, when he had to get back to camp.

Now when I look at that face in my mind's eye I can see the young boy of nine, ten – who had those dogged and obdurate eyes then, because he was preserving the inner core of himself, the sense of himself, against pressures from outside. Sitting in the Meikles Hotel Lounge, I did not see that young boy, only the young man, whose determination was making me uncomfortable.

His father was killed in the Trenches. His mother's sister's lover was killed too. The two women lived in a small country town on an allowance from the family and did small unimportant jobs when they could get them. What their energies went on was bringing him up. All over the house were photographs of the two dead young men.

When he enlisted to become a pilot he saw it as an escape from an atmosphere of war that suffocated him.

'Don't you see?' he insisted, 'I've never had my own thoughts. I've been thinking *their* thoughts all my life. My thoughts have been the thoughts of two women in mourning. And you are the same.'

I didn't like this. I argued. I joked. He wouldn't have it. We ordered more beer. Then more beer. We got tight and earnest.

He went on insisting. I *had* to see it, I had to, it was essential for him.

'You don't know what your own thoughts are. You've only thought your parents' thoughts. I've only thought my two mothers' thoughts. I've never had a feeling in my life I can say is *my* feeling.'

Soon, I was listening, not arguing. I was a good listener, all right.

I listened straining, because of the din of voices. And music.

I love you, yes I do, I love you,
It's a sin to tell a lie . . .
Millions of hearts have been bro-ken,
Just because these words were spo-ken . . .

'You're just like me,' he kept insisting. 'We're the same. Well all right then, when you have a feeling you know is only *your* feeling, just let me know!'

I don't know what we finally agreed on.

Goodby-ee . . . don't cry-ee,
Wipe the tears, wipe the tears
from your eye-ee.

Or, to put it another way,

We'll meet again
Don't know where,
Don't know when . . .

'Do you know what I'm going to do after the war? I'll tell you. I'm going to go right away to another country somewhere by myself and I'm going to stay there until I know what I feel. I mean, what *I* think . . . I mean, if I don't prang first.'

Dorothy Schwartz says next day, 'What were you doing with that RAF type in Meikles?'

In the provinces, never imagine anything can be kept secret.

'Jimmy rang me from the camp, he says to tell you you are wasting your time, he's not political.'

Good for him – I nearly say. When with – whatever his name

was – my life racketing and politicking around Salisbury seemed self-important, childish, quite mad, in fact. Now, back with one of the faithful, my three hours of earnest conversation seemed emotional, even hysterical, absurd.

'He thinks he's going to prang. His father bought it in the last war.'*

'Fair enough,' says Dorothy, giving judgement. 'Well, it's a great life if you don't weaken.'

Meanwhile, a Communist Party had started, as is described in *A Ripple from the Storm.* Since writing that book I have twice observed that process, the pike in a school of minnows, the pike being that perennial figure, the political fanatic. Then, his name was Frank Cooper, a cockney, but from north London, and I have met him several times, under different names. Their main characteristics are: charisma, a mysterious quality; then a secret and powerful pleasure in a possibly until then unsuspected capacity to dominate people; unscrupulousness; contempt for the people they are so easily manipulating. Once – London – this figure was mad, but really, quite mad, with very eminent people involved, some of them politically experienced, and what amazed me was that no one seemed to see he was mad. In politics it is easier to be mad unnoticed than anywhere else. My novel *The Good Terrorist* has a central character Alice, who is quite mad. A lot of people have not noticed that she is mad. What a nice girl, they say. That is because she is in a political context. If she were portrayed in an ordinary life, it would at once be obvious that she is mad. Political (and religious) movements and groups of an inspirational and revolutionary kind have room for any number of maniacs. Frank Cooper was not mad. He was the product of the hungry thirties in Britain, and his hatred of the middle and upper classes was expressed perfectly in Communism.

He was a corporal in the RAF, something to do with supply. He announced at a meeting that it was enough, all this running about and playing little games – but now we must be serious and start a Communist Party.

There was an inner core – never exactly designated one – in this loose and often changing crowd of left-wing people. The main people in it were Frank Cooper, Gottfried Lessing, Nathan Zelter, Dorothy Schwartz, and another RAF, a sergeant, Ken Graham.

* Prang = crash. He bought it = he crashed – RAF slang.

Gottfried Lessing and Nathan Zelter at once said there was no 'objective basis' for a Communist Party. Where was the black proletariat at this meeting, let alone in Southern Rhodesia? Ken went along with this. Frank Cooper sneered and said he insisted on a vote being taken. The atmosphere was such he won the vote. Nathan at once got up and withdrew, saying he disagreed, it was irresponsible. Frank Cooper sent after him contemptuous cold smiles and 'A good riddance to rubbish.' Gottfried said he would stay, though he agreed with Nathan. Ken said he would stay. Dorothy Schwartz supported Frank.

We told the Communist Party of South Africa, in Cape Town, that we intended to start a Communist Party of Southern Rhodesia, and they said they were against it, there was no 'objective' basis for one. Frank said, 'Oh, tell them we've started one anyway!' Later I found that the comrades there thought too little of us to include us in their mental map of Communist organizations: we rated a file in a cabinet.

This means I am perennially in trouble when asked about my time in the Communist Party. Emotionally, the right reply is that I was a Communist for perhaps two years, in Southern Rhodesia, from 1942 to 1944, but whether this is organizationally true is debatable. I joined the Communist Party in, I think, 1951, in London, for reasons which I still don't fully understand, but did not go to meetings and was already a 'dissident', though the word had not been invented.

The most interesting thing to me now is the language we used. For decades we have been laughing at phrases like capitalist hyenas, social democratic treachery, running dogs of Fascism, lackeys of the ruling class, and so on. They would fill a dictionary. Laugh – when this language was the weft and woof of indictments that sent millions of people to their deaths? For years the crudities of Communism were excused by the often unvoiced 'Well, look at their history, what can you expect?' Introduced to this vocabulary of contempt our first impulse was to laugh nervously, only to be fixed with the histrionic eyes of Frank Cooper. Histrionic, that's what I see now. We were playing a role. The play had been written by 'History' – the French Revolution, where the language was first used, and the Russian Revolution – and we were the puppets who mouthed the lines. We could not use it without laughing, despite Frank Cooper. We put the phrases between inverted commas, or exchanged glances while demurely mouthing them. Dorothy Schwartz was particularly good at this, pronouncing that such and such a public figure was a lickspittle of the ruling class with infantile left-wing disorders, while her eyes rolled gently, and her voice fell like an Anglican bishop's reaching the peroration of his sermon. Slowly, but within months, the abusive rhetoric was set aside.

When I used some of this experience in *The Good Terrorist*, I never have had more, or more interesting, letters from readers. Several were from people who had been in on the early phases of the Red Brigades in Italy, and they said that this type of rackety amateur politicking was how many groups began, and then 'the language took them over' and they became ruthless and efficient killing groups. 'The language took them over' occurred more than once. We should be careful of the company we keep – and the language we use. Regimes, whole countries, have been taken over by language spreading like a virus from minds whose substance is hatred and envy. When armies are teaching soldiers to kill, the instructors are careful to put hateful epithets into their mouths: easy to kill a degenerate gook or a black monkey. When torturers teach apprentices their trade, they learn from an ugly lexicon. When revolutionary groups plan coups, their opponents are moral defectives. When they burned witches, it was to the accompaniment of a litany of calumny.

If our group – it was not much more than that – had survived, instead of having as much chance of surviving as a camellia in the Karroo, we might have been taken over by the language.

In *A Proper Marriage* there is a passage: 'What she wanted, in short, was some sort of revenge: if the first political emotion of people like Martha is anger, the second is blind anarchy: if anyone had asked her in that moment to take a gun in her hand and go out to destroy those people who had been making a fool of her, she would have gone without a second thought. Luckily, however, there was no one to make any such demand.' When I wrote that I was throwing it away, just a little gloss on the situation. Now I read it with something like terror, and gratitude because 'There by the grace of God, I did not go.'

We say things like this: 'Stalin murdered nine million people in the forced collectivization of the peasants in the Ukraine.' 'Stalin murdered . . . millions during the Purges.' (Whatever is the figure considered accurate by the writer.) 'In China Mao murdered . . .' the millions of the Great Leap Forward, the Cultural Revolution . . . to mention just two of the bloodbaths there.

But these murders were carried out by young activists, dedicated members of the Communist Party. People like . . . for years I kept reassuring myself, No, no, I would not, I could not have done that. Can I imagine myself watching while millions of starving peasants, thrown off their land, or their food forcibly taken from them, took to the countryside, crowded the railway stations, died in crowds, masses, hordes? Would I really have said, 'You can't make an omelette without breaking eggs?' Well, to make myself feel I could not possibly have done anything similar, means saying, too, 'I am much better than all those hundreds of thousands of people, mostly young, who murdered, tortured, ill-treated, in the Soviet Union, in China, and elsewhere.' Why, how, can I think this? Believe this? In my time I have again and again observed masses of people being carried this way and that by emotion, with as much chance of saying No as fishes in a flood. Not only what we have observed tells us we are helpless against these tides. Experiments in universities confirm the same thing. The famous Milgram experiments, for one, tell us that a majority of people will follow orders to torture, to kill. Well, all right then, I murmur, but it was a *majority*, wasn't it? I wouldn't have *had* to be part of that majority, would I?

I could have been a pure soul, like Osip Mandelstam, like his wife Nadezhda? – but no, I have to face the fact that I and all my dear high-minded comrades, both those in that chimerical Communist Party in Southern Rhodesia, and many I have met since, some of whom are still drawing comforting certainties from the past certainties of Communism – all were of the stuff of those murderers with a clear conscience. We were lucky, that's all.

We were a real Communist group for, I think, eighteen months, not much more. When I say 'real', we had nothing in common with real Communist Parties in Communist countries or established Communist Parties in Europe. Ours was the authentic flame, the Spirit of Lenin lived in us, we lived and talked as if tomorrow we might face the firing squad. 'A Communist is a dead man on leave' – we would offer each other such phrases without irony. For a short time we did.

We had properly organized, chaired and minuted group meetings every day. We had Political Education classes at least twice a week. There were Medical Aid for Russia meetings, organizational and public, Friends of the Soviet Union meetings, The Left Club meetings, Race Relations.

There we sat, ten, fifteen or twenty of us, according to whether the RAF had been let out of camp, in an atmosphere of tense dedication, watching each other through cigarette smoke. We came straight from the offices, the Air Force camps, the holes and corners we all lived in, to our bleak and dusty office over a café. And there sat Frank Cooper, who seemed to come and go from camp as he liked.

'First of all, comrades, we have to take over all the progressive organizations in the town. Progressive – so called.' And he laughed his soundless contemptuous laugh. 'It is easy. Communists are always the best people for the job.' Here he might hold each pair of eyes, one after the other, in a gaze that managed to be dedicated, intimate, and impudent, all at the same time. For the female comrades it held a good dose of sexual innuendo as well. 'Remember, those fools don't bother to turn up at meetings. They don't care and we do. Any Communist worth their salt can take over an organization in one . . . month . . . flat!'

Gottfried and Ken were quiet, biding their time. They might interrupt with small points of order or fact, but knew they could not compete.

'Dorothy, you will be secretary of . . .' Let's say, the Democratic League – I forget. 'Bertha, you will be secretary of the Trade Union Social Club.' (I invent, I can't remember.) 'Comrade Tigger, you will take over the Beveridge Report.' There was a large and active organization studying the Beveridge proposals that later laid the basis for the Welfare State. I was almost at once elected on to the committee. Frank Cooper was right – unfortunately. Most citizens are too sane and well-balanced to want to spend hours of every week plotting or planning to take over an organization. But the ease with which we all became secretaries, chairmen, committee members in fact frightened us. We might joke about it, but we did not like it.

'But,' protested Dorothy Schwartz, 'all the women comrades are already secretaries and librarians of something.'

'Doesn't matter,' drawled Frank. 'The RAF comrades can't take part publicly in political activity, nor can the refugees, the good citizens of Southern Rhodesia wouldn't have *that*, so the girls will have to do it.'

The girls were Dorothy, Bertha Myers, who was a teacher, Phyllis Loveridge, another teacher, me, a few others. There were a couple of male Rhodesian trade unionists. All the RAF were like swallows or storks, and soon would be gone. When Gottfried remarked, 'I think you will have to agree, I was right when I said we have no basis here,' Frank sneered, 'Soon we will have recruited African cadres.'

The organization we called Race Relations – an anodyne title, we thought – attracted from the start ebullient, angry, or passionately partisan audiences, and the CID attended all the meetings. White trade unionists came to say the kaffirs should not be advanced too fast – in their own interests. The white trade unionists, for obvious reasons, were always the most impassioned opponents of black progress. From this organization we planned to attract African members, but in the end we got one, and he was a CID plant. Communism was too abstract and inhuman an idea to satisfy Africans – and in fact, when later there were Communist or Marxist regimes, they did not last long.

The short story, *Spies I have Known*, comes from this time.

People who know about political life will not be surprised to hear that, while some people worked very hard, others watched them work. For instance, Gottfried undertook to organize a big dance at Meikles Hotel to raise money for Medical Aid. When it was over and a great success, it occurred to me I had had posters and tickets printed, got them put up around the town, inserted advertisements, invited eminent sponsors, paid the band, had done all the work. Meanwhile Gottfried was being publicly congratulated on what a good job he had done. When I pointed this out he drawled that I had just learned an important principle of good organization.

There were other activities. One was the sale of the Communist newspaper from Cape Town, the *Guardian*. I was responsible for this. At one point I was selling 112 dozen copies every week. Dizzy with success I was not, for it did not take long to see there were unworthy reasons for these remarkable sales. I handed over several dozen to the RAF camps. Uncle Joe's picture was up everywhere, if the sentimentality with which he was apostrophized was self-conscious, on the edge of parody. Most of these copies were unsold, but left lying around they had a satisfyingly irritant effect on the camp bureaucrats. While I was reluctant to admit this could be a factor, Comrade Tigger was after all an attractive young woman, full of the earnestness that so often seems to guarantee dedication to other pleasure. My absurd high-mindedness made it at first hard for me to recognize that sex-starved, love-starved, homesick young men enjoyed selling the *Guardian* for this pretty Communist bint from town. Similarly, I sold stacks every week to the cheap cafés and restaurants where they were happy to have them on their counters, attracting custom – attracting our custom, for we went around in a group of sometimes up to twenty, trailing satellites. There were the sales in the Coloured quarter, where people actually read them. We also had individual subscribers.

And what were we selling? What the *Guardian* and other Communist or near-left newspapers said about the Soviet Union was untrue, though some people bought the paper precisely because of the 'political analyses'. But the articles and information about the situation of the Africans, the Coloured people, the Indians, were true. No other newspaper in Southern Africa had anything like it, for all were at best at the level of 'we must improve their conditions for the sake of our self-interest,' 'they don't appreciate what we do for them' or 'they only understand a good hiding.' What most people found upsetting about the *Guardian* was not the Soviet Union – after all our gallant ally under good old Uncle Joe – it was the attitude about the Africans, still called 'kaffirs' or 'munts' – 'natives' if someone tried to be polite.

What did we believe, what were the ideas that fuelled us? They were the same as those of Communists or near-Communists everywhere, not just the hectic fancies of a hotch-potch of young lunatics thrown together by war in Africa.

First, that, within ten years, well fifteen then, the whole world would be Communist, from free choice, because of the manifest superiorities of Communism. There would be no race prejudice, oppression of women, exploitation of labour – no snobbishness or contempt for others. This paradise would follow a brief period of resistance by reactionaries, only a minority, after all, because by then 'the State would have withered away'. This phrase, 'the withering away of the State', together with 'the contradictions of capitalism', were by far the most common source of the sardonic jokes with which the comrades everywhere indicated they were not lost to sanity.

Paradise, then, was on the world's agenda, and soon. Who would lead the world thither? Why, we would, people like us, Communists, the vanguard of the working class, destined by History for the role. Exactly the same mind-set as my parents, who believed they represented God's will, working by agency of the British Empire, for the good of the world. Or like the framers of the Atlantic Charter.

Secondly, that there was no way to paradise but by Revolution. We despised anyone who did not believe in Revolution – that is, with a few exceptions. (We would assure each other, voices full of the sincerity that goes with moral judgements, that such and such a reactionary was nevertheless *a good person.*) It was morally superior to believe in Revolution, and those who did not were, at the least, cowardly. We were united with each other by superiority of character, because we were revolutionaries and good. Our opponents were bad. People who did not believe in socialism were not credited with good intentions: a set of mind that continues to this day. It is satisfying to believe in the moral inferiority of opponents. That the people supporting the United Party in Southern Rhodesia or the Tories in Britain might actually believe their policies would be best for humankind was simply not admitted. So strong is this need to believe oneself better that, as recently as 1992, after all the storms of murder, torture, deliberate genocide committed by Communists, a female Red reproached me with: 'How can you turn your back on the Truth? *I thought you were a good person.*'

Thirdly, we were a part of a family that covered the world. 'A Communist can arrive in any country anywhere and at once be at home, with people who think the same, with the same ideals.' An enticing notion for people estranged from their families, or uprooted – and these days most are. I heard exactly the same from a Moslem friend. 'A Moslem can go anywhere in the world and at once be with people who think exactly the same: don't forget, the Koran is the mental and moral framework for every Moslem, and the stories and sacred and historical figures in it are shared by the Sheikh of Kuwait and the poor labourer digging the ditch in Indonesia.'

Fourthly, a Communist should always be better than everyone else, work harder, study more, look after people, always be ready to do the dirty work, both as a human responsibility and to attract people into the Communist Party, which embodied now, and would embody in the future, all the best qualities of humankind. This set of mind is religious. In the West, Christianity has shaped our thinking for 2,000 years. Poor humanity lives in a vale of tears and suffering (capitalism), but is saved by a Redeemer (Christ, Lenin, Stalin, Mao etc.), and after a period of pain and confusion (purgatory) there will be a Heaven where all conflict will cease. (The State will wither away, Justice will reign.)

It was a main article of faith that great men (or women) did not influence the tides of history. I have forgotten how this particular error was categorized. A left-wing deviation? A petit bourgeois distortion? Just to make sure the point is not lost, this was in the time of Stalin and of Hitler and of Mussolini. Not to mention Churchill.

We believed there would never again be nationalist wars or religious wars. Nationalism was obviously a thing of the past. So was religion. We used to congratulate each other: at least we can never have a religious war again, or a nationalist war.

We were supposed to believe – it was 'the line' – that an interest in people's emotions or motives was 'Freudian' and reactionary.

Only 'proletarian' literature was 'correct'.

We took it for granted that when the working class – or the blacks or any other disadvantaged people – took power, they would be inspired by only the purest and most disinterested ideals. Of all the absurd things we believed this was probably the worst. If anyone dared to murmur about 'human nature' we would reason patiently with them, explaining they had not understood the regenerative and transforming powers of Communism.

The most powerful idea, the one that underpinned all the others, taken for granted, and not even discussed, was that Capitalism was doomed, had been voted out by History itself. This frightful war was the creation of capitalism: capitalism spelled war, socialism was inherently peaceful. Capitalism had created the last war, and the great Depressions in Britain, in Europe, in America – the Depression had formed most of the people who came to the Left Club. When we had a lecture on the Depression, the speakers and most people in the audience spoke from a personal experience of unemployment and hard times, and passionate discussion went on until long after midnight in the spirit of what we now call *bearing witness*: 'I testify that this was how things were.' The books everyone had read included *The Grapes of Wrath, Love on the Dole, How Green was my Valley*, plays like *Waiting for Lefty* and Lillian Hellman's plays. We sang:

I built a railroad, now it's done
Buddy, will you spare a dime.

The physical results of the Depression in Britain were evident to anyone who had watched the RAF arrive in the Colony: the officers were a good foot or more taller than the Other Ranks, who were the product of a diet of bread and margarine and jam and strong tea. Capitalism was a killer and that was all that could be said about it.

We knew that anyone connected with business, of any kind, was morally inferior. 'Businessman' was a term of contempt. The portrait of the Wilcox family in E. M. Forster's *Howards End* as crude and philistine barbarians says it all. So does my picture of Richard in *The Golden Notebook*. I think this attitude was reinforced by the English aristocratic contempt for 'trade', which had percolated down to levels far from its beginnings. Yet 'business', trade, capitalism in short, was, in our canon, at times necessary and good. I do not remember that we made any attempt to reconcile, or even discuss, these 'contradictions'.

The most powerful ideas are those which are taken for granted. When people say now, But how *could* you have gone along with Communism, with the Soviet Union, knowing about the situation there? – what is forgotten is that in our minds there was no alternative to Communism, or socialism. Capitalism was dead, it was only a matter of time. The future was socialist, was Communist. Any 'errors' committed by the Soviet Union – the great exemplar – would be put right, they were only bumps along the socialist road.

We enter dubious and foggy regions with 'You knew what was going on'. In this country, Rhodesia, our contempt for the Press was total. Its attitudes towards the key question, treatment of the black population, were simply absurd. Five minutes with the *Rhodesia Herald* was enough to restore our faith in our ideas. When we used the phrase 'the lies of the capitalist press' we had good reason. When the people from Britain talked about 'the capitalist press' they meant newspapers that had supported the betrayal of the legitimate Spanish government, inertia as Hitler took power, equivocations or lies about Hitler's treatment of the Jews.

When people talked about the Purges or the Collectivization, and the resulting millions of deaths, we did not believe the figures. The capitalist press was concerned to blacken the nascent Communist paradise, and that was all there was to it.

Now I think all this is off the point. I have known now a good many people who have gone through the process, first devout Communist, then adjustments to various degrees of doubt, described by Arthur Koestler as 'coins dropping one by one out of your pocket' (interesting, that, coins equated with ideas), then sadness or depression, then loss of faith. This can take a long time. But why does it? – that's the point. There are people – but that goes with a certain kind of personality – who have a sudden reverse conversion, shedding Communist ideas (perhaps the right word is emotions) overnight. They are few. Most dawdled and drifted out of Communism, out of 'the Party'. For some people this was not painful. It was not for me. What I suffered from mostly was fear of epithets like 'turncoat' and 'renegade' – a powerful weapon indeed. But I was never committed with all of myself to Communism. I know this by comparing myself with some who were. The tragic figures were those very poor boys and girls who found in Communism a hope, a way of life, a family, a university – a future. Some came out of the poor families of London's East End, and joined the Young Communist League. For them Communism was everything, and when they lost their faith they were being deprived of everything that was best in life. Some died. Some had serious breakdowns. They were – but really – never the same again.

The question should be put like this: *if a person takes on a faith – political or religious – surrendering individuality in an inner act of submission to authority, then how long does it take to regain emotional* (I am deliberately not saying intellectual) *autonomy? There must be some psychological law that determines this, which has nothing or little to do with reason, with the rational level of a person.* In my case it took years to shed it all, and I was not committed with all of myself, as some people I knew were – and a few still are. That to me is the real question, yet to be answered. A person says: I left when the Soviet Union invaded Finland . . . because of the Hitler–Stalin Pact . . . because of the suppression of the rising in Berlin . . . because of the invasion of Hungary . . . because I learned the truth about the Purges and the Collectivization. But meanwhile they are still subject to this psychological law, whatever it is.

More influential in the long run than all these attitudes, some defined and debated, others implicit, was something persuasive that united them all, the atmosphere of sentiment created by the 1917 Revolution, and which has remained to this day, the Soviet Union as an idea, as the great exemplar. A generation, two, three, in the West, have absorbed an attitude towards the Soviet Union which seems able to survive any number of 'revelations'. Exactly at the time when Gorbachev was explaining to the whole world about the bankruptcy of the Soviet Union, literally and morally, a group of young people were demonstrating outside a theatre in London on the grounds that the play was 'anti-Soviet'. But the play's statements were mild compared to the truths emerging from the debates in their Alma Mater. When the Soviet Union invaded Afghanistan, there were many newspapers who would not criticize, and many did not begin to mention the Soviet atrocities in that country until the Soviet Union criticized itself. This phenomenon, the long-lasting, the pervasive myth of the Soviet Union's role as a beacon to all humankind, can be very well studied in the story of the Soviet invasion of Afghanistan and media attitudes to it.

In our little group, in 1943, 1944, the inner 'contradictions' – a word we used continually – began to tear it apart almost as soon as it was created. Frank Cooper was the destructive yeast. Gottfried and Ken, deciding that our level of political awareness was inadequate, said we should have more Political Education. Frank despised theory, and left, taking the RAF comrades with him – that is, the ones who had started with us. Another reason took a long time to see. Most of us had come to socialism by way of literature – night classes, personal and private adventures with books – at any rate, were soaked in the Great Tradition – which was European here, not merely British. In short, we found the language and slogans of Communism increasingly infantile, though we would not have said so. Within two years after the group began, most of its members were new, and soon the organization itself slowly fell apart. Which did not mean we did not think of ourselves as Communists. Frank Cooper had been posted back to England, he said because of his politics: perhaps, perhaps not. Some pilots had finished their training and had gone – new ones arrived. Many refugees, attracted because of the intellectual poverty of Salisbury, came to the Race Relations and Left Club meetings. In short, only a few of the original members, the founders, were left.

Now I wish I had photographs, but we were too busy and, in any case, too high-minded for such petit bourgeois activities. I can imagine it: 'I'd like to take a snap of us all, do you mind?' What sneers, what jeers. Besides, the photograph could fall into the hands of the CID. We were all paranoid – pleasurable, on the whole: it gives one importance to think the Secret Services are preoccupied with your doings.

No photographs, but I did see myself not long ago. At a big meeting in London, I forget what for, there she came, skipping down the aisle, a young woman vibrating with physical energy and the confidence that comes of feeling oneself in control, dark-eyed, dark-haired with a red mouth – bright lipstick having come back into fashion. She held a pile of leaflets, and was dealing them out as she came to the young people with her. She emanated the debater's combativeness. At a word she would face her antagonist, eyes fixed on an invisible prompt, and irrefutable facts and figures would be propelled from her mouth by the force of her pure conviction. Her sincerity was perfect. There she was. There I had been. She was a member of some Fascist group, and she had come to heckle and shout. No, no, I am not saying Communists are the same as Fascists, heaven forbid. We believed in the infinite perfectibility of humankind, the imminent triumph of kindness and love – our myth was the same as the religious one, so how could we be the same as racists, cynics and oppressors? When those Keepers of Accounts in the skies remark, 'Well, in *fact* the atrocities, the murders and the destruction caused by the Communists were much more than those of the Nazis and Fascists,' will They put on our side of the scale that weight engraved Good Intentions? An interesting little debate . . .

Our hearts were permanently swollen with compassion for the world. Any leisure moment between meetings or selling newspapers or 'working on' some 'contact' would find us sitting in a cheap café, talking about beautiful futures, our dreams fuelled by the sick rage we lived in because of the frightful war which we all believed could have been prevented. And in any case we were 'defending the bad against the worse'. Quite soon there wouldn't be any more war – like our parents, like my father, we believed this war had to be the last because war would have been seen – at last – to be so destructive. We would watch some ragged black child loitering on the pavement, looking in at the amazing riches of this eating place, and we would assure each other that quite soon, such children would not exist. We lived on heroic myths and fantasies. The Storming of the Bastille – only recently has it emerged that there were only seven people in it and they were quite well treated: we imagined ourselves swarming up grim walls, to release starving prisoners. The Storming of the Winter Palace – we identified with heroic revolutionaries, not a mob drinking themselves sick on the wines in the cellars. From inside Nazi Europe came stories of heroic resistance, brave words on the gallows, escapes to freedom in Switzerland, the feats of the French Resistance. Yugoslavia was a potent symbol: we knew that Tito, snubbed by the British Government, but recognized by Churchill, was fighting a heroic war as pure and as noble as the Battle of Britain. Of the myths that fed us, only the Long March has survived untarnished.

Something else began almost from the start, so low on our agenda we hardly noticed it. When I sold the *Guardian* around the Coloured quarter, once a week, I was for the afternoon immersed in 'toe-rag' poverty, streets and courts crammed with people who were listless, drunk, demoralized. They grabbed at the newspapers as if they were tickets to the Promised Land – America. A sick man, his eyes running pus, sits in the sun, grabs my skirt. 'Missus, missus, sit and pray with me, sit and pray.' But I did not believe in prayer. 'I don't think that's going to help you much,' I say, gracious, matey. 'But I'll tell my friend Mary who is on the Church Committee – she'll come and see you.' 'When will she come?' 'Soon.' 'Tell her to come quick-quick, I'm sick.'

In a small town all the members of the 'caring' organizations – a word not yet born – know each other. We might be Reds and Revolutionaries, tales told by the good citizens about us might make their flesh creep, but some of us were also members of the network of welfare workers. Informally, that is. When I left those poor and sorrowful streets I might spend a couple of hours ringing Welfare, various churches, the Education and Housing and Health departments. 'There's a woman with three children, her husband has left her, do you think you could . . .' 'Can do. What's the house number? Thanks a lot.' 'There's a Coloured child not going to school at Number 43 Selous Court.' 'Hey, is that you Tigger man? Leave it with me.'

When we sold the *Guardian* in the Coloured quarter we did not ask for money, and were criticized in the group.

'Since when do we stand for charity, comrades?'

'Oh for God's sake, have a heart, comrade, you make me sick sometimes.'

This comrade was more often than not Gottfried. He was always the embodiment of cold, cutting, Marxist logic, and his favourite phrase was 'And now let us analyse the situation.' His analyses at least exposed the bones of a situation, and even now, bombarded with information or rhetoric I find myself summoning the ghost of that voice, and I think: Well then, right, so do let's analyse the situation.

Another man, who was not among the founders, increasingly earned irritated dislike. He was newly from England, part of the bureaucracy that ran one of the RAF camps, a tall, thin, handsome young man, and all the women had a brief crush on him. He was the very essence of young working-class hero. In fact he was putting it on, as was the way then; he was middle-class. Like Gottfried he was always analysing situations, and in fact was made Political Education Officer for a time. He presented himself as a fanatic, Lenin's progeny, was serious, unsmiling, and sat apart, making tidy notes and consulting authorities, Lenin, Stalin. He would sit listening critically while Gottfried analysed something, and then deliver judgement, not necessarily in his favour. There was a meeting about the situation in South Africa – that is, an inner or secret meeting. It is I think worth recording that then, the early 1940s, the political situation in South Africa, the treatment of the Africans, Coloureds and Indians, was judged by us as so cruel that there was a revolutionary situation which must shortly explode in a bloodbath. We might spend whole evenings discussing how we could aid this process, 'when the moment comes'. The Chamber of Mines had put forward some proposal for its labour force. This comrade, John Miller his name was, sat silent long enough to get our attention, and then: 'In situations like this, comrades, it is enough to ask ourselves, What would the Chamber of Mines want? What would be their priority? Establish that fact, and then . . .' A pause, while the tension mounts. He smiles coolly, 'And then it goes without saying that our line must be the opposite.' Storms of applause. Yes, this was indeed the level of our political thinking.

But the storms of applause were already much less. In fact, this young hero had come in when things were falling apart, or at least changing. Now I see that if we had really been a Communist group, in a Communist country perhaps, then this man would have seriously challenged Gottfried. He would have been that forever recurring figure, the second in command (whether organizationally so or by virtue of personality), who splits the organization. Like Frank Cooper, he would have taken half the 'cadres' with him, formed a rival group, and calumniated those left behind. Ken Graham took no part in power struggles. His was the voice of moderation, that person who absorbs animosity and discord, who stabilizes a group, often using humour to do it. He was essentially a Labour Party personality, and when he got back to England, that is what he joined.

Long decades later I came to know a man with much experience of the process of government, and he said that most revolutionaries could be defused by offering them jobs. Nearly all are people of unused or under-used capacity. They do not realize that what they are suffering from is frustration. The job offered must be chosen carefully, without cynicism, giving room for this natural critic's talents for useful reform. If this idea had been put to me then I would have dismissed it with a string of contemptuous epithets, but now I wonder if it isn't true.

If there is one thing hard to be dispassionate about, looking back, it is the amount of contempt and dislike we projected on to everyone not actually one of us. 'Who is not with us is against us.' Religion again: it is rooted in the self-congratulations of religion. Communist, left-wing, revolutionary groups generally, legitimize envy. Nowhere is this more easily seen than in attitudes to art and literature.

In countries with a Communist Party, there is a framework – more, a formula – for the need to cut down, destroy, denigrate, established artists. If Thomas Mann or Proust is a lickspittle lackey of the ruling class, then that disposes of *him* – and the field is clear for the talents of the critic, who often aspires to be a writer. The writers (or painters) swept out of the way are always the ones still on the scene, the generation just before the critic's. The classics are safe – they may be venerated, for they are dead. This process has been at work in country after country in our time. When there is no Communist Party, no intellectually respectable place to put envy, and the need to slash and burn predecessors, then the denigration may take nationalist forms. He – or she – is fouling the nest, has turned his back on the homeland, because he lives abroad, or (feminist) is male, or (male) her writing is of interest only to women. Or it is a time when there is no organized or institutionalized place to put envy, and then the phenomenon may be seen for what it is. Sometimes you will see a review or critical piece in a newspaper or journal written by a new name, and it is radiant, incandescent, with hatred for the writer's elders. You know this person has just come out of university, has been given a job by an uncle, aunt, lover, friend, and is intoxicated with power: writers he or she has spent long years in university being taught to admire may now be slashed to the ground, demolished. Probably, later, this little critic will be ashamed, or embarrassed. The point is, there is always, in any culture where there are praised writers and artists, a sump or a well of hatred for them, and always people ready to do them down. Then, in Salisbury, Southern Rhodesia, the stage of demolishing the great was short, was mild indeed, compared to Britain, or the great exemplars like the Soviet Union. Certainly Gottfried and others exhorted us to admire only Mayakovsky and Gorki, only writers with a proletarian background, but the trouble was, they were talking to people who had been formed by literature, and not prepared to anathematize their spiritual parents.

A scene: we have been discussing proletarian literature. As we rise from our chairs drugged with rhetoric and cigarette smoke, it can be seen that Dorothy is smiling, and is about to address Gottfried in a way we all wait for. 'As for me, I am going to get an early night, and I might just *possibly* take *War and Peace* to bed with me.' Then, with a gentle but triumphant roll of her eyes, she departs.

I now rather admire the sleights of hand we used to admit writers to favour that we were instructed to despise. Lawrence? Well, he was a miner's son, wasn't he? Eliot? He was describing the decadence of the bourgeoisie. Yeats? He was Irish, an oppressed people. Virginia Woolf? She was a woman. Orwell? At that time he was being insulted by the Party, because he had told the truth about Spain. The trouble was some of us admired him. How did we get around this? I forget. But don't bother, Political Correctness, the offspring of Marxist dialectics, illustrates the ways of thought.

A scene: half a dozen of us are sitting around a table writing letters asking for money on behalf of the various organizations we run. We are all elated, laughing, inflated by our sneers at the people we are begging from, our 'respectable sponsors'. Because it is a small town, and there are never enough philanthropists to go around, we are swapping our sponsors like playing cards. 'You can have Councillor Smith for Medical Aid, if I can have MP Jones for Friends of.' 'Then I shall have Cabinet Minister Z.' 'Then I shall have Lawyer X.' Most of them appeared on the letterheads of all the organizations. 'We got a fiver out of him last time.' 'Then he can cough up another fiver. They are only doing it anyway to get their names on the letterheads.' What had our 'respectable sponsors' done to earn such contempt? They were successful, by definition. They were not young. Worst of all, they were not revolutionaries. People who believed in achieving socialism or even a just society by peaceful means were cowardly lackeys of the ruling class, at the very least.

Fifteen or so years later, when my name is on a letterhead as a respectable sponsor, I am in an office where I overhear a young woman, the treasurer, saying to the secretary (now a professor of resplendent respectability), a young man dressed in the left-wing uniform of the time, tight jeans, an over-sized sweater with a hole in the elbow, 'It's time we got some more money out of our respectable sponsors.' With the same sarcastic contempt.

Edward Upward, a British Communist, wrote a novel series that illustrates, like a little time capsule, not only his experiences, but ours – and a thousand other groups. The series title is *The Spiral Ascent*. In those days everyone still believed that we were living in a time when things could only get better. Humankind was bound for general prosperity and progress – if you were Red, then this by definition could only be achieved by the Communists. Volume One is *In the Thirties*. That is, when 'everyone' was a Communist, near-Communist, or was reacting violently against Communism. Volume Two is *The Rotten Elements*. The author's note says it aims 'to give an historically accurate picture of the policies and attitudes in the British Communist Party during the late 1940s'. You put the book down thinking you have been reading about the fate of nations, but it's a story of a tiny group of isolated people in a provincial town, whose every word, decision, action, is given the importance they would have in Moscow. Exactly: we are reading about the same psychological processes, the same group dynamics, that made and unmade the Communist Party of the Soviet Union. Heroes and traitors, splits and heresies, martyrs and plots and intrigues – it's all the same. Volume Three, not published till 1977, is *No Home But the Struggle*. The titles alone are like a little potted report of socialist thinking of that time.

14

I MARRIED GOTTFRIED LESSING in 1943, but only because in those days people could not have affairs, let alone live together, without attracting unpleasant comment. In his case it would be worse. He was an enemy alien, and risked being put back into the internment camp. To be a Communist when he was supposed not to be taking part in politics was bad enough, but to have a very public affair with a young woman who was good, insofar as she was a Southern Rhodesian citizen and therefore out of his reach as a German and an enemy alien, but bad, because she had so recently been unpleasantly divorced, was simply stupid. It was my revolutionary duty to marry him. I wish I could believe this was just one of our jokes, but probably not. We were having an affair because we were the only unlinked people in the group. Yet this affair was of no importance. Were we not dead men on leave? Were not 'personal matters' irrelevant to the struggle? We knew we were not well suited. We said, It doesn't matter, we will just get divorced when the war is over.

From the beginning he considered me unsuitable material as a Communist cadre. The trouble was fundamental – it was me, myself, my nature. What I liked best about myself, what I held fast to, he liked least. Any insight at all about another person was 'psychologizing', was – Freudian. Moscow had categorized Freud for ever as reactionary. If I woke up out of a dream that illuminated something for me, he hated it. He never dreamed at all, was scarcely able to believe there were people who did. Dreams and dreaming were reactionary. My interest in folk tales, legends, myths, fairy stories – he joked that if I were in the Soviet Union I could be shot for it. (There were people in 'The Party' who simply denied there were any shootings or torture in the Soviet Union; others couldn't see why anyone should bother to deny it. Rotten elements must be got rid of.) Folk tales, folklore – he would quote Lenin's 'the idiocy of village life'. In group meetings, when the agenda reached 'Criticism', he would coldly, correctly, take me apart for these and other petit bourgeois retrograde tendencies.

Gottfried Anton Nicolai Lessing was born in 1917 in St Petersburg, and fled from the Revolution in a train, with the family, a baby in the arms of his nurse, his nana, his other mother, back to Berlin. His great-great-grandfather, born Levy, made the family fortunes. He was one of those nineteenth-century traders who made fortunes and lost them, some in Russia. He built ships, railways, supplied all of Russia with horseshoe nails. Lenin himself commended the family as an example of well-used and fructifying capital. His children were many and large and weighted with clothes and furs and jewels, and they lived off, mostly, what he left, in enormous houses in Berlin. A photograph of them is like the early Forsytes, or Buddenbrooks. One, Gottfried's father, was an industrialist and speculator, like the originator of the fortune, but his heart was in his library. He married a daughter of a Russianized German family, who worked in his business in Moscow – that is, she was in advance of her time. The house on Nicolassee in Berlin was large, pleasant, but nothing like the semi-palaces of the second generation. Russian, German and French were spoken by the family and the innumerable visitors. The marriage was described by Gottfried thus: 'She gadded about and had parties, but he sat in the library reading history.' The two children, Irene and Gottfried, were rich young people, and they expected to continue rich, for the household took its cue from the mother, who decided that Hitler was a vulgar upstart, and no notice should be taken of him. Gottfried was at the university studying law. Suddenly the category of young men eligible for call-up under the Nuremberg Laws changed: Gottfried, as only part Jewish, had been exempt, but now, as a quarter Jewish, he was eligible to fight for Hitler. The family was one of the assimilated families, did not think of themselves as Jewish. Gottfried said that Hitler had made a Jew of him: it was a question of honour. He arrived in London, as a refugee, I think in 1937. He had little money, hardly enough to eat. On Sundays he was invited to a grand house near Park Lane by rich business friends of the Lessings, for lunch. He ate thin slices of dark brown beef, one slice of dark wet Yorkshire pudding, as many potatoes as he could, wet cabbage, a small piece of fruit tart and an ounce or so of hard cheese. He always went, because he was hungry. London, he drawled, might be pleasant for people with money. It was a Communist who spoke, but he had not discovered the disadvantages of poverty till he found himself in a bedsittingroom in London.

Meanwhile, in another part of the forest . . . The Lessings in Berlin had made good use of the 'au pair' system for the benefit of their daughter, Irene. One girl who spent time in the Berlin household was Margaret Morgan, the daughter of a self-made millionaire from Wales. Another girl 'au pair' was from Johannesburg, the daughter of a millionaire family – her father was a Jew from the Baltic, and he had made his money in the early days of Johannesburg out of timber and property. His daughter had brothers; the oldest Schneir son fell in love with the beautiful Welsh girl. He was very clever, literary, good-looking, but he was melancholic, or so he was described in those days before we became at ease with terms like schizophrenic, manic-depressive. Maggie married him, and tried to save him from his demons, but lost, and he threw himself into the sea from a ship travelling to South Africa. It was natural for her to look up Gottfried, whom she had met in his family's house. She was miserable, a very young widow. He was miserable and lonely. The Schneir son had made Margaret a Communist. She made Gottfried a Communist.

It took me years to see something obvious: the conditions were classically right for a conversion. This rich young man found himself penniless in an alien city. His belief in himself, his picture of himself and his family had been stamped on by Hitler's great black jackboots. He had been ill-nourished for months and did not know what his future would be, only that it would be bad. They fell in love. This was a passionate love. Certainly when I knew Gottfried nothing as intense had happened to him. Maggie was beautiful, with black hair in a ballerina chignon, dark eyes. She was full of uninhibited Welsh vitality. Nothing of the English in *her*, Gottfried might drawl.

The war had not yet begun. That wave of German refugees was given the choice of taking themselves off to Southern Rhodesia or to Canada. Gottfried chose Southern Rhodesia, and found himself in a raw, graceless little colonial town where his dark smooth good looks, his elegance, his sophistication, made him a butt for jokes. He looked like Conrad Veidt, all right in films but too much of a good thing in life. He became friends with a refugee from Vienna, a pretty woman who looked well in frilly blouses, scarves and jewellery. She had the ridged 'permed' hair of the time, and little fluffy ringlets. She was as worldly as he was, as much a city person. When they sat together in the lounge of the Grand Hotel (marginally more classy than Meikles) they were an elegant, but above all, alien couple. Later, the group called her The Merry Widow, or Gottfried's Countess. She had no money at all, having arrived as they all did with what they stood up in. She borrowed money and started the first dry cleaner's in Salisbury. She rented one of the small houses and let a room or two.

When war was declared Gottfried was put into an internment camp for six weeks. In Britain all Germans were interned, Nazis and anti-Nazis, often together. I hear the Isle of Man where the refugees were put was as good as a university, but this was hardly the same in Rhodesia. Gottfried had taken the precaution of making a friend of the man in the CID office who looked after his case. He dazzled him with talk of his rich life in Berlin and his mother, the Countess Schwanebach. She wasn't one, but never mind, the lie served. I can think of no other reason why he was let out of the camp almost as soon as he was put into it. Some Germans, anti-Nazis too, were inside through the whole war. Asked what the camp was like, Gottfried would smile and remark judicially, 'Not bad at all. It is not reasonable to expect an internment camp to be the same as a holiday camp.' When he was let out, a lawyer called Howe-Ely vouched for his good behaviour. He wanted a cheap lawyer for his foundering legal firm. Howe-Ely certainly got a bargain. He never paid Gottfried more than a mean minimum: when Gottfried arrived in the firm it consisted of a foolish old man, his stupid wife, and a typist – me. When he left it in 1949, the firm was in large smart offices, had several partners and was successful. There was a room full of secretaries and typists. Gottfried did it all.

When Gottfried was not working at Howe-Ely's he was helping his Viennese 'spot' dirty clothes before they went into the machines – a technique which must be obsolete. He was not in love with her, for he was in love with Margaret Morgan, but they were conducting an affair. They were trying to get her married. I was a romantic still and shocked at the cold-mindedness of it, but not for the last time Gottfried demolished me: 'You must learn to hold your tongue with things you know nothing about. All you colonial girls running about like chickens, you know nothing about life. Mizi [not her real name] is no longer of the youngest any more. She must have a husband. She must marry an officer from the RAF camp and then she will be provided for.' This is what happened. She married a Wing-Commander, a decent friendly fellow, like a young Labrador, who adored her, and she went back with him to England. And then?

All the refugees did well in Southern Rhodesia. They were, as the phrase now goes, successful immigrants. A couple of years ago I got a letter with a signature I tried to place. 'Do you remember . . . ?' Let's say, Nina. She was one of the smart intellectual refugee girls who came to the Left Club meetings: she could not be a Communist, she said, she was a democratic socialist: a term that then held a hundred overtones of political history. What arrived in my house was a large elderly woman, expensively dressed, too smart, covered with gold jewellery. She tried to remember when I asked if she ever thought of the old Left Club meetings. She had made a lot of money, she said, she had done well out of Southern Rhodesia, but she wasn't going to live under a black government. She was off to Australia.

Not long after I left Frank and the children I got sick. I agreed at once with the people who said sleeping so little and living on potato crisps and peanuts were responsible, but already I knew why I was ill. I needed to sleep and dream myself whole. I was full of division. I might be rushing around the town day and night, the embodiment of confidence and competence, but in my far too short sleep, staircases fell apart under me as I climbed, I sat examinations I had not studied for, I was due on stage, the curtain about to go up, but I had not learned my part. The flying dreams, so enjoyable, were grounding me in anxiety, for no sooner had I risen into the air than the knowledge I was flying brought me down again. It seemed that the moment I closed my eyes, I stood over ravines and gulfs where the ancient and unforgiving lizard, almost petrified, almost dead, stared with its dust-filmed cold eye. The farm had been sold, my parents were moving into town, and the house I had been brought up in was crumbling in my sleep, demolished by white ants and borers, the thatch sliding off the old rafters to lie in dirty heaps on earth blackened by a recent bush fire. Dreams have always been my friend, full of information, full of warnings. They insisted in a hundred ways that I was dangerously unhappy about the infants I had left, about my father – but what was new about that? – about my mother, and because I wanted so very much to have time to write, but could not see when that would happen.

I liked solitude, but had to fight for it against my poor lonely landlady, against my mother. The comrades came to visit me every day after work. The RAF men came any time when they could get out of camp. My landlady thought it nice I had so many visitors, but wondered what my mother thought of all those men in my room.

By then all the girls in and around the Group had been proposed to by all the RAF men. My high-mindedness continued to cripple my common sense. It never occurred to me or to the other women that poor young men from the bitter poverty of pre-war England might like the idea of marrying privileged colonial girls. When Gottfried pointed it out, I was shocked at his cynicism. And in fact, he was over-cynical. We were all awash with idealism and comradely feelings, we were in love most of the time, we lived in a never-never land full of possibilities. Cynicism, or 'realism', is often a poor guide to what is going on.

Gottfried came to see me. He said later that was when he first fancied me or, as he put it, saw me as a possible bed partner. Men who find it hard to break out of a shell of shyness enjoy being kind to sick girls in bed. He was avuncular and brought me ice-creams from the little cart that pedalled up and down the avenues, or a box of cream cakes from Pockets, the smart tea shop. 'Yes, and now you must eat,' he would say, handing me a teaspoon commanded from the landlady, and watching me spoon in ice-cream. It cannot be said that group life left much time for courting, or, as it was then put, 'going out together', and my being sick in bed led to the magistrates' court and a quick marriage, a scene described more or less in *A Ripple from the Storm*. It was with *Landlocked* I left autobiography behind. For one thing, there was a gap of years between *A Ripple from the Storm* and *Landlocked*. I wrote *The Golden Notebook* in that time, other books, short stories. I could not find in myself the right tone for that period, such a bad, slow, frustrated, blocked time. At last life came up with the psychological recipe for *Landlocked*, a melancholy book, pervaded with post-war disillusion. Even in *A Ripple from the Storm* direct experience was modified, because I did not put Gottfried into it. He was after all alive, and I was bringing up his son. I borrowed the husband of a woman friend in London with a different history and appearance, but he was the same psychological type. He was a poor boy, from a Berlin slum, the product of unemployment and the violent politics of the 1920s and 1930s, embittered with class hatred – a child of the First World War, in short. He was a Communist in his teens, one of the Germans who opposed Hitler. He was brought to England by a refugee organization. This poor boy, and Gottfried, the rich boy, were alike in that both were fanatics, what the Communist Party used to call 'the 150 per centers' – not with admiration. 'They always crack up, or they suddenly become 150 per cent anti-Communist.' The original of Anton Hesse neither cracked up nor became anti-Communist; the persona of stern, dedicated, unsmiling activist simply left him, it seemed, overnight. One year you met this man who would make speeches putting you right at the slightest hint of 'incorrectness', but the next he was affable, charming, social and said, 'I'm not interested in politics.' Meanwhile the rich boy had become part of the Communist ruling class in East Germany. When I met the original of Anton Hesse, in the early 1950s, it was like some dislocated dream, hearing Gottfried's words, seeing his reactions, in this cold, blue-eyed, tall, thin, blonde Communist ice-axe.

If I did not have time to write the novels and short stories that kept presenting themselves to me, I did write verses. Melancholy, the deep sadness of my dreams, set the beat, put words and phrases on to my tongue, and I would wake mumbling,

At evening strolling lovers pause outside the town
Where an acre or so of crosses lean in the sand . . .

Even as I wrote them down I distrusted them, because I feared the pleasures of sadness.

They were published in a magazine called *The New Rhodesia*, whose editor was N. H. Wilson, a man who had been in prison for embezzlement, but no one seemed to hold that against him. He was generally disliked for his impatient intelligence, his criticisms of this best of all possible colonies. *The New Rhodesia* was an idiosyncratic periodical, a weekly with few but influential readers. It was regarded as extremely right-wing, but wrong-headed because of its 'progressive' attitude towards the Africans: they should be better paid and educated. N. H. Wilson was one of the middle-aged and elderly men who befriended me, this abrasive, bright, pugnacious young woman so unlike most colonial girls. Probably they harboured romantic fantasies, but I thought they were old. Now I see they were lonely: to be intelligent, well-read, world-minded men was in that town a recipe for loneliness. They liked to talk to me, invited me to tea at Pockets, or to visit them in their offices, lent me books, used me – as I knew quite well – to find out what the Reds were thinking. Mr Wilson printed my verses, and, too, the spirited letters I so often dashed off in defence of the Soviet Union, Communism, socialism, or attacking the bad treatment of the Africans. He argued with me about my politics but conceded my right to them.

Another elderly friend was Max Danziger, Minister of Finance. I found his cool, ironical, not to say 'negative' approach to life a pleasant contrast to the fervours of the group. When we argued about politics he demolished me with quotations from the Greeks or the Romans, Adam Smith, or perhaps a clincher from Erasmus.

There was another, a magistrate, Frank Wisdom's friend, then Gottfried's, who married me successively to the two men, both times showing by an affable and heavy-lidded scepticism that he didn't expect much from these alliances.

I made of these worldly, world-weary, lonely men a composite: Mr Maynard in *The Children of Violence*.

A journalist on the *Herald* was another. He was in *Going Home*. I owed him a great deal. The vagaries of political thought can be illustrated by my derision, shared by my comrades, when he dismissed the Communist revolutionaries in China with 'There have always been warlords in China,' but now I wonder if it was so ridiculous after all.

When I was asked by the group what I thought I was doing, walking and talking with the Minister of Finance or that reactionary Wilson, or that magistrate, or that representative of the capitalist press, I said brightly that I was sounding out the enemy. The fact that they laughed showed how far we had drifted from the passionate certainties of only two or three years before.

Yet we still believed that the future of the world depended on us. It never occurred to us to ask what qualifications we had for changing the whole world, and for ever. Or, for that matter, what qualifications Lenin had. If we had been told, and had been prepared to believe it – unlikely – that we were the embodiments of envy, vindictiveness, ignorance, our attitude would have been the same as when people say that such and such a priest is delinquent or even criminal: he represents God, and his personal qualifications are irrelevant. We believed we personified the choices of History. The character of every one of us was as unlikeable as at any time in our lives: meanwhile we never stopped dreaming about utopias. Perhaps there is a connection.

Yet we were changing, and fast. Rather, most of us were. Gottfried did not change – or did not seem to. He became for a while a marker or monolith we measured ourselves against. Even the way he sat at a meeting, watchful, silent, made the claim. Nothing is simpler than to impress people by being silent, then coming in with a few decisive words. But you have to have the right personality, and Gottfried did. Did he discover this trick by accident? Shyness kept him quiet, and then, making himself speak, he saw the effect, and used it? Everyone was afraid of this cold, silent man, with his glittering lenses, which he turned on this speaker, then the next, allowing his flashing stare to speak his criticism for him. When people knew him, their attitude changed, became tolerant, or humorously affectionate. But he was too different from everyone else to keep his position of authority. The point was, he was an intellectual Communist, a description we may have heard, but did not understand. He had been introduced to Marxism by his Margaret as a system of thought, with no experience at all of hard times, and he maintained a purity of ideology, by reading the Marxist classics. Concepts, ideas, classifications, suited his temperament. An idea breeds other ideas, of the same kind and substance, and politics is perhaps the best place to see this. Lines of logic lead from a premise to successive intellectual positions, expressed, often, in dreadful cruelties. Arthur Koestler in *Darkness at Noon* explores the logic-chopping of Communism. It was no accident that tears might fill our eyes at the news that Stalin was in the habit of putting his hand on the earth, as a symbol of 'life itself'. (Or if he didn't, we believed he did.) Our minds were a muddle of ill-digested ideas, and somewhere or other we knew we needed Antaeus.

Gottfried was always in the right. His clear cold thoughts told him he was. We used to joke – 'we' being the colonials he despised – that he would have made an Inquisitor. When he knew, he took it as a compliment. Someone leaving the group in a rage of disappointment yelled at him that he was the kind of person who would shoot 100 people before breakfast for getting 'the line' wrong, and then eat a full meal and enjoy it. 'You are wrong,' drawled Gottfried, 'I would order someone else to shoot them.' We laughed: Gottfried's off again! For a long time I thought, Well, it is simple: if you are going to read Lenin and Stalin day and night, then political murder will not only be a duty, but heroic. But no one is simple, not even Gottfried, who tried so hard to seem so, to be all of a piece. Later I learned that his is a common type in revolutionary, or left-wing (and for all I know, right-wing) circles. These men find it hard to have simple, ordinary, easy friendships and loves, and they retreat behind walls of cold authority. Gottfried seemed armoured in arrogance. Very well, then, what is arrogance? Here I go off into the kind of speculations I spent so much of my time on then: is arrogance always a defence against shyness? Is shyness such a simple matter? I did not understand Gottfried. I do not know now, did not know then, what it was I was not understanding.

Decades later I was in Munich for the British Council, and after the lecture up came a charming elderly lady, who introduced herself as Gottfried's first girlfriend. This scene was not without its ironies, not least because we were surrounded by enthusiastic signature-seekers. And then, too, when a German and British person of the same age meet now, between us is the memory of the two wars, the thought we were enemies, and that our parents were enemies – a heavy, weary, painful incredulity – *how could it have happened?* – like a bruise that is invisible but which both know is there. People milled around while she painted a picture of that house on the Berlin lake, in a way that said how wonderful she had found it. She was a very young girl, awed by the Russian mother, an impulsive and generous hostess who still spoke Russian, by the scholarly father, and by the Russian nanny who dominated the family to the point of telling them what to wear and what to eat. Gottfried was twenty-years-old, glamorously good-looking and elegant. But, 'Did it ever occur to you to wonder, dear Mrs Lessing, that there was something about Gottfried . . . something . . . I don't know what to say.' 'Yes, it did, something – but what?' 'I was his first girlfriend. I have always wondered . . .' 'Yes, I know, but I don't understand . . .' 'Not like other people.' 'No, a strange man.' 'Something missing?' she suggested. But a man isn't a jigsaw puzzle. Do not imagine this was the exchange of two raffish old ladies. Our concern was something else, an unease people did feel with Gottfried. But who wants to be judged by what they were at twenty?

Our sexual life was sad. He was deeply puritanical and inhibited. I could have believed he was a virgin, but that was obviously impossible. Very well then, he didn't find me attractive? But he showed all the usual signs of doing so. There used to be a joke – and perhaps there still is – that some men react to ardent or even ordinary love-making in the same way a white boss or bwana reacts to over-familiarity from a black subordinate: don't get too close. A caress that stirred him too deeply, and not even a genital caress, made him defensive and angry. And yet he seemed to understand sensuality, for he would remark of a couple of refugees from Yugoslavia – obviously mismatched in other ways, she intelligent, he stupid, 'You have to understand there are couples who stay together because of the pleasures of the bed.' Yet as far as I was concerned he knew nothing of the pleasures of the bed, and this was long before I had wandered on to the wilder shores of love-making. It was all a mystery to me. I used to brood and ponder and think, trying to make sense of it all . . . What a lot of my time I did spend then, ruminating about Gottfried . . . but not miserably. For one thing, we were not going to stay married. But suppose I had never had a sexual partner? Suppose I had not already had enjoyable sex? I would have believed that Gottfried's unhappiness, my unhappiness, in bed, was all my fault. Women always think this kind of failure is their fault. But now I am haunted by thoughts of the girls – thousands of them – or millions? – married off to men they don't know and with whom they may be mismated. Everywhere, all over the world, silent misery, deserts of unhappiness . . .

Some day he'll come my way
The man I love . . .

Or perhaps, or even . . .

Night and day
I think of you . . .

Very well then, we were mismatched.

Our first home was yet another of the furnished rooms, this time in a house where the wife was extremely fat. There were a lot of children, some already adolescent. One day we were awakened by yells of laughter from the verandah. We found the woman sitting in a chair holding a just-born infant. She had not known she was pregnant. Her babe had gently slid to the floor while she was cooking bacon and eggs and drinking soda bicarb. for what she thought was indigestion. We all sat around a big table on the verandah, while the infant, dressed hastily in long since put away baby clothes, was passed around the family, from embrace to embrace. Her husband was delighted. So was she. This event was much admired in the group.

We had all decided that there must be more time for 'personal life'. It was 'counter-productive' to be at a meeting every night.

Meanwhile, we were trying to 'make contact' and to 'work on possible cadres' among the black people. One problem was that the 'line' – from Moscow of course – was that only a black proletariat could lead their people to freedom. Black nationalism was ritually cursed, in the usual rhetoric, 'lackeys', 'lickspittles', 'running dogs' and so on. We had doubts about the 'correctness' of this 'line', and there were heated debates about it. We had no contact with organized black groups, for the good reason there weren't any yet. Not in Salisbury, though in Bulawayo we were told there were informal and illegal black trade unions. One name we knew was Joshua Nkomo, described as an orator who drew crowds. The Bulawayo comrades were asked to make contact, but reported failure. Twenty years later I asked Joshua Nkomo about this and he said he didn't remember but probably thought we were government spies.

The only black man we were continuously in contact with was Charles Mzingele who for years had been the old Left Book Club's 'token' African. There, in those meetings, he had gently and humorously repeated that Britain, because of the entrenched clause in the Constitution that gave the colony independence, was responsible for the bad treatment of the natives, yet no one ever reminded her of this dereliction of duty. With us he did the same. For him, this was the nub of the situation. If Britain could be made aware, she would tell the Southern Rhodesian government to behave. He usually came alone to our meetings, but sometimes brought a friend. They went off with selections from our pamphlets and books, refusing the glossy offerings from the Soviet Union, but gratefully accepting anything at all with information about the situation of their people. They disliked Communism. They found the choppings and disingenuities of 'the line' irrelevant to them. The Nazi–Soviet Pact was incomprehensible. He did not understand why the Communists abused the Labour Party – socialists, like them. Claims that this war was 'for democracy' they listened to with polite smiles, sighing perhaps, and shaking their heads. When pressed they said they could not see that the treatment of their conquered peoples by the Nazis was any worse than what they, the natives, suffered in Southern Rhodesia.

Charles Mzingele is now regarded as an Uncle Tom by people who have always, all their lives, been part of an accepted structure of opinions, have never, ever suffered for what they thought. (Incidentally Uncle Tom was rather an admirable character, not an Uncle Tom at all, but never mind.) Oh, Charles Mzingele! – they sneer now. Charles had thought his own way into opposition to the whites at a time when there was a sullen or angry opposition, but very little information. By himself, often without any allies at all, he kept up a continual sniping, lobbying, letter-writing to newspapers, to Members of Parliament, government commissions. At the time we met him he was middle-aged, tired, and unhappy because he was a Roman Catholic, devout and a churchgoer, and a couple of priests had come to his home to tell him that if he persisted in trying to form a trade union he would be excommunicated. It was a union of office workers, but he dreamed of a trade union of mine workers. He did not do too well with his office workers, for the messengers and office boys were an elite, better paid than most, and not keen to risk their jobs by seditious activity.

When Charles or a friend or two came to a meeting, we put aside whatever agenda there was and talked about their interests. They came to the Left Club office, or any office we were using, because he was under surveillance by the CID and at that time none of us had a flat or a house black people could be invited to. Meetings were not easy. The African sense of time meant it was no use saying six o'clock and expecting him to be there. If we said four o'clock, meaning six, he might arrive at six or unexpectedly at four. And then, there was the curfew. All blacks who did not actually live in the town had to be back in the Location, some way out of the white town, by nine o'clock. Charles used a bicycle, was always afraid it would be stolen – it had been, more than once. It had to be brought up the stairs into the office. We joked Charles' bicycle should be given a vote. We might be in the middle of talk and argument when Charles and his friend got up, having kept an eye on the time, made apologies, and left in their characteristic way, smiling, patient, but stubborn. When they had gone we might burst out in frustrated rage, hating so much what we were part of, or sit depressed, because of our impotence, hardly able to look at each other. We knew that these men would, once they reached the Location borders, separate and go carefully to their homes where they would hide whatever books and papers we had given them, because of the Location police, a particularly nasty class of black man who used a careless and jocular brutality which made them easily recognizable even when they were in ordinary clothes. When Charles remarked he had more than once been beaten up by these men, or told how they had invaded his little house and torn up his pamphlets, knocking him about in front of his wife and children, and we were indignant, he found us funny, like sheltered children being told of the wicked world. 'Yes, that is so, that is how things are with us,' he might remark, patiently, smiling.

He was always asking us to remind the British Parliament of its obligations. We were always sending off letters, copies of minutes, notes of resolutions, relevant parts of the Constitution to Members of Parliament in Westminster supposed to be 'good' on colonial issues, but if we got a reply at all, it would be a polite refusal. It was the war, we told Charles, people in Britain did not have time for anything but the war. 'They did not have time for us before the war,' he might remark, smiling as usual. It goes without saying that no member of the Southern Rhodesian Parliament was interested. Charles Mzingele was an agitator, and that was the end of it.

He did, however, have an ally in Gladys Maasdorp, the Lady Mayoress of Salisbury. When we dropped in to her office we might find him, and friends, drinking tea and talking. By then we were all members of the Labour Party. Her attitude to us, the Reds, was that we would grow out of it. She was a remarkable woman. As we would say now, she was a role model for me and the other women. She had been a child, then a girl, in Graaf Reinet in the Cape, as isolated as Olive Schreiner, as she said when we asked how it had been. She had read her way, while being part of the old-fashioned, racist society of that time, into socialism and feminism and equality for all races. She was most solidly married, with children. That a woman with opinions extreme for that time and place could be voted Mayor was a tribute to her personal qualities. She had a prominent position in the Labour Party, but despised it, because it was not socialist, and its attitudes to the Africans were no better than the United Party. She knew that there would be no chance of getting Africans accepted as members, but proposed a branch of Africans. In the same way as white trade unions had for decades blocked the advance of the blacks, saying they could only join the official trade unions if they earned the same wages – but the whites earned thirty times as much as the blacks, and would continue to do so since 'preserving White civilization' meant keeping a big gap between even the poorest-paid whites and the best-paid blacks – so the Labour Party rejected the idea of an African Branch because it was undemocratic. I have written about the great battle that followed in *A Ripple from the Storm*. The reason Mrs Maasdorp wanted us in the Labour Party was that we could vote on her side about the African Branch.

Imagine the scene: the usual dusty, dreary office, with filing cabinets and a plain deal table behind which sat Mrs Maasdorp, a large, solid, calm woman, while opposite her, crowded on a dozen chairs, the Comrades being told what to do. We all laughed – the incongruousness of it, the unexpectedness, for it had not occurred to us we would be welcomed into the Labour Party. Besides, most of us were not even citizens. I wasn't, for a start: marrying Gottfried had made me an enemy alien. (This made me so angry I simply decided to forget it. I was supposed to report at the CID once a week, but never went. Quite soon Gottfried didn't either – easy-going Colonial days and ways.) But what about the RAF? What about the refugees? Never mind, that democrat Mrs Maasdorp, manipulating an extraordinary variety of rules and regulations, got us all in as members. Meanwhile she spelled out her terms to us. What we did outside the Labour Party was our affair, but if she saw any sign of Communist dirty tricks she would personally throw us out. Meanwhile the whole country, and I do not exaggerate, rocked with rumour about the Reds who were inciting the natives to rise and drive the whites into the sea. I had been brought up with this phrase. Interesting, since there is no sea for hundreds of miles. The same habit of mind describes England 'set in a silver sea'.

Chief among our enemies was one Charles Olly, a town councillor with Mrs Maasdorp, and an ancient enemy of hers. He was a little fat ugly man, in a striped businessman's suit, despised by people like Max Danziger as a vulgar little upstart. Just like the Lessings in Berlin, and Hitler. Charles Olly was full of bullying self-confidence, because he was in the right, and continually wrote letters to the newspapers: 'Citizens of Southern Rhodesia. It is time you woke up to what is going on. Agitators and kaffir-lovers are at work in the Location, inciting Revolution. There are foreigners and Communists among them. The natives are not ready for political activity. They have only just come down from the trees . . .' and so on.

Our short time in the Labour Party was all enjoyment. Meetings, conferences, intrigues, discussions filled our days and nights. We were full of the elation that goes with incongruity, with contrasts bordering on farce, always dramatic in the colony anyway, but particularly now in wartime with its influxes of violently contrasted people. 'Only in Southern Rhodesia could such a thing happen' we locals were reminded by the people from outside.

Our activities, organized by Mrs Maasdorp, succeeded in splitting the Labour Party, which was the only possible alternative government. Reproached for this, she answered that a real alternative government would be a socialist one, and the Labour Party was indistinguishable from the United Party, full of careerists.

Perhaps Mrs Maasdorp was lonely, like so many of the other 'old' people I knew then. She did not share her politics with her husband, we knew. Who was likely to understand her struggle, in solitude, as a girl, to inform herself about the world? Charles Mzingele, for one. They were real friends – who could only meet in her dingy little office, nowhere else. She liked me, I think, because she saw her own youth in me. 'It takes a long time for some of us to grow up,' she might remark, severely. She also liked Gottfried, while despising his politics. What they had in common was their ability, and that they were both authoritarian by nature. They, too, sat for hours at that old deal table, talking. She wanted to know about the Labour movement in Germany, but he had not been political before he left, and got out of it with an airy, 'I'm sorry to say I have nothing to tell you about Social Democrats.' She was curious about his family history, which not only she, but all of us, listened to as if to a saga. 'Why bother to read novels?' she commented. All of us young people she had recruited into her Labour Party were mad about literature, about poetry, and neither she nor

Gottfried shared this. 'Sorry, can't understand poetry, it's no good expecting me to,' she would announce, while Gottfried backed her up, 'Anything worth saying can be condensed to a paragraph.' 'I've got a blank spot, that's all there is to it,' and she would defy us with her innocent blue gaze, with the satisfaction that usually accompanies these feats of philistinism. Or they would read each other a verse or two pretending awe or ignorant admiration. Few poems stand up to being intoned in emphatic rhythms, with sarcastic intention.

'"Turning and turning in the widening gyre . . ." – how would you pronounce that, Gottfried? "The falcon cannot hear the falconer . . ." But what does it *mean*?'

'"Things fall apart . . .",' Gottfried orates. '"The centre cannot hold . . ." What is falling apart? What centre? That's what I mean. It's so imprecise.'

Charles Mzingele loved poetry. We gave him poets he hadn't heard of, and supplied him with suitable bits of Shelley to inspire him and his fellow conspirators. He quoted 'Tyger, Tyger burning bright . . .'

'But Charles,' Mrs Maasdorp said, with severity. 'There aren't any tigers in Africa.'

'There aren't any lambs in Salisbury,' he might say. 'Only down in the Eastern Districts I think? "Little lamb who made thee? Dost thou know who made thee?"' And his eyes filled with tears.

At which Gottfried and Mrs Maasdorp, both atheists on principle, regarded him with that politically diagnostic look which probes for possible future apostasy.

'But what *good* does that do, Charles?'

> *'"Cruelty has a Human Heart,*
> *And Jealousy a Human Face,*
> *Terror the Human form Divine,*
> *And Secrecy the Human dress . . .",'*

recited Charles, smiling, but sighing too. 'This has been my experience, Mrs Maasdorp. I am sorry but I have to say that, I am sorry.'

Mrs Maasdorp had another natural ally, Jimmy Lister, a Scotsman from Umtali, kingpin of a branch of the Union of Railway Workers – white. He came from the 1930 battles on the Clyde against the forces of Capital, and was a passionate socialist. Not a Communist. He had achieved the impossible, again by force of personality, of getting his branch to support the African Branch. When asked by us how he had done it, for we needed to know how it was done, 'I just let them have it, that's all it is. I said I was ashamed of them as working men, they needn't expect me to run things for them if they turned their backs on basic socialist principles.' And he recited,

'For a' that, and a' that,
It's coming yet for a' that,
That Man to Man the world o'er,
Should brothers be for a' that.'

to show us how he had shamed his fellow white workers with Burns.

He did not have time for airy-fairy poetry, said he.

A scene in Mrs Maasdorp's little office, which is crammed with people. Three RAF, pilots in training, are reciting Byron, Shelley, Keats, Blake, at Jimmy Lister, while Charles Mzingele – sitting at the same table, and this is only possible in these wildly revolutionary circles – listens and sighs and laughs, 'Oh yes, I like that, I think it is true.' Jimmy Lister, a little fighting man, his pugnacious chin lifted, waits until the three privileged youngsters, serene in their education, stop and gaze at him, waiting for his capitulation. But he isn't going to give in. 'Och, weel, that's fair enough, I dare say, but give me Burns, Burns every time.' And he recites, 'Wee, sleeket, cow'ring timrous beastie', the poem about the mouse, right to its end. 'There now, you match that, match that if you can.'

Meanwhile there sits Gottfried in his pale and elegant linen suit, there sits Mrs Maasdorp beside him, like a well-kept housewife, and both look ironical, for they know they must make allowances for people's weaknesses.

Jimmy Lister had a wife who didn't approve of him and his principles. He spent his working life with men who might vote for him but who didn't approve of him. He was not the only one, far from it, who found with us a temporary alleviation for their isolation. Later he blotted his copybook politically, I forget exactly what he did – I think supported the 'reactionary' part of the Labour Party – and was reviled for it with the vitriol the Left uses for heretics.

Jack Allen, the old miner from the Rand who was dying of lung disease from the mine dust, was Mrs Maasdorp's closest friend. He lived on the edge of the Coloured quarter, in a tiny house always full of black children, brown children, Charles Mzingele and his friends, RAF let out of camp for the afternoon, any of us who had half an hour to spare. He was the generation after Granny Fisher, and what he remembered was not luridly-lit canvas and tin streets and drunkenness, but the big confrontations in Johannesburg between capitalists and workers – white workers, that is. And poverty – the kind I saw with Stanley, the Griffiths' chauffeur.

One of the waves of young men sent to the Colony to learn how to be pilots were students from Cambridge whose interrupted studies would be continued when the war was over. Three of them were our special friends. One was working-class, like D. H. Lawrence, for that was how I placed him, in a context entirely literary. And that was how he then saw himself. One was upper middle-class and later had a fine career in the Federation of British Industry. One had been to Harrow, and he said that if you survived an English public school you could survive anything that ever happened to you, but he had not survived. He had been most cruelly bullied. He drank far too much. Two had been friends in Cambridge: all three were good friends now. For me they were a confirmation and a promise. Every young person dreams: Oh, if only I had real friends, someone to talk to. And here they were. These three, who after all were around for only a few months, changed me, gave me confidence, because they brought England close, as somewhere I might really be one day – and soon, the very second the war ended. Once there I would . . . for one thing, I would talk . . . What a pleasure it was, talking with these three, talk not as debate, argument, confrontation, rhetoric, accusation, but, simply, a natural and friendly exchange. Talk for fun. They had arrived well after the first ardours and certainties of the group had evaporated, and in any case, unconditional enthusiasm was hardly their style. It is this style that interests me now. They were the very essence of 'We are defending the bad against the worse', and of '*Well, what can you expect?*' This was our style too, but in them degrees more intense. If anything went wrong – parts for aeroplane engines sent to the wrong RAF camp, or some blunder had ensured a food shortage, if a speech in London seemed more than usually fatuous, if the Second Front was again postponed – then we laughed, sneered, jeered, joked, shrugged our shoulders. A gentle, almost tolerant cynicism was general among the RAF who all knew they defended the bad against the worse. Perhaps no country can drag itself through a decade of such dreadful poverty – the result of World War One – as the 1930s in Britain, and then expect its population to embrace a new patriotic war with unmixed fervour? These three shared the general mood of the RAF, but there was something else: they were from Cambridge. Cambridge was the nursery of the famous spies. No, I am absolutely not saying these three were or could be spies, but their particular tone or style was the product of that university, then. We even called it 'The Cambridge Style' because of them and other Cambridge RAF. This disbelief in their own country – *our* disbelief – was a kind of poison. This level of cynicism is nearly always inverted or betrayed idealism.

Which brings me back to: why do we expect so much? Why are we so bitterly surprised when we – our country – the world – lurches into yet another muddle or catastrophe? Who promised us better? When were we promised better? Why is it that so many people in our time have felt all the emotions of betrayed children?

We might – I think now – have quite reasonably chosen to see things like this. Britain, ruled by feeble and incompetent men, colluded with France to permit Nazism and Fascism to win in Spain, and allowed Hitler to become powerful although he had announced openly, from the beginning, exactly what he intended to do in *Mein Kampf*. Churchill, who saw what was happening, was ridiculed and kept out in the cold, and when at last he took over government Britain was unarmed and unprepared for war. In spite of this, she got herself together, fought the Battle of Britain in the air, and the Battle of the Atlantic on the sea, stood up to Hitler when France collapsed. Then, as well as sending armies to North Africa, achieved the surely extraordinary feat of sending – in spite of U-boats – hundreds of thousands, perhaps millions, of men to Australia, Canada, Kenya, South Africa and Southern Rhodesia, to train pilots, an achievement which surely has never been paralleled? We were fighting a sea war and a land war in the Mediterranean. Surely we might have allowed ourselves to be proud of this? We might, if our set of mind had allowed it.

Another emotion paralysed us. It was that none of this need have happened, it could have been prevented. When we were in the cinema, watching bombs fall on to cities, or saw or read of ships being sunk, aeroplanes destroyed, tanks exploding, we felt a sick and paralysing rage: because of this waste of resources, of wealth. When we saw a stick of bombs falling we were thinking, *That could build and equip a hospital.* Or a tank exploded: *There goes a library.* We could have transformed the world with the wealth we were squandering in war. Now I would like to know if this emotion, so strong in the Second World War, was new: were we the first generation to feel it? Did they feel like this in earlier wars?

Now a different and deadly disbelief afflicts us: we are not intelligent enough – the human race – to make a new world or even prevent the old one from being destroyed. This is a continuation of that old cynicism, the *What can you expect?* which was the other side of our unashamedly naïve dreams.

Gottfried enjoyed the company of the Cambridge men – to a point. With them he could discuss history and ideas, but found their allusive, throwaway joking style dismaying, for he could not join in. Serious matters should be discussed seriously. He was made even more lonely by my easy friendship with them, and criticized the way we talked about books. He might not enjoy literature, but insisted on the party line. Where was our serious Communist approach to literature? And indeed, where was it? For years in the Soviet Union it was common to speak one language during formal Communist occasions, but use a different one for ordinary purposes. Long ago our minuscule group had developed the same habit. I had never been able to take Socialist Realism seriously. I give myself credit now for standing up to Gottfried, Nathan, Frank Cooper – anyone else espousing 'the line' – but this did not mean I could not use the language. If I had actually been in a Communist country, would I have stood up to the tyrannies of 'the line'? I like to think I would. I can give myself the benefit of the doubt over this, as over the question, Would I have really gone out with the activists and hounded the peasants to death during the Collectivization in Russia in the late 1920s? And yet I know how few individuals withstand a prevailing mood, or atmosphere or 'line'. Our little group – in a Communist country – I expect would have conformed, at least for a time, agonizing and anguishing over literary choices and definitions that could mean imprisonment or even death.

In Salisbury, Southern Rhodesia – as in thousands of other towns at that time – our group had latent possibilities. It could easily have become a literary society. Everyone confessed they hoped to write poetry, a novel, short stories. And what else might we have become? Soon, we would find out . . . it is a commonplace of sociology and psychology that a group anywhere, no matter what its first inspiration, political, literary, even criminal, tends in the end to become religious: 'religious' interpreted broadly. But our group had not had time yet to do more than dwindle into debate and speculation. We were too diverse, there was too much potential for schism. By the time the three Cambridge RAF came we had already lost two waves of RAF – that is, of the permanent RAF, the ground crew, mostly working-class. They did not sever themselves from us. They might drop in for a beer, or to get books or pamphlets. They had their own group in camp. Why did they not invite the three Cambridge RAF, and other Communists among the pilots, to join their group? It was a question of class, but they were not going to say so: they hedged and made excuses. They saw the pilots, as they did us in town, as luxurious and privileged. When I pointed out to one that Gottfried and I lived in furnished rooms and then in a one-roomed flat, he laughed at me. Just like Charles Mzingele laughing at us, as if patting a child on the head. *'You see, that's how things are with us.'*

This period, when the Cambridge RAF were with us, a time with its own flavour and taste, went to make up the Mashopi parts of *The Golden Notebook*, which I have just re-read. There is no doubt fiction makes a better job of the truth.

15

I HAD LEFT HOWE-ELY'S because Gottfried and I thought it too much of a good thing to spend working hours together as well, and I became a junior secretary in a legal firm, Winterton, Holmes, and Hill. Winterton and Holmes having gone to fight up north, Mr Hill was holding the fort. These were large light airy offices, unlike dingy and dusty Howe-Ely's. The senior secretary was Mary, whose real name has dissolved into her literary name. Mary was from Britain and regarded all colonial girls as lazy and incompetent compared with women trained as she had been. She typed with two fingers and a thumb faster than anyone I have known, long legal documents without a single mistake. I was content to be junior and to earn so little because I didn't want to waste vital energies on earning a living. I did ordinary undemanding letters, easy documents, and then took over the books, double entry. I was surprised it was so easy. But my main task was the debtors. They had a special cabinet for their cards, and in the store room stacked with files theirs took most room. Here it was again, the world of real poverty, the toe-rag world. Most debtors were white, some debts were years old. Most had begun in the Depression. The debtors came into the office all day. Some were men who drank and their wives had left them. They stood glaring at me with furious reddened eyes, or were ashamed and would not look at me. They did not have the money, they said, Mr Barbour or Mr Hemensley or whoever it was could say what they liked. Some men had been brought low by illness. The women had babies in their arms or children pulling them by the hand, the tired, just-coping women of real poverty. Often the names on the cards were Coetzee, or Van der Hout, or Van Huizen, or Pretorius or Van Heerden, and they were poor relations of the big Afrikaans families down south. When the Coloured women came with the children from the Coloured quarter, they might greet me, and then Mary was shocked. All brought with them the atmosphere of cheap hotels, or shacks in the bush somewhere, or slummy courtyards. If they did not pay up, I gave them grace for one, two, three weeks, but then had to appeal to Mr Hill who said, 'Why are they so stupid? They only have to pay more legal costs.' Then I telephoned the creditors and asked if they really wanted to sue these poor wretches, and they were always irritated, and the conversation usually ended with, 'Oh do what you like – but why do I have to carry these people?' or, 'Get a garnishee order.' This was a court order deducting weekly sums from the wage packet. The cost of the order was paid by the debtor. Then, one could be sure, the garnisheed one would change his or her job or go to another town. This wasn't Slump-time, it was war-time, there was work. Mary disapproved of these smelly shabby people cluttering up her nice clean office. She thought they should all be punished, let's say, by imprisonment for life, anything that would get them out of the sight of decent people. I was amazed by the uselessness of it all. I rang Mr Barbour, I rang my old friend Mr Hemensley, I rang up men of property all over town and suggested it might be more sensible to cancel the debts, since no one was likely to get more than a few shillings back, but they were every one of them shocked. My proposals threatened to open the door to anarchy. To repay what one owes is a question of principle. I gave up. When principles are invoked, common sense flies out of the window.

Meanwhile, on the shelves were files with well-known names on them, debts that had ceased to be, because the culpable ones had gone bankrupt. When I pointed these out to Mary she would turn away, with the look of one whose ideas were not going to be upset by an inconvenient fact or two.

Gottfried and I had moved again, this time to a flat that had a large room useful for the casual, dropping-in-and-out life of that time, and for study groups, or informal discussions that seemed to happen most evenings. I used to go to bed wondering at the number of people I had been with in the day, yet I was someone who thought of herself as solitary. I longed to be alone. I wasn't asking much – even an hour occasionally would have been enough.

This was the time of the bombing of German towns. I would find Gottfried sitting on his bed, his head in his hands, a newspaper cutting in his hand, or listening to the radio. Or lying silent in a dark room, his cigarette brightening as he drew in deep breaths, illuminating a chest of drawers, a shabby curtain, the radio. I dared not switch the light on.

'Well,' I would say, 'some time the war will end.'

'Yes, the news is not of the most pleasant.' Or, 'They deserve a good hiding, and they are getting one I'm glad to say.'

If, in the cinema, the newsreel showed bombs falling into German cities, he would say bravely, 'That's right, give it to them.' Or, in company, and we were discussing the Second Front or the bombing of Germany, he would judiciously light a cigarette, allow his lips to curl over the smoke, and say, 'Yes, if you make wrong decisions you have to pay for it.'

He slept badly, and whimpered in his sleep, or cried out. Just once I woke him, saying, 'You have been having a nightmare,' but he was angry, saying, 'It is not so. You must not say these things.' He once refused to speak to me for a week because I joked about the Unconscious, and it was not even his. So, when I woke him from bad dreams I did not say, 'You have been having a bad dream.' We might smoke a companionable cigarette. Or try to make love, out of a kind of good faith, for we did try sometimes, as if we believed our incompatibility was a temporary misfortune. We might lay awake and chat about the people in the group, but he judged most of them harshly, and I was afraid of that cold drawl. He was having such a bad time, poor Gottfried, in a country he despised, surrounded by 'so-called Communists' and raw colonials. Whom did he respect? I think no one, apart from Mrs Maasdorp and Hans Sen, the Red Cross representative. While he quite enjoyed talking to the three RAF from Cambridge, he thought them lightweight and middle-class. It might seem incredible now but then there was nothing unusual in a person from a privileged background criticizing others from the same background for not being working-class.

Gottfried resented the three RAF too, because I flirted with them. I was in love with two of them – that needs definition. Surely one should not use the same word for the yearnings of lust and those of love? One I was romantically in love with as much as I have been with anyone – but I except one or two. There was something classic about this love. He was going to leave the Colony soon to pilot bombers and would be in danger, love's most potent aphrodisiac. I was married and about five years older than he was. Our meetings were always in public and contained by group necessities.

I flirted because *it was my right* and, too, from a most ancient and savage female imperative: Gottfried could not make love, so what right had he – and so forth? And besides, what about all those marriageable girls who – as in any political group of any kind – dreamed of winning the favours of the Leader?

But the really bad thing, the worst, was that he was working for an old skinflint who never gave him a word of praise while he built up his legal firm for him.

What this unhappy man needed then was something so simple, so obvious – decades later I am intelligent about him. Back in that luxurious childhood of his, where his socialite mother partied and entertained, where his father maintained the family fortune and read in the library, where his elegant sister led the life of a rich young thing, was, too, the nanny, the Russian nyanka, the loving genius of the family, who had kissed and scolded the mother and then the mother's little children. Like an English family of that class, it was the nanny who brought up the children and loved them. In Chekhov's *Uncle Vanya* there is a wonderful scene where a fretful and peevish invalid savant sits late at night with his young wife who makes sensible remarks to him, but in comes the old nanny and treats him like a little child, kissing and stroking him like a child, and the poor pedant dissolves into trusting love and allows himself to be taken off to bed.

What Gottfried got was a patient young woman, kind enough, but too much of a raw female to allow herself to treat her man as a baby, even for a few hours of the dark. And, after all, he probably would not have tolerated it, any more than he would admit to a bad dream or to say that he wept secretly because of German cities being bombed.

Scene after scene from that time, of those night-time hours . . . the bedroom filling with cigarette smoke, visible in the light from the window, the barking of dogs, from the gardens the scent of shrubs. Silence, for there was little traffic. From Gottfried an acid smell I know now is anxiety. What I thought was, Funny, even his smell is alien to me. Perhaps this is one of the basic things children should be taught: if someone smells like this, it means so and so, and an acid smell means they are scared or unhappy.

At that time I was certainly running around like a chicken. Every day had to accommodate people who could not be allowed to know about each other, they would be so shocked. But that has been true most of my life.

Fifteen years later, in London, exasperated by the compartmentalization of my life, I invited two young people working for the Soviet Trade Organization, whom I met at a party, to dinner to meet an American. He was currently committed to Freud, being in that stage of the journey taken by so many in our time, disciple of Marx, then Freud, then shaman. (Amazing how many former Marxists are earning their living as holy men.) My intention was to break down barriers, broaden minds. I cooked a good meal, but not a mouthful was eaten, for these ideological enemies took one look at each other, squared up for the fight, at once started shouting key phrases from their respective faiths, and half an hour later hastened down the stairs together screaming abuse. They had forgotten about their hostess.

A day, for that short time – 1944, '45 – went like this. As soon as I woke I rushed into the bathroom, because once Gottfried got there he took what seemed hours. I used to dress in five minutes. No time for breakfast. Off I went on my bicycle to the office. There Mary would already be, a reproach to me and all less-than-perfect secretaries. At lunchtime I might meet Dora, to find out how things went in Fife Avenue. Offices shut at four. I might drop down to Mrs Maasdorp, where I filed newspaper cuttings for her. She used to say that in her lifetime the News had got steadily worse. As a girl she would not have believed what she read now and took for granted. She said that most books and articles on the state of the world described a bad if not fatal situation, and all ended with a list of prescriptions for better everyone knew would not be used. 'No use denying the state of affairs is bad and worsening, but if we all . . .' She described such books and articles as 'The but-if-we-alls', thus: 'If you have half an hour to spare perhaps you'd like to drop down and file some of the but-if-we-alls.'

From Mrs Maasdorp's office I might go to Jack Allen.

The police could be there. More than once I heard something like this:

'What are all these kaffirs doing here?'

'They're visiting me. Do sit down, lad, make yourself easy.'

'But Mr Allen, if everyone behaved like you then the blacks would arise and cut our throats.'

'But everyone doesn't behave like me. I wish they did.'

'But, man, you're a bad example. And you're putting ideas into their heads.'

'Don't imagine I'm giving them any ideas they don't have already.'

The frail, thin, dying old man, with his brave blue eyes, his oxygen tank at his elbow, smiled at the big healthy policeman, whose face was screwed up with suspicion and the effort of taking in difficult thoughts.

'But they're backward compared to us, you know that . . . They're just down from the trees, and their brains are smaller than ours.'

These perennial truths of racism Jack Allen greeted with a laugh, and called for the woman he lived with to bring tea and biscuits. And if some black children came running up, hesitating as they saw the white policeman, then running off, terrified, he would say reproachfully, 'You see that? I hope you're not proud of that, children afraid of you?'

'They'd better be scared if they know what's good for them. Oh hell, man, Mr Allen, I don't know what to do with you, it's against the law, you know that.'

'I don't think you'd find a law on the statute book forbidding black children to visit a white man.'

'Well, I'll let you off this time, but I'm not saying I will next time. *We* know it's not just black kids you have here.'

I then might do what Gottfried dismissed as 'social work' – that is sorting out problems with the Welfare departments. Then, perhaps my mother, with whom I had to watch every one of my words, since if I mentioned Jack Allen, or Mrs Maasdorp, her already great unhappiness would be confirmed and deepened. 'But how *can* you know such people?' 'Mrs Maasdorp is Mayor, mother. She is Lady Mayoress of Salisbury.' 'Maybe, but she cares more about the natives than she does about her own people.' 'And how is Daddy?' 'You know how he is. I'm simply at my wit's end.' 'And have you heard from Harry?' 'No.' 'Don't worry so much, he'll be all right.' 'Perhaps he will. I pray to God for him every night.'

Then, perhaps a meeting. After that, perhaps, another. Or it might be the day for distributing the 112 dozen *Guardians*. This meant bicycling around the poor streets, then from café to lunch-room, or meeting the RAF comrades for tea somewhere. There I heard news from the camps. Perhaps I met Athen Gouliamis, who had been a newspaper seller on the streets of Athens. He went about in a group with other Communists, because the authorities put together ELAS and ELAM, the Communists and their opponents, God knows why, and this meant that the two factions had to watch each other as narrowly as they might back in Greece where they were killing each other. More than once, at night, when we took Athen and his friends to the bus stop, and saw the enemy group there, we drove them back to the camp. When they had to return to Greece they came to us and said they would almost certainly be killed, though they would try to reach the partisans. Complicated systems of signals were arranged; this kind of envelope would mean this, that phrase would mean that, and in fact about a year after they left, we heard they were dead. Sometimes I am glad that such and such an old friend, a Communist, died before they could know how shameful a failure Communism turned out to be. Athen is one of them. There were saintly Communists. Few of the characters in my novels are unchanged from life, but Athen Gouliamis, under his own name, is in *Children of Violence*. Unchanged. As he was. A small, a very small tribute, to one of the best people I have known.

A large group of us might eat together in some café before going to hear a Left Club lecture, or coming back to our flat for a committee meeting that had become so informal Gottfried refused to call it Communist. We drank white wine from the Cape, or from Portuguese East, while we talked. My drinking had become civilized. Now I was surprised I had survived what Gottfried and the other Europeans called the barbarous drinking habits of the Rhodesians. That heavy, hard, grinding kind of drinking obliterates the variety and fun of drink. Now, getting tight – not very often because we worked so hard – was a treat, and full of the most unexpected revelations. Gottfried, for instance, was a different person tight, all Russian, whereas sober he was German. Drunk, he wept at gypsy music, sober, he said music was only fit to stir up cheap emotions. Drunk, he permitted himself to flirt in a fatherly way with the girls who yearned for him, but sober, he talked about marriageable girls who only wanted to catch husbands. 'Our marriageable girls.' Sober, he judged, he criticized, he categorized. Drunk, he danced cossack dances with strangers. Brecht wrote a play about a landowner who was unpleasant when sober, but delightful when drunk, so his peasants conspired to keep him drunk all the time.

On the evening of VE Day, 8 May 1945, it seemed the whole town was dancing. There were Victory Dances at Meikles Hotel and in the Grand Hotel and the Sports Club and everywhere were crowds in blue-grey uniforms and in dance dresses, and everyone shouted and sang, and then ran shouting and singing into the streets. We were all drinking. I remember exactly how I was, that is not from outside, not – medium shot – a young woman with an emotional flushed face, in a day dress, walking slowly along a corridor in the Grand Hotel, but from inside, what I was feeling and thinking. I am alone. The music is quietening behind me. 'Run Rabbit Run', 'We'll Hang Out the Washing on the Siegfried Line', 'Lilli Marlene', and, of course, 'There'll be Bluebirds over the White Cliffs of Dover'. Among the dancers running past me are most of the people I know and it seems every RAF in the colony. I am angry and I am sick. Nothing new about any of these emotions, it seems that is what I had been feeling all my life. In my mind, I am with the crowds celebrating in London and in Paris, but I am also in Germany, and that is because of Gottfried, who for weeks now has spent hours of every day by the wireless, listening to the fighting in Europe, to the reports of the millions of refugees, the general chaos. Like the end of World War One, only much worse. I was not feeling the pleasure in sorrow, that poisoned well, all that had been burned out of me. Earlier I had been to see my father, a very sick old man in bed, and he said, 'Now I suppose they'll start preparing for the next.'

But it was the end, after those long years of war. Well, not the end of the war, actually, only the war in Europe.

I walked back home through excited crowds, mostly white, it goes without saying, because of the curfew, and found Gottfried sitting with Simon Pines, drinking wine. He was a refugee from Lithuania and, like Gottfried, like all our refugee friends, tonight he was in his mind in Europe, not Britain. To go into that room was jumping into cold water after hot. Gottfried did not know what had happened to his mother, his father, his sister. Simon had relatives who had been overrun by the Germans, by the Russians. He expected them to be dead. Usually he was full of aggressive restless energy, but not tonight. Simon and I 'got on' – it was because both had country childhoods. We would amuse each other and ourselves swapping details. He would say: 'Do you mean to say you fed the sour milk to the chickens? We ate ours. Sour milk and potatoes – I'll cook some for you, one of these days.' I can look back and *feel* myself inside the sombre anxious mood of that night: look back and *see* the three of us, Gottfried, like Conrad Veidt in his elegant suit, his hand with the blue stone loose around his wine glass, his polished black hair. Then Simon, a stout brown man in his khaki uniform – a bear in khaki, as he said himself, and me, standing over the hotplates, cooking supper for them and for anyone else who would drop in, when they were tired of running around the streets singing 'The White Cliffs'.

Our new flat was peculiarly constructed, in a building called Leander House on Jameson Avenue, now Samora Machel, on the site where the Jameson Hotel stands. It was a two-storeyed building. Our flat consisted of a large room, entered from the very wide corridor that bisected the ground floor. Another corridor, outside the room, went from nowhere to nowhere, for it paralleled the outer wall of the building on Jameson Avenue and the wall of the big room. It was neither a corridor nor a room. It had in it a large wardrobe, a chest of drawers, and quite soon there would be a pram and a cot, all lined up side by side against the wall. This long space, or corridor, turned a corner to become a narrow room where on a marble slab stood two electric hotplates and a little refrigerator. The bathroom was an extension of this kitchen. The inner wall of the big room opened with two windows on to a small court, and opposite, across the courtyard, was another flat, exactly like ours, where lived a mother and a daughter.

The move to this flat had been an outward sign of a change in the style and pace of our lives. The Communist group had quietly died, as an organization, though we all considered ourselves Communists. The principles of sound organization as enunciated by Gottfried were proving themselves: the time was over when every one of us was secretary, librarian or chairman to half a dozen organizations: other people did this work. The Cambridge RAF had gone to fly their aeroplanes in the war over Europe. Our men, 'the Boys up North', had quietly come home, if they were alive. The Group, always fluid, had few of its original members. Dorothy Schwartz had gone down south to Johannesburg, to work for the South African Communist Party. The RAF were still in the country. Our friends were not necessarily political, but homesick for conversation and needed books to borrow. This was the time when I might cook supper for up to fifteen or twenty people on the two hotplates, eggs, bacon, sausages and tomatoes, or great stews, or braised chicken or duck. Gottfried had said he could no longer stand English cooking, and a woman friend of his had instructed me in the use of spices and herbs and garlic. In the flat we kept supplies of Castle beer and wine. There was always clearing up in the morning. We did not have a servant. Many nights people slept on our floor or in the bath: men from the camps who had missed the last bus, or people who lived a good way out. Gottfried did not like this huggermugger living, the informality. This was not a question of principle, but his temperament. In those days I was a natural bohemian. 'You're just a boheem, my dear.'

'In the provinces you are a bohemian if you drink wine and do not keep a black servant.'

We were working very hard, Gottfried and I. Several days a week he got up at five to go out to the Tobacco Auctions where he worked as a clerk until it was time for his work for Howe-Ely. He got paid a clerk's wage in both places. 'One should not speak ill of the dead.' I would speak even iller of Howe-Ely if I thought it would do any good. But now, having observed successive waves of refugees, I know that employers will always take the opportunity to pay a pittance to refugees with professional skills, and describe themselves as charitable.

So Gottfried had two jobs. He was learning Russian. For recreation he read the history of Byzantium, which fascinated him, and spent some evenings with Hans Sen, a Swiss. Hans spoke, I think, twenty-five languages but read and understood as many more. Like Gottfried he thought of himself as an exile from civilization. He was a Roman Catholic, Gottfried a Communist. When I joked that the two faiths had a good deal in common, Gottfried was huffy with me for several days.

I had left the lawyer's office and earned three or four times as much typing for Hansard and government commissions. I worked for Mr Lamb, an old man whom people pointed out as having been one of 'Milner's Kindergarten' – that is, young men groomed for power and influence by a famous English liberal planning civilized ways and thought for all of Southern Africa. He was a shorthand writer. The system then – made obsolete by technology – was that two or three shorthand writers took turns to go into the Parliament

House, for ten minutes, fifteen minutes, twenty minutes, rushing out to dictate as much as they could to waiting typists before their turn came around again. The requisite in the typists was speed and the ability to use words reduced sometimes to no more than a letter.

Min of Agric: T H M is ooo, t sbjt o cattle food w b o t agen tmrw.

This ruined my typing for ever. Extreme speed, yes. I was the sole typist for three government commissions. Conditions of the recruitment of native labour. The proposed Kariba dam. The control of sleeping sickness – by the deliberate shooting out of all game in areas hundreds of miles across. The first one enabled me to know when the government was lying, when asked questions about, for instance, a practice of deliberately kidnapping Africans walking down from Nyasaland to get farm work so as to take them to the Rand. The second was a violent clash between experts, some of whom maintained the Kariba Dam was impossible because all the water would disappear into rifts in the earth and be lost, to go forever gurgling through subterranean caves and aquifers. The third, the deliberate shooting of hundreds of thousands of head of game, explains why so much of the bush is now without animals. The policy – to control tsetse fly in this manner – failed.

I was also writing *The Grass is Singing*, which is now a tidy little book on a shelf. It began three times the length it is now, and was a satire. The central character was an idealistic young Englishman of the kind who so often arrived in the Colony, only to be appalled at what he found. Since most were escaping the Slump and extreme poverty in Britain, to leave again was impossible, and they adapted themselves to the local mores: it was acknowledged that these new converts to White Civilization (like all converts) tended to be more extreme than the old-timers. But supposing one of them arrived and did not either at once leave again, or conform? I would have that classic theme for a comedy, the innocent idealist in conflict with corruption, or, as it is put, realism. Like a Western. The trouble was, I did not have the experience to write it, and it was heavy and clumsy. In its first shape I sent it to London. No airmail then. It went by ship. If it was not sunk, that took six weeks. Then as long as the publisher took to read it. Then back by ship. Six months, a year, more. Later I kept a sub-plot and threw out the rest. I sent short stories to magazines in London. The same process.

Some were later published in *This Was the Old Chief's Country*. I sent poems: these came back with encouraging letters or rejection slips softened by 'Do send us some more.' This exercise in patience was valuable. Nor can one take the pronouncements of editors so seriously when rejected short stories were later printed and praised. I also wrote plays. I had been in love, besotted, with the theatre since I saw, aged nine, *Oedipus Rex* put on by some sixth form on Government House lawn. A play I wrote at this time was later put on at the Cambridge Playhouse, when it was already dated, for things were changing so fast in Africa. The local Reps put on *Dangerous Corner*, and *Blithe Spirit*. When they did *They Came to a City* – now a forgotten play by Priestley – they had an emotional, grateful audience. We needed news of better times. Another lamp in the dark was the film Laurence Olivier made of *Henry V*. We – a group of about twenty – saw it in a large cinema where there were perhaps five other people.

Several times a week, Gottfried and I drove out to the suburb where my parents lived now to sit by my father's bed. He was dying. But he had been dying for a long time. We sat with him while my mother went off to visit someone or, as she still put it, pay calls. Those afternoons at my father's bedside, they were a horror, and still come back in dreams. His own personality, his own real self had dissolved long ago into the illness. As far as I was concerned my father had died long ago. I was always trying to talk to *him*, call him back, make him respond, be my father, be anything even for a moment, that wasn't this self-pitying, peevish, dream-sodden old man talking about his war.

Gottfried was kind – but correct. With my mother I was dutiful – and correct. It must have been a nightmare for her, this polite daughter, using a cold foreigner – a *German* – as a shield. They did not like Gottfried, the Countess von Schwanebach or not. It was not because he was Jewish – or part Jewish. (This kind of distinction would seem to them silly.) I don't remember ever hearing an anti-Semitic remark from them. Besides, we were just beginning to understand what was happening to the Jews in Europe. Only beginning. We had not 'taken it in'. Who is 'we' here? The view of my parents and people like them was much less dark than that of people on the Left. We – the Left – prided ourselves that we had been for years pressuring our government (British) and governments generally to tell the truth about Hitler's treatment of the Jews. It turned out we were not much better informed than my parents. 'It's terrible what Hitler is doing to the Jews.'

Our view went something like this. Hitler had begun by killing off all his German opponents – Communists, socialists, Catholics, Protestants. (It is only recently that these brave people have been remembered.) Then he began persecuting, and killing, Jews, gypsies, homosexuals and the mentally defective. They were being used as slave labour, and so badly treated they died. The reality about the death camps had not begun to 'sink in'. The point is, if your mind is not 'set' to take something in, facts are rejected. Our view – the Left's – was in fact as conventional as the general view. We were seeing the war in terms of allies and enemies, theatres of war, the great battles that were the turning points, the Battle of Britain, the Battle of the Atlantic, North Africa, Stalingrad, sieges like Leningrad, defeats like Dunkirk . . . Yes, refugees. Yes, the cruel killings of opponents. But now it seems to me frightening that what later generations will surely see as the worst thing, the characteristic thing about that time, *our* time, we hardly saw at all. The most significant thing, with its implications for all our futures – left out of this map of the world. The deliberate murders of millions of people – systematic murder. Systematic torture. Gas chambers. Concentration camps. Laagers. Genocide. Ethnic cleansing. Our mental map of the world was still innocent. This was a war of the bad against the worse. It was a good war, in spite of everything.

I rather doubt now whether that is how it would seem to that mythical watcher in the skies.

16

THAT WAR INVOLVED the whole world, and so many different kinds of people and experience. What could they possibly have had in common? A soldier fighting his way up Italy, or in Burma, or in Stalingrad – the real combatants. Refugees. Prisoners of war. Civilians overrun or occupied. And then people thousands of miles away from fighting, watching from a safe distance? Yes, there was one thing. It was that populations were being stirred and swirled around, and people were colliding who never otherwise would have met. When I look back now that is what strikes me first: the improbable clashes of people, and at once I feel an ebullience, an exhilaration, an energy. Is that what I felt then? Yes, often. But memory is a great maker of comedy. Decades after an event which was painful or even frightening, it may seem merely absurd. I have to remind myself that arguments or events described here humorously might end in physical violence. I say incredulous, Is it really possible that my good friend Mark assaulted my good friend Abe for calling him a typical intellectual? (What were they really fighting about, one is tempted to ask?) Or that at a lecture about Lysenko the Soviet scientist, Stalin's protégé, and the inheritance of acquired characteristics (the 'line' required adherence to the proposition), a group of 'Trots' challenged the 'line' as a result of which there was a brawl outside the hall that ended in two young men being taken off to Casualty? Or that Jane (a Communist from London) relinquished politics altogether rather than sit in the same room with Marie (from Cape Town) who, she accused, suffered from racial prejudice because she said that all the Afrikaner families in South Africa had black 'blood' in them? The word objected to, not the fact.

Ripeness may be all, mellowing you into a shrug of the shoulders and a smile, but the raw rub of the time itself was the engine of events. Even then the conjunction of Kurt and Esther seemed to us a paradigm of surreal war. And even visiting them was a reminder, for while Esther's garden might be a little paradise, from it you could see the tall fence of the nearby RAF camp. No one could look at that fence without a constriction of the heart. It was a fence of a kind we had never seen. Our fences were low and straggly, and the wire might be looped around convenient trees (shielded by strips cut from rubber tyres) or roughly-cut posts. This fence meant business. When your eye was drawn to it you had to think of all the other air force camps in various parts of Africa, the swarming grey-blue-uniformed men, who were mostly unwillingly here, behind fences and under guard. You had to think of war.

This new suburb, one of many hastily flung out around Salisbury to stretch a town developing fast because of war, was imposed on the veld in a rapid cross-hatching of streets, with a narrow badly-tarmacked road leading to it off the main road to Umtali. This suburb seemed as makeshift, as temporary, as the camps, but in 1956 when I drove past, the cheap little houses had been enclosed by gardens. At the time they were built, the box-like houses stood dotted in rows inside squares of wire fencing that enclosed raw and ravaged earth. But almost from the moment you turned off the main road, there, among all the wastes of fences and builders' rubble was a garden that displayed itself as a crammed exuberance of colour held upright inside its wire ties like a bouquet. This was Esther's and Kurt's house. As we walked up to the house, on a bright red brick path already subdued by portulaca and thyme, we were crowded by roses, plumbago, cannas, jasmine, oleander. The verandah just ahead was barely visible. On its steps stood pots full of plants, and from the verandah rafters hung curtains of ferns. The houses were all the same, two rooms, the front one the sitting room with the bedroom behind, and a minute kitchen behind that, off a miniature verandah. The front verandah was wide and shady, like a third room. The sitting room was furnished adequately, with native-made table and chairs. On the floor were reed mats. Cretonne curtains said *England, English.* Vases of flowers stood decorously on most surfaces. On the walls were English watercolours, and reproductions of Pieter Breughel from the art galleries in Vienna.

The garden was entirely Esther's, made in the early mornings before going to work and in the evenings after work. When we went to visit, and stood on the brick path, we saw a disturbance in the depths of the garden, and then Esther emerged upwards from green waves where flowers tossed, and she said, 'Oh how nice of you to come, please come in.' She stepped carefully out, smiling, and accepted a hand extended to steady her, while indicating it was the last thing she needed. She went up the steps in front of us to the verandah calling, 'Shilling – tea please.' Already Kurt's voice reproved, 'Esther my dear one, his correct name, please.'

'So sorry,' she said airily. 'I always forget.'

Standing on her verandah among the dangling green fronds, she was a slight, upright, pretty Englishwoman, in a sensible dress and heavy gardening gloves. She walked lightly into the room, saying, 'Look, darling, how nice, we have visitors.'

Kurt sat brooding in a chair too small for him. He was a large man, heavy rather than fat, with a skin so dark it seemed greenish or bronze. A bronze man, all long sloping planes, and his face was heavy too, with long cheeks, a flattish thick nose, and small dark intent eyes below heavy ridges. His hair was cropped short, and showed the clumsy modelling of his skull. He was an ugly man, yet there was a fascination about him. He had been born and bred in Vienna, and it had been agreed among us, with his indifferent consent, that he must have Mongol ancestors. In those days people said, 'So-and-so must have Mongol blood,' or did if they were not 'progressive'. Because we progressives had not yet been rescued by the word 'genes' from the word 'blood', discussions were always full of pitfalls. Not for Esther, uninterested in politics. Contemplating her husband with her cool but affectionate smile, she would muse, 'But when you think of how many centuries the Mongols were invading your part of the world, of course there must be a lot of Mongol blood. Like us and the Vikings.'

'Esther, I beg you, not that word *blood*.'

'But why ever not?'

'Hitler,' he groaned, fixing her with his mournful gaze, so full of history. A gaze, however, which she always bravely confronted 'But I am not Hitler, am I?' she might observe.

She came from an English country town. If you were trying to find her opposite, her antithesis, it must be Kurt: of his, it must be Esther. These two beings, one so light and brisk and commonsensical, one so heavy and bound and tormented, made it inevitable that anyone seeing them together must exclaim at the incomprehensible choices of nature. She was a teacher and poor, because she sent most of what she earned Home to her mother, who was an invalid. He had not been able to get a better job than as an official in the Public Works Department. He was a Doctor of Philosophy. That had been his preparation for the harshness of Hitler's Europe. Had there been no war he would certainly have spent his life in the ambiance of some university or newspaper, talking in cafés. He had in fact spent most of his time, since the age of about twelve, talking. In short, he was an intellectual, a word at that time even more emotive than usual.

If there was one thing we raw Southern Rhodesians found most fascinating, it was that these refugees were all the time, day and night, political, ideological. Of course we – particularly in the farming districts – talked about politics, but we had not thought that the vagaries of the government and the Company spelled Politics. The intellectual allegiances of these immigrants were so important to them they told us what they were before we knew anything else about them. 'You see, I am a Freudian.' 'I am a Marxist-Leninist.' 'Reich!' 'Jung!' They never stopped arguing, discussing, quarrelling. They conducted silent and contemptuous vendettas or passionate ones.

Gottfried said Kurt was only an intellectual. Kurt's real education had been, he said, the conversations in the cafés of Vienna and then later when he lived in a commune or ideal collective in Vienna, run according to the ideology of one of the psychological geniuses who succeeded to Freud. His talk, rather, his monologues, always returned to the years he had spent in the commune. He would sit leaning forward in his chair, as if in the grip of a need to reach for an idea still imperfectly glimpsed by him, but he was held down in this ridiculously flimsy and irrelevant chair by the weight of his bones, his body, the burdens of material life. His thoughts were nowhere near us, our level – how could they be? His mind was pursuing some truth that he would one day capture, to have and to hold, once and for all, brought back wriggling at the end of a pin: 'There! I *told* you, didn't I?' He could not sit still, but jigged and jiggled as he sat, one foot tapping, fingers tapping on the chair arm.

'You *must* understand! We succeeded! That's the point! For years we lived the ideal life, the life of comrades! *Real* comrades!' And here he sent an accusing look at Gottfried or any other Communist exemplar who happened to be around. 'We shared everything, everything! We did not own anything, only our clothes. We were allowed a jacket and a pair of trousers and two shirts and a jersey and some underclothes. That was all. We shared our food, our money and our books.' And he would give us all a calmly triumphant stare.

'Hard on the girls, I am sure,' Esther might comment. She would be sewing or knitting, but set down her work to smile at him. She never spoke to him except in tones of the gentlest respect. The word has to be love. Other people might find him exasperating or – simply – impossible, but she knew this Caliban embodied something wonderful, and that is why she married him. She called him Kurt, the *r* sounding. Most of us unregenerate colonials called him Curt, just as they called Gottfried Godfrey.

'If you can call me Koort, showing you respect me and my language, then why don't you give this man here his proper name?' He meant, the cook. 'His name is Mfundisi. Say it!'

'It says Shilling on his identity document.'

'But it is not his real name. You simply do not understand the importance of a human being's name. You people have taken their names away and you call them any sort of rubbish. Sixpence. Tickey. Shilling. Blackbird.'

'But they also get called by very beautiful Biblical names,' said Esther, approving of these not on religious but on aesthetic grounds.

'But they didn't choose these names. Esther, you really must understand this matter. It is important.'

'I didn't choose Esther. I wasn't asked.'

'You people are simply not forgivable. History will not forgive you.'

'But I didn't take away their names. I wasn't here. I only came here a few months before you did.'

And about the Commune. 'You do not understand, Esther. Sometimes you say things full of delicacy and understanding, and then I think Yes, she understands . . . but you don't understand. You see, the women believed in the ideal life. Often they inspired us when we weakened. They kept us from temptation. None of them cared about petty things like dresses and lipstick.'

Esther took up her sewing again, with a guilty but determined look. She did not believe in idle hands. Our hands held cigarettes. Through clouds of cigarette smoke we contemplated the ideal life, which could not be found in mundane Salisbury, but had to be in Europe, in Vienna, where we all wished we had been, with other noble creatures, living a noble life.

'It was wonderful,' intoned Kurt, gazing at where the bulldozers were clearing yet more bush for the spreading suburb – smashed trees, torn earth, a rubble of stones. 'We proved that it was possible, the life without property, without ownership.'

Here Gottfried might remark, 'Petit-bourgeois idealism.'

'And without jealousy. We forbade ourselves jealousy.'

Here Esther raised her eyebrows.

'Yes, yes, it is true. Then one of the girls had a baby and that was the end. Everyone began to quarrel and suddenly there were couples in their own rooms with locked doors. It was horrible, horrible.'

'But Koort,' said Esther gently, 'anyone could have told you it would go on nicely until the first baby. Surely your Professor Fischel knew that?'

'And why should he know it? He was a great man. He was interested only in the real possibilities of the human being. We abolished property – we had put an end! – and then just one child, hoppla! That was it. A child's birth is not the affair of its parents, a child is born to the whole world, the human community.' He stopped to check his audience. Some eyes shone with emotion. Others most definitely did not. 'We loved each other. We trusted each other and then phut! Just one small baby and – the end!'

Now, the real possibilities of human beings were what we were all dedicated to.

'Wait a minute, Curt,' demanded Jane, one of the girls. 'You are seriously telling us that you all slept with each other and no one was jealous?'

'Yes, that *is* what Koort is telling us,' remarked Esther, threading a needle, or counting stitches.

'You mean, if two men were in love with one girl, or two girls with one man, it was all sweetness and light?'

Kurt jerked his leg up and down, tapped irritably with his great clumsy fingers, stared at the wall, or at his invisible ideal. 'You don't understand. We all loved each other. When something like that happened we were *kind* to each other.'

Meanwhile the Comrades shared patient and ironical smiles: the future was on their side.

'All the same, there must have been a heartache or two,' mused Esther.

'Life,' announced Kurt, 'must have pain. It is our birth pangs that propel us into true humanity.'

We all contemplated this remark, from our ideological standpoints.

Esther said, 'I think it was greatly to your credit that you all made it work for so long.'

'Yes, and it was at least three years. And don't forget, the war . . .'

'That wasn't very long,' said Jane, a neophyte. 'I dare say I could submerge myself in the Ideal for a year or two. You must have been jolly well pleased with yourselves. Were you all having analysis?'

'Naturally we were. That was the whole point. Professor Fischel was analysing all of us.'

'You mean, one person analysed all of you? How many in the commune?'

'There were eight of us. You see, a great many people wanted to join, but not everyone was suitable.'

A silence.

'I really do feel it was a most impressive achievement,' Esther insisted, giving us all a look. 'But I think Professor Fischel should have made some allowances for property feelings over the baby.'

'Human nature!' said Jane.

'A tragedy,' mourned Kurt. 'It was the death of a dream, when that baby was born.'

'From the way you talk, it could have been yours,' said Jane.

'No it could not,' said Kurt. 'I am never irresponsible in these matters.'

Neither Kurt nor Esther came to meetings, Esther because she thought political meetings childish, Kurt because he thought these meetings childish. But he was always arriving at our flat. I was learning that some people cannot stand being alone, even for an hour. Not Esther, who was always happy reading or gardening. But Kurt invaded my precious people-free mornings. If another person similarly afflicted turned up, I might leave them to each other, shut the door on them, and work standing up, using the marble kitchen slab as a desk.

But didn't Kurt have to go to work himself? – you might ask. At the Public Works Department, his work was mostly done on the roads, seeing what condition they were in, or supervising repairs, or looking at parade grounds, or drains, or other public amenities. He was not often in his office.

'It is not my real destiny, the Public Works Department,' he might announce, drinking tea in his own way, which was to interrupt his tormented gaze after the unreachable long enough to gulp down half a cup at a time, when he would sit looking at the cup with gloomy distaste until I refilled it. 'No, I was not meant for this kind of petty occupation and I cannot pretend to be what I am not.'

'But what do they say in your office?'

'It is not that I don't do any work at all. I do enough. Nobody works hard in the Civil Service, and I am only conforming to your mores. Besides, my boss enjoys despising me. Esther tells me I imagine this, but I say to her, No, Esther, a kind heart such as yours, and your protected English upbringing, has not been the best preparation for life. You do not understand evil. I see through that man, and he knows it. I take him a file about something or other, roads or such nonsense, and I hand him the file and he has started sniggering before he has even opened it. "What time did you get to bed last night?" he asks, and then he laughs as if he has made a real joke. "You will not believe this," I tell him, "since you live in a country where everybody is asleep by ten o'clock, but there are people in the world who despise sleep. They prefer to spend the night talking. Or even thinking." Then he slips open the file and sneers. "Do this again. The columns aren't correct. Then bring it back and we can *talk* about it." And he shuts the file, full of triumph. Once he threw it at me. And so I spend my days. When I think of what I have become, I am full of despair.'

By the time the war ended, the thousands of refugees had stopped being needy and sad and desperate but were businessmen and farmers and exporters and importers and builders and ran transport firms and played in orchestras. A talented lot. A credit to the Colony. But Kurt stayed in Public Works. 'I have no intention of wasting my life making money,' he would say, while Esther gazed fondly at him. 'For one thing, a lifetime is not long enough to read the books one should read. Have you read . . . ?'

Often, I hadn't.

'You are very ignorant. Anyone who lives in the twentieth century and who has not read all of Freud and Jung and Adler and Klein and Reich is ignorant of the main influences of our time.'

'How about Marx?'

'That is ridiculous.'

'At the very least he would appear to be having a certain influence.'

'Merely a passing phenomenon.'

'And Freud and Jung and the rest aren't passing phenomena?'

'I simply can't understand why you don't see it,' he fretted.

Later, as was only proper, Esther became a headmistress. Esther wanted him to leave Public Works and devote himself to his book. In fact he continually worked on this book, which was all in his head, so he said.

'Why publish it when it is already perfect? For fools to misunderstand?'

'But Koort, you must expect a really good book to be misunderstood.'

'This is more than just a good book. This is a *real* book. There are only half a dozen real books.'

'And what are they?' I wanted to know.

'First of all, *Don Quixote*.'

'And then?'

'*Hamlet*. A man whose heart is too great for the pettiness of his surroundings. And then, *The Man Without Qualities*. Of course you have never heard of it.'

'Sorry to disappoint you.'

'Then you couldn't have understood it. And then Stendhal's *Love*. But you are too young to understand that.'

'I can't argue about that, can I?'

'No, you can't. But time will cure a good deal that is wrong with you.'

The list of the real books changed, often.

By this time the manuscript of *The Grass is Singing* – final version – had been read by a few people, Kurt and Esther included. She disliked it because it did not give a hopeful picture of race relations, Kurt because it was written in English: he agreed with Joseph Conrad that it is a language unsuitable for novels, and only French has the necessary clarity. No one had a good word for it. Comrades and friends continually arrived to tell me how to rewrite it. I believe the worst enemy of any beginning writer is the group of loving friends. Most hope to be writers, and the story which the poor aspirant has entrusted to them is nothing like what they would attempt themselves. Everybody, these days, is likely to be an ideologue of some kind. It goes without saying that the Comrades disapproved, and for the same reason as the non-political Esther.

I have met several people who kept in their minds perfect books that could never be written down, that must be kept well out of sight of the vulgar and our contaminating thoughts.

One deserves commemoration. He was brought up in Spain by the Jesuits, destined for the priesthood. He became a Marxist, as pure and undeviating as he had been a seminarist. He married an English girl, and persuaded her to share a caravan and poverty with him in the West Country. Four children. London he told her would spoil her essential self. They lived on social security. He spent his time discussing points of doctrine in the company of other revolutionaries in the bars and pubs of the charming towns of Somerset and Devon, or in London, with his literary peers. Meanwhile, he was working – in his mind – on his book. He would fix you with stern eyes and say, 'Why write it down? Why compromise?' He lost his faith in Marxism, became an anarchist and left his wife in her caravan with the four children. He went to Paris and became a *clochard*. There in a café he found another middle-class English girl, who was easily persuaded her life was a sham. Together they lived the life of beggars, which, it seems, is pretty hard work. Every morning you meet with other *clochards* to find out where the free snacks will be in the exhibitions and trade displays, what charities are giving free meals, and how best to exchange goods stolen from the shops in a way that will not attract the police. He agonized at the amount of vital energy that was going into all this, and his girl agonized with him. But all the time he was working on his invisible book. Then the girl announced she was pregnant, and proposed to bring the child up as a *clochard*. This was logical, since he had taught her to believe it was the only honest way to live. Leaving integrity behind, he took her off to the country in England, put her in a tumbledown cottage, where soon there were four children. They lived on social security. We knew some ideology would come to his rescue. Yes, it turned out that women and children are the poisoners of integrity. He abandoned this family too and became attached in an administrative capacity to one of the innumerable organizations bringing advice to bits of the former European empires. His boss was a black woman, a feminist bringing up four children whose fathers had abandoned her and them. He moved in with her: he had found his destiny. Sometimes he visits Europe. And how is his masterpiece getting along? He fixes you with a calm unfractured gaze, full of contempt for the venalities of literary life. 'And why should I write it down? It is safe where it is.'

Esther tried in all kinds of tactful ways to make it easy for Kurt to write his book.

'Esther insists I should put my book down on paper,' Kurt accused her.

'I, for one, would like to read it,' said Esther.

'It is strange that a sensitive soul like yours cannot understand such a simple thing.'

Kurt never wrote his book, and would not discuss it with us. He might describe its airy and phantasmagoric mansions, but not in a way that invited comment. But there was one person he did tell everything. Kurt did not drive a car. Probably he was the only white person in the Colony who did not. He said, truthfully, that he would be a bad driver, because it would be impossible for him to keep his mind on the road. The Public Works Department allotted him a driver. According to his pass, or situpa, this man's name was Joshua, but Kurt insisted on being told his real name, which was Muesaemura, and on using it.

They spent a good deal of time together, driving around in Salisbury, or even making trips on the various roads out of Salisbury. Not all these trips were necessary. The two used to stop for long philosophical discussions in the bush, well out of sight of the road, because of course both would lose their job if discovered in this seditious activity. As it was, Kurt had been warned that his attitude towards 'the munts' was unacceptable.

'I regard it as fundamentally insulting to you, calling you Musa,' Kurt insisted.

'But my mother calls me Musa. So does my sister. And all my friends. So why not you, Baas?'

'Don't call me Baas. It is an insult. And I keep explaining to you, if people shorten or distort a name it means a fundamental lack of respect.'

'But you are my boss. A white man. Mlungu. So what you call

me can't be more respectful than the name my mother gives me.'

('So you see,' Kurt might say, reporting such conversations. 'He has an innate sense of logic.' 'Perhaps you ought to tell Charles Olly? He'd be convinced on the spot.')

'And what does your father call you? It is significant that you never mention your father.'

'I didn't have a father.'

'But that is not possible, Musaemura.'

'How, not possible? It is possible for us kaffirs. We often have no fathers. My mother was the sister of the cook who worked in the house of a Member of Parliament. She lived illegally in his room at the back of the house. She got pregnant by a friend of the cook. A husband? That's too much to ask. She had a place to live in. Not a legal place, but a husband *and* a place to live in is too much to ask of life. For a kaffir woman.'

Musa was about twenty-five years old. He was very tall, thin, restless, and eaten with violent energy. His life had been colourful, precarious, sometimes criminal. 'But surely that goes without saying,' Esther said. 'You cannot expect good behaviour from badly treated people.' Musa was clever, a good mechanic, and valued by the Department. 'He's not a bad munt,' Kurt's boss might remark. 'But you'd better keep your eye on him, I'm warning you.'

'That man,' Kurt brooded to us, 'he should be Prime Minister. It's a crime he's just a driver. I keep telling him so.'

'And what does he say?'

'He says, "Baas, you know a black man can never be Prime Minister. Our heads are the wrong shape. We do not have enough brains stuffed in our skulls to be Prime Minister." He calls me Baas. He does it to annoy me. I tell him, "All right, I concede your right to insult me in this way. Not me, as an individual, but as a representative of the oppressing race. It is your right to behave unkindly. I support your fundamental right to negative emotions and destructive impulses, as an oppressed people."'

'But do you support the right of the whites to be negative and destructive? They fined me £2 last week in the Magistrate's Court for being drunk. That is half my month's wages.'

'You get drunk because of the frustrations of your life,' said Kurt.

'Yes, that is true, Baas.'

'I would make you Prime Minister. You couldn't be worse than what we've got now.'

Independence, a black President, a black Prime Minister, were thirty-four years in the future.

'I would be much worse for the whites. I would be terrible.'

'No, I believe you to be a noble and magnanimous soul.'

They were, as usual, sitting under a shady tree in the bush. Kurt asked Esther always to provide vacuum flasks of tea and plenty of sandwiches because Musa did not get enough to eat. Kurt made a sandwich last a long time, and watched Musa eat up sandwiches, cold meat, cheese, fruit. 'And I'm going to take this back to my friends,' Musa would say, wrapping up the remains. 'You have too much money, Baas, and my friends are hungry. And I'm not going to be noble and magnanimous just to please you, Baas. What is magnanimous? Do I know that word? It is something to the advantage of the whites, I know that.'

'If you are no better than we are, why bother to have blacks in power instead of us?'

'Because we have the right to be shits in our country, if we feel like being shits.'

'I see that one has to accept that you will insist on going through all the stages of stupidity instead of learning from our mistakes.'

'Learning from you to be noble? And mag-nani-mous-damn-fool?'

'Learning that it is better in the end to be just kind and magnanimous. In the long run.'

'You whites have done all right out of being shits.'

'You are wrong. Not in the long run.'

'And why should I care about any long runs? I don't want to be Prime Minister just to please you, Baas. I want my own business as a transport driver with my brother working for me. When he gets out of prison. But I have no money. You could have your own business if you wanted, because you have enough money. You could have a transport company. But you are stupider than me.'

'At some things I am.'

'You can't even change a tyre when we get a puncture. You don't know where the sparking plugs are. If we had a breakdown and I wasn't there you'd sit and wait for someone to come.'

'I'm no good with my hands,' admitted Kurt, believing that he was humbling himself.

'Then what are you good at, Baas? When we go to a job I have to tell you what is wrong. What about that burst pipe yesterday on the parade ground?'

'I understand the real meaning of the intellectual movements of our time.'

'If that was any use, why are you working for the Public Works Department?'

There, in the bush, to the sound of doves, the go-away birds, the louries, while ants sent out their scouts for crumbs and ran frantically over their four legs, black and white, stretched out companionably side by side, Kurt told Musa all about his book, the perfect book, forever unfolding in his head like a cumulus cloud on a hot day.

From time to time Musa might enquire, 'But what is so new about that?' (The Unconscious. Repression. Incest. The Id. Good and Bad Breasts. Even the Collective Unconscious.) 'I knew that when I was so high.' And he held out his hand about five inches above the earth where ants were carrying away cake crumbs larger than themselves. 'Except that I didn't have a father so I couldn't kill him. You have to be a white man to have a father so that you can hate him. If I had a father I would have more food because I wouldn't have to give all my food to my mother and my sister and my friends. No, all that is for rich whites. You have money. You have fathers.'

And he sat grinning, while Kurt defended himself.

'Musa is exactly like a friend of mine in the Commune. His name was Wolfgang. He was a sceptic by temperament. It was he who destroyed our Commune because he insisted he was the father of the baby. That was after we had decided the baby would have no father or mother but belong to us all.'

'And was he?'

'The baby did look like him.'

'Have you told Musa about your Commune?'

'I tell him everything.'

'What does he say?'

'He says he thinks we were backward because with his people all children have more than one father and mother. As a matter of course.'

This was before everyone started to talk about the extended family, so much better in every way than anything we enjoy in Europe.

'Musaemura thinks we are very primitive. He remembers biting the breasts of his mother because she did not have enough milk. He says good breasts have milk. Bad breasts do not have milk.'

'How can he remember?' enquired Esther.

'He was four. They breastfeed longer than we do.'

'Four! Well I think that is primitive.'

'No, no, Esther. I tell him, human beings are primitive, the whites and the blacks, as bad as each other. He says he would rather be primitive like the whites, with more money. He wants us to lend him money to start a transport firm.'

'But we don't have the money.'

'As far as he is concerned all white people are rich. I am afraid he has very large ambitions. He wants to have a trucking and transport firm, on the same scale as Hamish Van Doren's.'

'But Van Doren is a Member of Parliament.'

'He made his money out of a transport firm.'

'It's not going to do Musa much good to be like Van Doren. He's a first class crook,' said Gottfried.

Musa had become something like a touchstone. If we wanted to know what black thought was on any subject – if one may put it like that (and who knows? – by the time this book is published) then we asked Kurt to ask Musa. Thought, that is to say, that was not political, not informed. For correct black thought we asked Charles Mzingele. (Charles was shocked by Musa – by the new generation – disapproved of this anarchic disrespect for serious matters. Musa thought Charles was a sentimental old fool.) We invited Musa to come to discussion groups, but he refused. 'Baas,' said he to Kurt, 'you and me in the bush, that's all right, only the ants and the chameleons can see us, but when I start going in and out of a big building in Jameson Avenue, then that's big trouble. How many laws would I be breaking?'

'I don't think any. Provided you got home in time for the curfew.'

'There'd be a law. There always is.'

'Provided you didn't want to marry one of the girls,' said Kurt, attempting a joke, 'then who could object?'

'How do you know I wouldn't want to marry one of your girls, Baas?'

'Oh dear,' said Esther, when this conversation was reported. 'No, really, that is going too far. I hope you told him so?'

'I will convey your opinion to him, Esther, and then I'll tell you what he says.'

'If that damned crook comes to our meetings, then you can count me out,' said Simon.

'Now, now, comrade, do we see here some little tinge of colour prejudice? Do we ask the white people who come to the meetings if they've been in trouble with the police?'

'Colour prejudice my foot. If he did get enough to start his transport firm, he'd suck everyone dry.'

'He'd say he was only following our example.'

'He's not following my example,' said Simon. 'I'll be in Israel.'

17

ALL THROUGH THE WAR, I had been saying that the very moment it ended, I would leave. What nonsense. My adolescent dream was intact still. The Gobi, the Kalahari, or footloose and fancy free, I would go wandering around the shores of the Mediterranean, all wine and roses and tender nights. As soon as the war did end, and it slowly seeped in to us what terrible damage had been done everywhere, the dream ended, just like that – gone.

The news out of Europe, out of Germany, Russia, was all destruction, misery, death, concentration camps, refugees, lost children. We have been watching it, listening to it, from one part of the world or another, ever since. But then, everyone believed it was the last time humanity would be stupid enough to permit such suffering, such waste. A long way ahead of us all, decades, was the suspicion, now becoming general, that humanity is not in control of what happens, that we are helpless. The Korean War, five years away, that horrid 'little' war, was the same kind of shock as the wars at this moment going on in former Yugoslavia are to Europe's idea of itself. *It was simply impossible* that after all the horror of the Second World War humanity could permit Korea.

The news Gottfried got from Germany unfroze the picture he had painted for us of the Countess at her parties, the all-loving nanny, the scholarly father, the sister whose fiancé, drawled Gottfried, was 'the eternal student and would never amount to anything'. When the war began Klaus and Irene were out of Germany. They returned to Germany, and to mortal danger because it was their duty as Communists. The mother behaved as if the war was not happening. Although there was a death sentence for listening to the BBC she listened openly. The family said she was mad. She got away with it, even though at one point there were SS billeted in the house. Irene hid Klaus, who was a Jew, from the Nazis, once even in the house. Klaus's father was a doctor: he got Klaus's name off the death lists by doing an abortion for a Nazi high-up. When the hungry time came at the war's end, Irene, a small slight woman, walked miles out into the countryside two or three times a week and carried back to Berlin, on her back, a sack of potatoes bought from the farms. This food kept the family alive. The eternal student later became a high official in East Germany.

News of the Nazi death camps came first through Hans Sen, who had heard rumours as Red Cross representative, but had only believed the half of it. He and Gottfried would sit together, in our flat, or in a café, in silence. Or, exchanging comments that sounded as if they were dripping acid on to their own flesh. They might have opposed viewpoints on everything, but shared something it seemed they were not aware of, an irony, a scepticism about the claims of civilization that was in the tones of their voices, if not in the sense of their words. (It is this scepticism that separates Europe from America.) Sometimes ten or a dozen people sat around our room until late, or even all night, listening to the News, discussing it. And meanwhile the war in the Pacific was being fought, cities in Japan were bombed into dust and ruins. What we saw in our mind's eye about Japan was not accurate, because of ignorance. By the end of the war few people were talking about 'good Germans'. There is a book, unjustly overlooked, *The Forgotten Soldier*, by one Guy Sajer. It gives a picture of what it was like to be a very young, ignorant soldier fighting on the Russian Front, retreating back, back, as the Russians advanced, sometimes with nothing to eat for days, in winter, in thin useless clothes and boots. He was half French, half German. It was only chance he fought with the Germans. Most of the young men he fought with were killed. At the end of his years of being a soldier for Hitler, Guy Sajer (a name which can be, was, pronounced either the French or the German way) was invited to join the armies of Occupation by a French officer anxious to save the life of this half-starved ghost of a soldier, still not much more than twenty. From that book comes a dark and angry pain, a comprehension, you can't call it 'protest' – for the layers of experience it inhabits are deeper than that. When I read the book years later I recognized the mood of the refugees who were friends at that time in Salisbury, Southern Rhodesia. It was also my father's. His reminiscences about the old war had become a defence against thinking about this one, which distressed him so much my mother hid the newspapers. But she could not prevent him listening to the radio, breathing in gasps, eyes staring as he mentally saw the bombs falling into Japanese towns, and watched the retreat of the Japanese, island by island. 'It's all very well . . .' he kept muttering. 'But whose fault was it? We didn't start the war, they did.' As if he had been accused. 'Why did they have to do it, why did they bomb Pearl Harbor? Why? And now look at what is happening.' And every day he said, as if this were the first time he knew it, that Harry was very deaf because of the gunfire in the Mediterranean: he was in a famous ear hospital in England. 'But it needn't have happened,' he might whisper, his bony hand tightening around my wrist. 'If they had listened to Churchill . . .' And then, soon, 'In the trenches we used to . . .' I left the house with a sprightly wave to the RAF, my mother's protégés, sitting about on the flight of stone steps up to the verandah, playing with my mother's little dog. The war had ended before their training was finished and so they would for ever feel that their youth had kept them from the real experience of life. As I drove off they would be saying to each other that I was the daughter who had married a German and was inciting the kaffirs to rise etc. If they had stayed in the Colony they would have been assimilated at once.

When the atom bombs were dropped on Hiroshima and Nagasaki they did not seem to us so much worse than the pulverizing of Tokyo and Osaka and Dresden and Coventry. What we felt was, Thank God the war is over. Only later did we learn the war would have ended anyway, and very soon.

Gottfried had decided he wanted to stay in England, only partly because I intended to live in London. He would not be the first Lessing to become British: a cousin had recently been Member of Parliament. Other members of the talented clan had lived and worked in Britain. He intended to sit it out in Salisbury until he got British nationality. Now, with hindsight, his anxiety seems unnecessary: he would have got British nationality almost as a matter of course. For one thing he was by now well known in the legal fraternity. But after ten years of exile from his country he was feeling and thinking as a refugee, that is, as someone who has been rejected. And then, he was a Communist, and he had spent years afraid he would be sent back to the internment camp for taking part in politics, when he was not supposed to. He did not feel he was bound to be accepted as a British citizen. He wanted me to wait with him until he was British before we divorced. A divorce would be a bad mark, he said, and might tip the balance against him. I agreed to wait. We believed it might take a couple of years. Meanwhile, we might as well have a baby. It was in this light-hearted way we informed our friends. 'We're going to fit in a baby now, because we've got nothing better to do.' Our discussions were on the surface the essence of practicality and common sense. After all, we had never intended to stay married: we had been forced into marriage by circumstances. When divorced, nothing but a formality, just like getting married, we would be friends, and we would both be responsible for the child. Later on we both would be too busy to 'fit in' a baby. How did he reconcile his Communism, which remained intact, with his being a consultant for some part of British industry – a job he knew he would be good at? He didn't reconcile it. During those years before he left Rhodesia, people were coming to him for advice on how to set up businesses, make money, invest – all that went on in one part of his mind, parallel to the thoughts of Comrade Gottfried. Suppose he had in fact stayed in England and become a business consultant? Would he have had a reverse conversion and become, like his parents, a good liberal? I doubt it. He would have been one of the people – and I have met many – who kept the Communist structure of ideas intact in one part of his mind, while living the life of a conventional citizen.

I got pregnant around about Christmas of 1945, after trying for about a month, to general applause – that is, of the Comrades. The first baby born into a circle of young people just married, soon to be married, hoping to be, is a lucky baby, embodying the hopes of them all. Gottfried was delighted. He said it would be nice to have a little chap crawling round and peeing in his nappies. Note that this baby was going to be a male. The women joked that they hoped Gottfried was going to take his share of changing the nappies. Feminism was not born in the 1960s.

My father said, 'Why leave two babies and then have another?' My mother was fiercely, miserably, accusing.

Now I think that our rational and intelligent discussions had nothing to do with it. Just as in 1939, when I got pregnant with John, and then, so soon, with Jean, I believe it was Mother Nature making up for the millions of the dead. Here was this healthy and fertile young woman, she would do nicely. Besides, I wanted another baby. I yearned for one.

From this period two memories, both getting heavier with significance as time goes on. One was when the Russians decided to hang, publicly, on a platform like a theatre, German officers who had committed war crimes. There were protests about this barbarity from everywhere. That is to say, after five years of war, crammed with every kind of atrocity, the world still could react to a public hanging. A return to the Middle Ages, to savagery, and so on. And, too – a note that had not been heard since 1941, for the Russians-as-savages had been in abeyance: 'The Russians with typical brutality and indifference to public opinion . . .' A similar event now would hardly be noticed. We have supped full of horrors. Who would protest? As I write unspeakable horrors are going on in what was Yugoslavia. Yes, we do note that they are. Are we shocked? – I think not.

The other incident: Simon Pines gave a lecture at the Left Club on the collapse of Hitler and National Socialism. At the end he remarked that quite soon Germany would again be our ally. There were two or three hundred people there and the uproar was such we joked it was lucky he wasn't lynched. Even more remarkably, he let drop, quite casually, that we were supposed to be afraid of Russia but he was a thousand times more afraid of America. Sometimes people who brood about politics develop a feeling for probabilities. It seemed a crackpot thing, to be hostile to America. Why did he say it? Where did it come from? It was an attitude which was to become even stronger on the Left, but that is not my theme, which is that in 1945 it was an astonishing thing to hear.

The Cold War was in the wings, ready to walk on stage.

Meantime, I wanted a break before becoming subject to a baby's needs twenty-four hours a day. It was not that I did not know what having a baby meant. I longed, I ached to be out of Salisbury, out in the big world, particularly as the RAF would soon be gone – in the event, it took much longer, because of the shortage of ships to take them home. I was certainly aware of the contradictions of my position, that was the trouble. I had chosen to have a baby, when for the first time in my life I would be free to do as I liked.

Now I wonder about a certain set of mind, particularly when I see it in young people: I never had a moment's doubt that I would be able to – not only survive – but cope competently with any circumstances I found myself in, or any obstacles I set up for myself. This expressed itself in a quiet obstinacy, even exaltation, even relish, when I contemplated possible futures. I do not think this is a simple thing at all. Why is it that one young woman or man is floored and sometimes for ever by what seems to be something quite minor, while another, without one moment's self-doubt, survives oceans of difficulties?

Gottfried approved of my going for a holiday. He was always a generous man. For instance, when he was earning only Howe-Ely's pitiful wage he bought me a new typewriter. He encouraged me to write, though he never liked anything I wrote. He contributed towards my going to Cape Town.

Again, the five days' train trip. I was with five other people, but I do not remember them, only watching the Karroo go past, the mountains of the Hex River Valley, trying to fix in my mind scene after scene of that barren solitary splendour which is so particularly South African. I watched for the little stations along the railway lines, where the gangers' wives and children came to the wire fences of the dry dusty little gardens to wave as the train shrieked past to the big city down south. I knew how those children, those women, those girls in their grown-up dresses and hair permed for love thought and felt, as the train disappeared. All through the war I listened to the farewell yells and wails of the train leaving Salisbury, taking people, out, out, *out* and away. For a generation, more, up and down Southern Africa the sad wails of a train meant loss, meant goodbyes, meant being left behind. I tell you, the roar of a jet is not at all the same thing.

The Cape Town docks and the ships in them no longer made one think of submarines and torpedoes and convoys, but freedom, escape. And I was three months' pregnant.

The *Guardian* newspaper had said of course their so successful representative from Salisbury could work there for a few weeks. The *Guardian* had been started just before the war by two efficient middle-class Englishwomen who had become Communist in the way people did in the 1930s. The formula for this Communist paper was the same as for the *Daily Worker* in London. The subscribers should be made to feel it was 'their' paper, and that they financed it. Every issue had a feature where the sums of money raised last week were announced, with exhortations, on behalf of the working class, to do better. The *Guardian*, like the *Daily Worker*, was perennially accused of 'Red Gold' – that is, taking money from Moscow. Everyone I ever met in the Party laughed heartily at the idea of Red Gold, just as they did at Reds Under the Beds. In fact there were Reds under beds and there was, too, Moscow Gold – only last week I read (again) how the *Daily Worker* was financed from Moscow. Whether the *Guardian* took Red Gold from Moscow I do not know. Given the characters of the two formidable ladies who ran it, I think it unlikely. The *Guardian* had flourished all through the war because of the popularity of the Soviet Union, but now sales were dwindling. For instance, no longer were 112 dozen copies sold in Salisbury, only a couple of dozen, to the faithful.

I was set to work in the subscription department, a room filled with desks where sat people, some volunteers, writing letters to backsliding subscribers, thus, 'We see you have not renewed your sub this month. We feel this must be an oversight. Remember, it is on people like you that the future of South Africa depends. With comradely greetings.'

There were Indians and Coloured people working in that room with the whites, an impossible thing in Southern Rhodesia. This was a lowly and unimportant layer of the newspaper. Far above us were the high-ups, the stars, though Carina Baldry, a founder, came every day to dictate to me admonitory letters. I was surprised that inside this socialist newspaper, considered so subversive, with its very existence at an angle to everything else in the country, was a strong hierarchy, everyone watchful of their privileges. In short, 'come the Revolution' – a phrase used with sardonic relish – that is, if it had ever in fact come, here was an essential bit of Communist machinery all ready to fit into a power structure.

The newspaper had lines out into the populations of the poor – not the blacks, at that time not so plentiful around Cape Town, but to the Coloureds and Indians. Several times I was taken out by car to some miserable suburb of Cape Town, to a factory where very poor Coloured girls, or Indians, made cheap goods. Here was the familiar smell of poverty, of deprivation, of desperation. The girls, in their cheap dresses and bright bits of jewellery, came crowding around to offer their coins for a copy of the *Guardian*, with its heady messages of freedom, but what they wanted was information: housing, benefits, grants. Was it true a shoe factory was starting in Woodstock? Was there work for them as maids or nannies? This was like selling the *Guardian* in Salisbury, where we reckoned to spend three or four times as long on 'welfare work' as on selling the paper, and where we had long ago understood that the newspaper was for them a kind of talisman for a better life. These *Guardian* salesmen and women offered advice. Yes, we will ring the doctor, yes, we will ring Housing, yes there is a TB clinic, but no, it is not within our competence to make your bosses raise wages.

Between the miserable levels of these lives, and those led by the Comrades, mostly white and middle-class, lay gulfs.

The Communist Party of South Africa was at that time legal. It had flourished during the war and was in 1946 enjoying a mood of excited confidence. There was an election when I was there, I think for town councillors, and the Party had put up candidates. There were noisy and ebullient meetings most evenings, and the Nationalists, and groups from the Nazi organizations, like the Ossewa Brandwag, came to heckle. There were always scuffles and fights. The atmosphere of elation, or hard recklessness, had in it a jocular self-parody, apparent in phrases like 'Come the Revolution' and 'See you on the barricades.' At that time I was too inexperienced to understand it, but have learned since that this is the atmosphere of 'revolutionary' groups inside a democracy who are not in immediate danger, and a good part of it comes from the pleasure of cocking a snook at authority. No revolutionary group inside a repressive or brutal regime can afford this elation, this intoxication. This is because they are not thumbing their noses at indulgent Daddy; they face torture and death. When the Communist Party of South Africa was banned three years later the atmosphere of privileged children being given their heads evaporated overnight, the people who were there for the ride disappeared, and those who stuck it out had a very hard time.

In the 1970s in the States, the 'revolutionary' groups were pervaded with the thumbing-the-nose, cocking-the-snook mentality. (I got to know what was going on, by chance.) For instance, one young woman, famed through the world as a dangerous terrorist, got her kicks by living almost opposite a police station in New York, 'You can't even recognize me although my face is on your posters.' Another, who looked like Patty Hearst, used to speed on motorways, so that she would be stopped by the police, who said 'You're Patty Hearst.' Oh what fun, what thrills.

In 1946 in Cape Town the Communists behaved as if there was no tomorrow. In their offices I saw a filing cabinet and in it a file marked the Communist Party of Southern Rhodesia, an entity already dead and buried. I told the ebullient Comrades that there was no Communist Party of Southern Rhodesia. They had been right all along: there was no basis for one, and it was a pity we hadn't listened to them. I also remarked that it was foolish to have the names of 'secret' members of the South African Communist Party in files anyone could see simply by opening a drawer. They joked it didn't make any difference because it went without saying there were informers among them. I was pointed out among the comrades as an example of paranoia: 'our little neighbour from up North' was displaying all the backwardness to be expected of a British colony. It was salutary to move from Salisbury where I was a big fish in a tiny pool, to Cape Town where I was a minnow among great and glittering fish – who, in London, would seem very small fry indeed.

It is a rule that perhaps we don't attend to well enough, that even a visiting minnow may have useful contributions to make to a situation, simply because not part of the communal mind-set. It is impossible not to be absorbed into the ways of thinking of a group, or a party.

I had not been in Cape Town more than a day or two when I knew I was being 'looked over', just as when I first arrived in Salisbury. There was a new girl in town. I was determined to have a love affair on this trip: I felt one was owed me. (By whom, or by what?) At a riotous election meeting I met an Afrikaner trade unionist, with whom I flirted. Later that same evening, presented himself a man who I saw at first glance was the perfect candidate for a love affair in every way different from anything possible in Southern Rhodesia.

And now, the dilemma familiar to all writers, which is, how much to say, what to leave out. The trouble is that children, grand-children, the writers of theses, the professors, would really like it best if their Writer was born at let's say the age of fifty, in a silk dress or nice suit, in the act of accepting a Literary Prize, with a gracious smile. 'A sixty-year-old smiling public man'. Or woman. Quite agreeable to have an eminent relative, but why can't they shut up about that disreputable youth? 'Oh God, what would they say, if their Catullus came their way.' The trouble is, just as much, that old Catullus quite often disowns young Catullus. How many times some old friend with whom I have shared this or that adventure has said, 'What are you talking about? I am sure I never . . .'

At that time, in Cape Town, everyone would have known who I meant after half a dozen words, 'An Afrikaans artist, tall, colourful, dressed like an artist, behaving like an artist, but mocking bohemia by being always slightly over the top in everything.' Now? Probably no one.

I put him in *In Pursuit of the English.*

He was a Communist, but then who wasn't? The love affair was only possible because of my naïvety. Having ditched my first opportunity, the trade unionist (reproachful, but all's fair in love) with a ruthlessness that now I marvel at, I went off with this artist, let's say René, in his car, to where we sat on the high slopes above Cape Town and watched cloud pour over an edge of hills like milk into a dish. We necked. With much pleasure and promise. He then deposited me in his studio in District Six, and said he had to go off on a business trip for four days. I gave notice to the *Guardian* and enjoyed the studio, which smelled wonderfully of paint and turpentine, and had his pictures and reproductions of masterpieces on the walls and, too, a collection of old plates that to me was a revelation of possibilities, like a treasure ship from Far Cathay. It was in an old, little, white house. There was a flavour of danger in going out and coming in. Gangs of what were then known as Skolly-boys frightened everybody in the neighbourhood. In fact, René was with another woman. He juggled girls, women, with a laughing and unscrupulous relish. He always had women after him. This was because he adored women. He loved everything about us, how we look, smell, feel, sound – are. Such men are irresistible (particularly where few and far between, such as in England), are immediately recognized by women, and treasured. (It is not possible to fake this love of women.) Nor should we expect from such men fidelity or other domestic or civic virtues. René at that stage of his life was full of nostalgic and tender regrets – my guess is for the first time in his life. He had successfully evaded marriage until now – he was forty, but had got a girl pregnant and had to marry her. He was attracted to me. He did not love me: such men are insulated against the passions. Sentiment – that's their food. He was cool enough to take me on a drive in a car where his soon-to-be wife, pregnant, like me, sat in the front seat, smiling. She was charming, funny, an Afrikaans country girl, the perfect wife for him. Did I see that this was one of his women? I did not. Quite soon, probably within a year or so (ripeness is all), it would have taken me no longer than it needed to settle myself in the back seat to know they were lovers and that she was pregnant. (How dare you! How can you be so insensitive! You are a monster!) But then, I wouldn't have had an affair with him.

The studio was not supposed to be lived in: or perhaps that was an excuse to conceal yet another girl, I don't know. When he came back from his trip with whoever it was, we were in a little white house with trees in the garden lent to him by a friend. For a few days, not long. (I say this because now they say I lived in this house for a year.) Lovemaking, again, was of the robust kind, but frequent and prolonged. We laughed all the time. He was appalled at my ideas about food and taught me several dishes, as hearty as his lovemaking, the pork stews, the lamb stews, the spiced minces and meat cakes of the Afrikaans cuisine, which owes a good deal to the Malays. I was putting on weight. He loved it. He loved me being pregnant. He would sit with his big hand on my swelling belly, as if listening to the infant. Then he rushed to the easel and made sketches. He identified with Renoir. Probably Renoir was like this, always intoxicated with the generosity of nature. From this period in his life must have resulted dozens of sketches of pregnant women, of his wife, of me. I like to think that my then Renoir-like body is on a wall somewhere. Not my face. In René's sketches faces tended to be a just-indicated cheek, or disappeared behind a fall of hair. While I was certainly smiling as he worked, enjoying his enjoyment, the smile would include ingredients he could not like. How he hated any suggestion that a woman might also be a detached observer. When not eating and making love, he drove off, me with him, on what he referred to as My Revolutionary Duty. The big blocks of flats where the Coloured community lived were his special responsibility, because, like them, he spoke Afrikaans, and because they loved him. He would set up a soap box – this was deliberate parody – climb on to it, holding a loudspeaker, and begin. He was a figure that had to draw attention – then. He was tall, bony, with long blond locks and loose coloured clothes. I adored these clothes – nothing like them could have been worn then in Southern Rhodesia. Or, for that matter, by most people in

South Africa. At once the windows and balconies of the Coloured flats filled with laughing and shouting people. I had to laugh too, not understanding one word of Afrikaans. He was like Till Eulenspiegel or the Baron Münchhausen, a magician from a world where it was natural to laugh, insult government and authority, where poverty was just a big joke. The Communist Party, the Comrades, relied on him to bring in votes.

Then he might drive me to visit friends, for instance Jack Cope, the writer, and his wife Lesley. He was tall, dark and handsome. She was slim and blonde and beautiful. Both were then Party members. They took soap boxes to Market Square and made revolutionary speeches. They were a wild success, this beautiful couple, particularly Lesley, who was like a princess. The Comrades got sardonic pleasure out of the improbability of this very English middle-class couple on their soapbox. So did the Copes. So did everyone.

They lived in a little flat at Seapoint, where from their windows they looked down at the sea sweeping in and out, not far from where I had been in that seedy hotel with the strings of coloured lights six years before, with John. I could not prevent myself sinking into the rich and morbid and pleasurable melancholy. I knew this meant I was demoralized.

René, as we drove away, might say, 'Hey, what's got into you? I don't like to see a girl of mine with a long face. What's got you down? You're thinking that we have to part? Wait, let's just get home and into bed and I'll make you happy again.'

René might love women but at any sign of cleverness he got restless. He liked Tigger making jokes, though. It was, let us say, at least interesting, to move from Gottfried, to whom it had never occurred that women are better stupid, to René, who would say, 'Ah, hell, man, but why did God give girls brains? It spoils everything, I tell you.'

I was not the only woman with brains he was afflicted with. He either just had had, or was currently having, an affair with the woman who would later be Gottfried's second wife. Ilse was a refugee from Germany, a Communist, but known throughout the Left generally as an energetic clever woman – and very brave. She would have to be brave to marry an Indian in that South Africa. She defied bureaucracy and the petty persecutions of the colour bar with a bravura that made people speak of her with admiration, and in that particular tone earned by people who cannot fit into the space allotted to them.

It was time, he said, to go for a fishing trip. Only later I saw he wanted to get out of Cape Town where everyone disapproved of him for having an affair when he was engaged to be married. Off we drove along that coast as beautiful as anything in the world, which in those days was mostly wild and empty of people, the sea crashing in and flinging up spouts of spray on rocks, on empty white beaches. Then we drove a little way inland, across fields, through vineyards, to a country store where the Coloured people who ran it looked me over, smiling. 'What is this new girl like?' they were thinking. René rented a tiny house or shack, one room only, about a hundred yards from the shore. In it was a vast old-fashioned bed, oil lamps, a table, two chairs. Outside, under a tree, a brick fireplace. The sea roared, crashed, groaned, battering all that coast, and the earth under our feet seemed to tremble. At once we set out through the low shrubs that smelled of salt to the shore, to get our meal for that evening. He took serious fishing tackle to a pool among sharp rocks where the water boiled and swirled, and burst in spray all around him as he waded and jumped to a rock in the middle where he stood with the sea racing around his knees. He was roaring, laughing, yelling with exultation as he cast in the line, again, again – until up leaped a slippery wriggling fish, which he flung wide and high off on to the rocks at the sea's edge. Then, soaked, smelling of sea, he embraced me with the same exultation, and ran with his arm about me up through the salt bushes to the little shack, where he gutted the dying fish, and had it on the bars of the grill within a minute of leaving the sea. Wine. Bread that tasted salty. Hanepoort grapes. Dried fruit crusted with sugar. We went to bed with the flame of the oil lamp turned low. Shadows and the sound of the sea filled the room. We made love and listened to the sea, made love and listened, our bodies as slippery as fishes, and then he slept and I lay awake, listening. The waves crashed and roared, each seemed to assault the land and then drag it out with the retreating wash, all night, all night . . . as if the little house was deep under the sea.

I have been tempted often enough to write about this paradise of physical pleasure, painting out the truth, which was that I was depressed. Not the 'real' depression, or the pleasurable melancholies of 'See how sorrowful I am.' I imagined a story where the individualities of the lovers dissolve into the sounds of the sea, wind creeping about in the bushes, lamplight on a lean bony back like a boy's, where a drift of tiny gold freckles on glistening white skin speaks direct to the heart, a truer story than that told by the ravaged face . . . a woman's brown silky knee, her young hand lying abandoned in sheets that smell of the smoke from the fire. I was *almost* able to sink myself into being found a treasure of pleasure – and narcissism does have its part to play, how much you learn only when you are old. Did there have to be a worm in this apple? After all, I did not care that René patronized me, for I would soon be gone – and that was why I was sick with misery, because I had to go back at all. *If* things had been different, I could have stayed here in Cape Town, and, probably, within months, post-war-time chaos or not, I could have got myself on to a boat and then off to London or Paris. It was certainly not that I yearned to stay with this man – who I did not yet know was about to get married. Nor did I feel that it was my destiny to write to erring subscribers for the *Guardian* newspaper.

The train journey, five days. Not down to the coast, to the sea, freedom, love, but back again away from what I was sure must be my real future, through the mountains, through the Karroo, slow, slow, stopping at every little siding, wailing past the gangers' cottages and the waving women and children, the wheels thundering, I'm going back, I am going back – while the train filled with dust and I lay on a bunk as unhappy as I have ever been in my life. I was confronting that emotional or psychological double helix coiled at the roots of my nature. It was not a question of 'How stupid to be pregnant again when I didn't have to be.' The trouble was, I had to be. My early childhood said so. Not a question of, 'If I had not married Gottfried Lessing out of a revolutionary romanticism and recklessness I now think stupid, I would not now be travelling back and away from everything I want to travel towards.' What is the use of ever saying, I should have done this, should have done that? The point was, nothing else could have happened given my nature and circumstances. And I was quite able, thanks to Kurt and others equipped with the vocabularies of sophistication, to apply to myself the suitable labels and even epithets. And there was another thing that made it inevitable I should be on that train travelling the wrong way: my father's illness, his long, slow, dying.

It is the strangest thing that one may be the picture of physical well-being, while inwardly unhappy. When I got home I was five months' pregnant, brown and vibrant with good health. Gottfried was pleased I had had such a good time, but disconcerted, for I had left him my usual shape and now . . . Gottfried could not have been more kind about my being pregnant, but he was not a man to enjoy the process. He had been having an affair with one of the marriageable girls. Since she is now a pillar of society, better say no more.

Gottfried and I resumed our unhappy but kindly marriage.

There is not much in this history that I am proud of, but Gottfried and I, born to disconcert and shock each other a dozen times a day, behaved well, we really did.

A scene: in one of the what seem now innumerable little bedrooms we inhabited is a chest of drawers where ten pairs of perfectly rolled socks, graduated by colour, lie side by side. In the drawer next to it are exactly ordered pants and vests. In the long drawer are three stacks, not a quarter of an inch out of place, of laundered shirts, plain, coloured, striped. In the wardrobe hang his white, cream and coffee-coloured linen and cotton suits, not a crease or a grain of dust to be seen.

I am looking at this perfection with incredulity and despair. Gottfried is looking at pulled-out drawers where spill stockings, bras, pants, jerseys, a salad of colour. In my wardrobe are crammed dresses and slacks. At the sight of his face remorse engulfs me. 'Oh Gottfried,' I cry, 'I am so sorry, I really will try.' And I impulsively embrace him. He stands stiffly in my arms. 'I am glad to hear it,' he says, cold. 'But I take the liberty of doubting it.' He is angry. Much worse, he is miserable, he is discouraged. 'For God's sake, Gottfried, it's only clothes.' 'That is not my opinion,' he says, turning away.

And then he is in his bed, and I am in mine. We both stare into the dark. He is about to say something, and I brace myself. But then he says, slow, judicious, humorous, 'This kind of incompatibility is more of a misfortune than a crime.'

I am quick to laugh, from relief. He laughs, because I have.

We smoke. The smoke that drifts around the room is visible in the shafts of light that fall here from the upstairs flat opposite. Music too, dance music from the flat across the court, dance music and cigarette smoke, never one without the other. The flat across the court always swirls out music, day and night, even at three. four in the morning. Good times . . .

Some day he'll come along
The man I love
And when he comes my way
I'll do my best to make him stay . . .

The voices of a mother and daughter who had danced and partied their way through the war, and assorted RAF reverberate around our room.

Somebody loves me, I wonder who –
I wonder who she can be.
Somebody loves me, I wish I knew . . .

'Well,' drawls Gottfried, 'they say it is love that makes the world go around.'

'You wouldn't think so, from the evidence,' I say.

Gottfried clears his throat, in the way that means he is thinking of making a joke. 'That is a negative thought, comrade.'

'Oh dear, oh dear, I'm sorry.'

He lights another cigarette. So do I.

From across the court they are singing.

They asked me how I knew,
My true love was true . . .

We hum along with the crowd opposite, good times by proxy.

I of course replied,
Something here inside,
Cannot be denied.
Now laughing friends deride
tears I cannot hide
I just smile and say
When a lovely flame dies
Smoke gets in my eyes.

'It cannot be denied,' drawls Gottfried, 'that at least the last line is true.'

18

WHEN I CAME BACK from Cape Town I was thinking, It won't be long, then I'll be off . . . yet we already knew that nothing was going to happen quickly. In fact I was in one of those periods in a life where nothing can move. It is a log-jam, a marsh, a quicksand, you have weights on your feet. It was certainly not the last time I have been held fast by circumstances. There is nothing to do but sit it out. During the war we had joked, 'Once there was a hundred years war . . .' There wasn't much joking in that time after the war. If war creates out of itself a breeze of general elation, then post-war is flat greyness and depression. How was it possible that these terrible things could happen at all, is what people are secretly thinking, and in rather the same way as when, after a long bad personal experience, you simply want to go to sleep. If I said, It was only three and a half years, what's that? – it would be dishonest. We didn't know it was going to be three and a half years. We did know that all the offices dealing with passports, visas, naturalization, repatriation, were stuffed floor to ceiling with appeals and applications. Gottfried as a lawyer knew people in the relevant departments who told him how things were moving – slowly. Besides, Howe-Ely's, like every lawyer's office, was busy with former refugees wanting to become British, or trying to find out what had happened to their families. That was when the news of what had happened in Germany's concentration camps became real. At first it was hard to assimilate, to 'take it in'. Now, when we hear that yet another dictator is exterminating thousands, hundreds of thousands, or millions of people, we have heard it all before. Hitler, Stalin, Mao, Pol Pot, Khomeini, Saddam Hussein, there seems no end to them. Osip Mandelstam said it, for us all,

My animal, my age, who will ever be able
to look into your eyes?

* * *

Cruel and feeble you'll look back
with the smile of a half-wit:
an animal that could run once,
staring at his own tracks.

During those years I worked for Mr Lamb and earned pretty good money. I enjoyed it – politics by the back door. No one brought up as I was, with the BBC News as a high point in a day, and continual talk on verandahs about the government, can ever really shed politics. Now I can say to myself, Suppose you had never, ever, read a newspaper or listened to the News, never had anything to do with politics, what difference would that have made to you or to the world? It is no good, I find politics forever fascinating. Besides, Mr Lamb was an old man and loved telling and re-telling stories from his past as one of South Africa's ambitious young men. Personalities . . . the clash and the colour . . . the intrigues, the battles between Capital and Labour – this was the history I knew from Jack Allen and Mrs Maasdorp, but from the other political perspective. 'What an old reactionary,' they admonished me, when I told them what I was hearing during my afternoons of typing. 'What capitalist lies.' And Mr Lamb was saying, 'My dear, one has to remember, that as Terence said, "So many men, so many opinions . . ." and now, would you set the machine to single-spacing for the next bit, it's Max Danziger and his plans to ruin the country with his next Budget. He is probably the most conceited man in Southern Africa. When he is speaking I always hear him saying, "O happy Rome, born when I was Consul." I am sure you are familiar with Cicero? Shall we begin? "I shall introduce my Budget with a quotation from Francis Bacon, 'He that will not apply new remedies must expect new evils: for time is the greatest innovator'." Well, if he's going to quote Francis Bacon – no, no, don't type this my dear – if it's Bacon, then Danziger might do better to remember that he said that the remedy may be worse than the disease . . . Shall we start again? "I shall introduce my Budget . . ."'

'When they start quoting Latin,' says Jack Allen, 'then you know they're trying to get away with something.'

In the mornings I was learning to write. This was when I wrote and re-wrote *The Grass is Singing*, short stories, and, always poems. I see my verse-writing now as the equivalent of throwing quieteners out from the back of the snow-travelling sledge to the wolves of melancholy.

Some short stories were published in a magazine in Johannesburg called *The Democrat*, others in *Trek*. Mostly I wrote and tore up, wrote and tore up.

In October of 1946, exactly to the day, as with the other two babies, I went to the Lady Chancellor Nursing Home. I had no expectations, since I had been wrong twice before, but was in my usual state of pleasurable excitement, invigorated with the need to paint the entire flat or walk twenty miles. This time I did know all this energy announced imminent birth.

The Lady Chancellor, as usual, was too busy to have time for women not actually in the labour ward. I was grateful for this. I was in the room where I had been with John. Alone. It was morning and the yells of hungry babies assaulted my backbone which is where, if you are on that wavelength, you hear babies crying. I was listening for that first unmistakable pain also felt in the backbone, meaning that real labour is beginning. The pains were mild. I wandered about the room, till a nurse popped her head around the door to enquire, Did I want a bath yet, and if so, would I just get on with it. I was happy to do so, said yes, and was in a hot bath for over an hour. No pains at all. Back in the room I sat in a chair and even slept a little. I woke and admonished myself that this was no way to go on, and at once there was a useful pain. Wait a minute, I thought, wait, what is this? In short, I found I was controlling the pains. When I was tired I slumped with slack muscles in the chair. Then, recovered, I stood up, and walked about, saying, Now I shall have a pain . . . and did. I have never read of this in any book, and if I had had a nurse there distracting my attention would not have discovered it. Is that what that black ward maid had meant, when she told that fat, slack, slummocky woman having her third child (who I now resembled, and did not care), 'Now you are a real woman'? There was no similarity between me, now, confident and in control, and that tense pain-racked girl of my first labour. All the time I was wondering, when are the real pains going to begin? With the second child I had waited for the 'real' pains but they didn't begin, not until the very end. All this went on for some time, all morning, and from time to time a nurse rushed in to say, 'Would you like a cup of tea?' or, 'It's the Christmas rush,' or, 'Do hang on to it a bit, we'll have a bed empty soon.' Not until two o'clock, without any serious pains at all until then, did the cold pain slice my backbone, and I put my finger on the bell. I was already shouting for chloroform though I had said this time I wouldn't have it. Present was the Matron and, soon, Dr Rosen who . . . and in no time I was awake again and there was the matron, announcing that it was a fine boy. Again, I was taking it for granted that whether girl, or boy, it would be healthy, strong, and whole. That the Matron disapproved of me so openly made it easy for me to demand to see the afterbirth, when it plopped out. She was shocked. I appealed to Dr Rosen, who joked that after all I had made it. The afterbirth was held, by her, in a kidney pan, for about five seconds a yard or so from my face, accusation and disgust on hers. Long enough to see it was like raw liver. I then demanded to hold the baby and knew she would say it would be time enough when I was in bed and 'tidied up'. This time I was not weeping with rage and frustration, because I had an ally, a young nurse from England, who was not a Truby King product. She brought the baby and was close beside me, protecting me when the Matron came in and stood waiting for me to hand the baby over to her.

No woman who has had more than one baby can subscribe to the doctrine that character is made and not born. When you take an infant in your arms the first time, you are holding what the human being is, its real nature, and whatever else is later done to it, that is the bedrock, the basis, the foundation. This babe was unlike brave and battling John, unlike sweet and confiding Jean, he was a sleepy but amiable and interested infant. I managed to see more of him than the other two. He also saw more of his father. Gottfried thought the colonial custom of getting drunk with the boys barbarous, and came often and brought anyone among our friends who happened to be around. While with the other two I felt as if I were entertaining Frank's friends, now these were my friends too, and every visiting hour was a party. Gottfried simply ordered the Matron to let the nurse bring the baby in. She did as she was told. Gottfried said he knew how to deal with a bullying woman. So Peter spent a part of every day being dandled by a variety of people, beginning as he would go on at home. Not only was this the group's first baby, but a baby born soon after a war brings a feeling of hope and of resurrection. I was out of that nursing home six days earlier than the other two times. Gottfried simply informed Dr Rosen that I was going home. With a sweet, smiling, bitterness, the Matron watched me, or rather Gottfried, go off to the car.

The large room that was so used to hosting so many people easily accommodated a baby, who had a cot in the corridor-space. But really, he spent his time with us. This being the third, the layette was minimal and sensible. A hot climate demands no more than nappies, a couple of dozen – which dried in a couple of hours on the clothes line outside the flats – vests and jackets. My mother was distressed, believing that this was parsimony and evidence of bad feeling towards the baby. Long before he was born his teddy bear was waiting for him in his cot. And for whom was this bear bought? 'But what's the point?' cried my mother. 'He won't want it for a couple of years yet.'

This time, having a new baby was easygoing and pleasant, for I was holding Dr Spock, blessed as a deliverer by all us women then, as a shield against the visiting nurse. She disapproved of me as much as I did her, but it was ideological. She was a rather nice Scotswoman. She did not deny that the baby throve, that the scales she never moved a yard without were in my favour and not hers. 'But you'll ruin him,' she cried when I said the baby was fed on demand and there was no timetable. 'His character? Have you thought of that!' To which I replied that my character had been formed by Dr Truby King, and surely she couldn't regard that as a recommendation? Soon she came no more. Not only I and Gottfried enjoyed the baby, it seemed everyone I knew made excuses to drop in to watch him have his bath or play with him. From the camps came our RAF friends, who had been deprived of any ordinary family life for three or four years, and knew they would have to wait for places on the crowded ships, and it might be months, or years, before they got home. Very young men who normally would not have time for an infant competed to hold him, to fill the bath half full of water and swim the babe up and down, a steady hand under his head, while he kicked and crowed.

I was in love with this baby. The mists cleared from my eyes when I sent a photograph to a friend with a letter demanding, Wasn't this the most beautiful baby she had ever seen? She sent the photograph back, with the suggestion I should actually look at it, demanding, 'He's just like every other baby. Have you gone completely crazy?' Well, yes, but women do, and it was only for a couple of months.

Most afternoons I drove out to my father, with the baby. He propped himself on his elbow, with difficulty, took the baby's hand and examined it as only people can who are matching new and shining flesh against Death. 'It's my hand,' he announced, while the small fingers tightened around a bony digit. 'It's mine. Isn't it?' he would enquire, uncertain now, while he peered at me under his white eyebrows. I knew what he was asking, for his question was not really about mortality – continuance – inheritance – death, but about fatality, the secret grinding of the wheels. 'There's absolutely nothing to be done about it,' I might have said, to this old sick man. 'Don't you see?' But who was it who had taught me to see, long ago, and probably even before I was born? In my mind, far too often, was that long, long, five-day trip from Cape Town to Salisbury, in the dusty coupé, while the train wheels drummed, That's how it is, that's how it is, that's how it is . . .

Or, he might demand, 'What did you do it for?', leaning forward to stare into my face. Really he might never have seen it before. Dying people often do see things they have never seen before. The intent, close, intelligent gaze of a dying old man or old woman is focused on your face in a question, but what is the question? Perhaps it is, 'Why is it I have never in my life actually seen your face before? Why have I never given myself time to actually look, to *see* anything properly?' And then, with a sigh, he collapsed back against the pillows, and let his head slide sideways and lay looking at the vigorous little creature, kicking there beside him. Just as if he had never seen a baby before. Sometimes he was too drugged to wake properly, or he woke and soon fell back into sleep. But it seemed he knew I was there, for if I began to gather myself and the baby so we might steal away, I saw his dark eyes burning out at me under the shelf of his white brows, and he gestured, Stay. And I sat for an hour, or two or three, until it was time to take the baby home to bed. And when I got back to the flat, there might be half a dozen people there waiting for me, for the baby.

About then the Cold War began. Suddenly. From one week to the next, we became pariahs. Certainly from one month to the next. A salutary experience. For several years we might have been Reds, kaffir-lovers and so on, but we were quite popular, because of Uncle Joe, because of our gallant ally. One week, walking down a street, I might be stopped several times – 'Hey, hang on, what's the hurry?' – from people wanting a chat, but more often, for me to telephone an official, organize an interview, write an article, right some wrong – in short, social work. In a small town, people may be as revolutionary as they like, but the fact they are continually involved in one way or the other with the processes of government or administration means they have influence and even friendship with people nominally enemies. 'She's a real right-winger, but her heart's in the right place. Ask her to . . .' 'He's a Fascist, but you can always appeal to him on a question of principle.' And now suddenly old friends and acquaintances crossed to the other side of the street when they saw us coming. That was when I learned, right at its beginning, what the Cold War meant, at a grass-roots or pavement level. Later, when I met Americans who had suffered under McCarthy, it was the same story. This does not have only political relevance. From time to time in Britain there is a scandal, and the culprit or victim tells the newspapers, 'I had hundreds of friends, they didn't mind drinking my champagne and coming to my parties, then this happened and I found I only had two real friends.' No, I don't want to push this too far, because, since we were a group, we could not be isolated as individuals. Increasingly we were, as a group. The experience was not without its benefits. People who lived in Communist countries where every fourth or sixth person was an informer or a spy developed ways of judging people unknown to those living in democracies. You learn to know who will stick by you, in serious trouble. This is not a bitter knowledge: though it may be to begin with. When I was a prohibited immigrant, for decades, in Southern Rhodesia, no white person had a good word to say for me. My name was mud. Time passes, and the same people who said all those things write letters, greet me with smiles, invite me to give lectures and offer eternal friendship. It is the way of the world. And we all do something of the kind.

Now a little speculation . . . what would have happened if there had not been a Cold War? I think this would. Within a few weeks of the start of the Cold War all the 'progressive' organizations collapsed, first of course, Medical Aid for Russia and the Friends of the Soviet Union. No loss, either of these. But Race Relations went too, which supplied information – facts, figures, ideas – to people who got them from no other source. Of all our organizations this one was the most attacked, threatened, continually mentioned in the newspapers as dangerous. The good citizens of Southern Rhodesia knew that all ideas to do with improving the lot of 'the munts' were Communist. Now that Communism, or, rather, the Soviet Union, had conveniently become an enemy, it was easy to shut a door on any dawning notions of progress. There is no way now, in the last decade of the century, to convey – convincingly – the stupidity, the idiocy, of the average white person's ideas about the blacks. Anything I may say must sound exaggerated. Is it enough to say that Mr Charles Olly, who would adorn every public speech with remarks like, They are only baboons with small brains, they are an inferior race, they are just down from the trees – was Mayor of Salisbury? If I were to say that as late as the 1970s a man known for his liberal views on race questions made his servant cycle in seven miles every morning to make him his early morning cup of tea, at six, and that this kind of thing was taken for granted? The point is, the meetings, lectures, pamphlets of Race Relations, which would now strike everyone, black or white, as wretchedly feeble, were found incendiary then.

The Left Club, despite its name, had lectures on all kinds of subjects: and some were far from 'left', for there were not enough people with Left views to maintain a weekly lecture. A hundred, two hundred people came to these lectures. These people, who regularly attended, would have been a leaven with enough influence to change thinking, but instead they lost interest in ideas, became reactionary, or bitter and disillusioned. Soon the University would come into being. It has always been a narrow-minded and provincial place, where the few genuine liberally minded people have had to fight every kind of sectarian politics – at the time of writing, Political Correctness. If . . . but 'ifs' are not worth anything when not authenticated by events. But a stratum of informed and open-minded people might, perhaps, have prevented that war, that stupid, unnecessary, bitter and infinitely damaging war that went on for ten years, and resulted in a Communist government, too extreme for the natural temper and style of the black people. That Cold War froze very much more than attitudes towards Communism and the Soviet Union, just as the springing into existence in 1942 of a 'Communist' group – it really does have to go into inverted commas – caused all kinds of useful yeasts and ferments to start working.

Because of the increasingly poisonous atmosphere of the Cold War even more people found refuge in our flat. Not all had called themselves Communists, far from it: they were people with 'progressive' ideas. I put that into inverted commas because a good many of the ideas then called progressive turned out to have ambiguous results. From now until he left the Colony in 1949 Gottfried spent even more time with exactly the people he would have been friendly with had he in fact become a business consultant in a democratic country – the lawyers and civil servants he worked with and who admired what he had done for Howe-Ely's – the firm, and for the old man. His close friend remained Hans Sen, the Roman Catholic, who might come and share the bohemian goings-on of our evenings. He was an ugly man, and he hated women – or said he did. All of us women treated him with affection, not to say tenderness, like a child who doesn't know why he is having a tantrum. When he went with us on trips to Macheke, he would stand in the middle of an empty space of vlei, and announce that he would build a tower there with no ingress, but he would let down baskets and we could send up books and wine and food. Wouldn't he get lonely? – we teased him, and he conceded he might let one of us – that is, women – in to clean up, me, or Gottfried's marriageable girl or in fact any woman who happened to be there. He said that being representative of the Red Cross and knowing what went on everywhere was enough to make anyone a humanity-hater, and we, the human race, did not deserve to live. He would look at the small baby, the centre of attention, and frown and say that if he had known what this world was like he wouldn't have agreed to be born. Athen Gouliamis, once a poor boy on the streets of Athens, with many brothers and sisters, would sit on the bed beside the baby, and gently stroke the energetic arms and legs. That little olive-skinned man, with his burning black eyes, stern, fierce, could not prevent himself laughing and smiling as he played with the baby. The group of Communist Greeks would come into town especially to be with him for an hour or two. 'It keeps us sane, do you understand?' Athen might tell us. For they all took it absolutely for granted that they would be dead inside a year or two. The war might have ended in Europe and the East but they were all waiting to begin their war, against what they always called the Fascist dogs.

Simon Pines' relationship with the baby was admonitory and

instructional. He stood by cot, pram, or the bed on which the infant lay, regarded it – like Athen – with the eye of a man brought up where children competed to survive at all, and then lectured me, or Gottfried, on how to equip him for the battle of the world. Simon did not reach Palestine to help the making of Israel. This large, strong, stout man, who boasted he had never ever been ill, got malaria, and was so shocked and frightened that he turned his face to the wall, like a black person who has been cursed by a shaman – and died. We could not believe it. Sometimes I still can't.

I was fighting hard for time to write. Not against the baby, who was amiable and easy, but against baby-lovers and Kurt, and others who needed an attentive ear. It had not occurred to me then that people with experience of unhappiness attract their like. Years later, a line in John Osborne's play, *Epitaph for George Dillon*, enlightened me: *She is an emotional soup-kitchen*. Since then I have certainly done more than my share of listening, but I do not deceive myself: I know that this victim who apparently depends on you for understanding will – if you withhold it – simply go off and find someone else.

Very few people – perhaps one in fifty? – respect women's privacy. If you say, 'I spend my mornings writing' that will not prevent the furtive knock on the door, and then a moment later, the guilty, embarrassed, smiling face appearing around the edge of the door. 'I've just dropped in for a *second*.' The trouble is, a good part of the intruded-upon one conspires with the intruder . . . particularly if you are a novelist. The young woman who came most often was a compulsive talker. Once Marie started she talked, and talked, certainly not to me, but to some auditor invisible to me and perhaps to her, her eyes fixed in a stare at nothing. She was a small woman, or girl, with thin white freckled arms and legs. Her hair was dank and fine and dark. Her eyes were like prunes in a small pale lightly freckled face. She worked in one of the factories that began with the war: secondary industry benefited from war-time restrictions, just as it did later with sanctions during the Unilateral Declaration of Independence under Smith. She worked long hours for bad pay. She had been thrown out, or had run away from a home where her father had sex with her – had since she was a child, like her brothers. I had never met anyone like this, though it was generally known that in the rural districts of South Africa incest went on among the poor Afrikaner families. I listened to her with that exhilaration that comes from improbability, the sharp clash of different substances – she talked about what surely most people would say were appalling experiences as if saying her father smacked her and her brothers stole her hair-ribbons. When she said that her mother did not stand up for her, but supported her husband and sons, she might have been saying it was a pity her mother was a bad housekeeper. She now had a baby of her own, and a man who beat her, but she wasn't going to marry him for she wanted a man who would be good to her, after all her troubles. For one hour, two hours, three hours, she talked, eyes fixed on space. Then, she got briskly to her feet, smoothed down her cheap flowered dress, old-fashioned, which might have been worn by her Boer granny, and said, 'Thanks ever so, Mrs Lessing, and thanks for the tea,' and wandered out. In those days there was a periodical called *True Love Stories*, which had less love than gloomy melodrama, just on the edge of pornography, and when love did triumph it was only after murder, rape, threat, imprisonment, larceny and blackmail. I imagined myself writing up this woman's story and submitting it: 'Dear Mrs Lessing, Thank you for your interesting little contribution, but we do feel you have gone over the limits of what our readers would accept as probable.'

Sometimes I left Kurt and Marie together, and from the passage outside where I fed the baby or rewrote *The Grass is Singing* heard the two streams of talk running on, one about the intellectual and emotional dilemmas of his commune in Vienna, one about incest in the Orange Free State. Neither heard one word the other said, and it could not last, for both needed a listener.

My brother came home from his war. He was even more deaf, though he had had an operation done by a world-famous specialist. My mother was deaf. My father was deaf. It was a house where the family shouted at each other, among my mother's non-deaf guests. Harry was slow, smiling, and seemed to be behind some glass wall. He was in a state of war-shock but did not know it, or at least did not talk about it until decades later. Out of his glamorous naval uniform he was a good-looking, polite, courteous man. My mother now had two grown-up children armoured with politeness, helping her with her dying husband, sitting with him, sharing night vigils, but never giving her what she needed. Harry and I hardly talked at all: at no other time in our lives did we have less in common. He did not like Gottfried. He drove out to the farm and returned to report the house was a heap of rotting grass among borer-eaten rafters and bits of linoleum. He did not seem upset by this. Soon a bush fire swept across the hill and nothing was left.

Harry and I met most afternoons on either side of my father's sickbed. We urged mother to go for a drive, go visiting, go anywhere, to get a rest from the drudgery of it. Or he drove her off to the park and I stayed.

My father kept saying, 'Why don't they put me out of my misery?' muttering it angrily, clutching my hand, or roughly and bitterly stroking the baby. He said it when Harry was there, too. Harry, as it now would be put, was not a man in touch with his emotions – at least, he was not then, though when we got to know each other, later, both of us elderly, he was very different. When my father demanded to be given a lethal dose of something, Harry put a polite enquiry to me and to our mother, 'What do you think? Is that what he would wish?' This phrase, what he, she, would wish, did wish, might wish, would have wished, appears at a certain point in the death process, to be seen as an ultimate in hypocrisy, or, perhaps, an admirable anaesthetizing of grief by a deliberate use of banality. My mother was distressed, and, too, angry. She had had to think about it, forced into by his demands, 'You'd do it for a dog.' But she knew very well it was much more complicated than it seemed. Sometimes when dying people say it is all too much, give me a dose, they mean exactly what they say, and sometimes they are saying, This is all unbearable, and you don't understand what I am suffering – a demand that these crudely healthy and alive people around the bed should really, but really, share in what they are experiencing. We might see our father as a parody or travesty of his real self, a sick and querulous old man, instead of the vigorous one we remembered, but he was all the time inside there, unchanged, and he did not identify with that rotting body. When he said, 'Why don't you put me out of my misery?' he was saying, really, 'Why am I saddled with this body – it isn't me at all?' Or I think he was. And we would hear painful dialogues between my mother and him. She had a simple, not to say practical view of the afterlife. 'Don't you see? We'll meet over there, it will be beautiful and we'll just go on from here.' 'I don't want to go on from here,' he might say. 'What do I want to go on from here for? Am I going to be saddled with *this*?' Meaning his illness, or illnesses, his swollen pulpy white legs, which he looked at with horror, his swollen white stomach. 'No, no, Michael, don't you see? We'll have new bodies, it says so in the Bible.'

'Well, it seems a pretty funny business to me.'

We wanted him dead because of the strain of it, and because we thought how terrible it must be for him to be in this state, but really it was terrible for us to think of him knowing he was in this state.

One morning a man appeared at the flat where I was bathing the baby to say my father was dying in the hospital, and if I wanted to see him, I should come at once. I did not go. Partly, I did not believe it, because for years I had been summoned peremptorily by my mother, or by her sense of the dramatic, to an imminent death-bed. But I did not want to be there when he died. I sat bathing the baby, and was full of emotion like a howl, or a scream. I wanted to kill someone or something, but who? I could have torn out my hair with both fists or raked my cheeks with my nails. A general ward of the Salisbury Hospital would hardly have put up with this kind of thing. I went on bathing the baby.

My father had been afraid all his life of being buried alive, and had made my mother promise to cut both wrists in case he came to life again deep under the earth. When I did see him those thin and bloodless wrists had pale wounds across them. He was not at all as if 'asleep' or 'dreaming' or any of the other lies people use. He was, quite simply, not there at all. He had gone. I've now seen quite a few people die, and then dead, and all of them have taken themselves off.

We buried him. I sat with my mother in the car driving out to the cemetery and we talked about insurance and wills. I thought it was all quite horrible and tried to put my arms around my mother and I said, 'Poor Mother.' She shook me off, with a look that rejected me. I was being false and she was rejecting the falseness. We went on talking about the insurance policies, like a scene from Balzac, or from Samuel Butler.

I was so angry . . . oh, I was so angry. I could not see what this funeral service had to do with my father's death, or with my father, and I knew what he would have thought of it. When I saw the Death Certificate I wanted to scratch out what they had put for Cause of Death, heart failure I think it was, and put instead, the First World War. Long decades later I say to myself, this kind of anger is immaturity, and it's time you grew out of it. But any music from the First World War, or scenes in a film, or those old photographs or shots from the Trenches that we see again . . . again – and up springs the anger as fresh as it ever was. *But whose anger is it?*

My mother was lonely. Coming into town, at last 'getting off the farm', she had believed that her social self would be given room to enjoy itself. She had been in cold storage for twenty years on the farm. Nursing her husband had given her little time for a social life. She knew that our flat, every evening, was full of people. She might say wistfully she had heard that we knew interesting people. For the purposes of the daydream she chose to forget they were likely to be Reds, Communists, kaffir-lovers. I lived in dread of her meeting them. More than once she arrived, and her face lit up with pleasure as she came into the room and saw so many faces. She sat drinking a cup of tea, and slowly her usual look of brave disappointment came back. I used to wonder, Who, *who?* – could she be asked to meet she might like, or rather, approve of? There was Esther, middle-class, English. But Esther was married to Kurt. There were the Loveridges, English, middle-class, teachers. But it was not possible to ask the Loveridges to come by themselves: this kind of formality was not our way. Besides, anyone might drop in. Suppose Charles Mzingele did? He was certainly the most interesting person we knew.

Quite soon, my mother arrived one morning to announce calamity. She was pale, she was distraught. She said my brother was going to get married. 'But what's wrong? What's the matter? Would you like a cup of tea?'

'She's completely unsuitable. It's a tragedy.'

The word unsuitable should have alerted me, but as usual we went on in a muddle of misunderstanding. The girl was Monica Allan, and she had been going round and about with Harry for some months. She was beautiful. She was kind. She was clever. She had a rich father. She was also the female champion swimmer for Mashonaland, surely a good mark?

I remember my unease: I really did think that my father's long, long dying had unhinged her. The trouble was, I had spent years now in the company of people of all classes, and even colours, if not many of these, and from many parts of the world. Most of us believed that quite soon there would be an end to class, race prejudice and other discreditable emotions. I had become lost in a never-never land of beautiful thoughts and had simply forgotten about the real world. What *could* be wrong with Monica? Well, she was not English middle-class. Her father was one of the best farmers in the country, his farm was visited by people from all over Southern Africa – but he was not middle-class. And he was Scotch.

My mother actually advanced on me, beside herself, infinitely distressed, hands out in appeal. 'You've got to do something. You've got to stop it. He won't listen to me, he never does, no one ever does.'

If she was shocked, then so was I. It must have been years since I had behaved with anything other than a cool politeness, but suddenly I was shouting, 'Leave them alone. Don't interfere. Don't spoil things for them.' She fell back, stammering. 'But what do you mean . . . can't you see . . . ?' 'Leave them alone,' I ordered, doing the right thing in the wrong way. Those psychological doctors who have claimed I should have 'stood up to' my mother were proved wrong by this scene. She turned and blundered out, vague, not seeing where she was going. She stood under the tree in the garden – we had moved again – sent towards me long, puzzled, hurt – but above all, uncomprehending – looks, and then went to her car where she sat, limp, for a long time.

When I did see, what I saw was intolerable, almost worse than my father dying as he did. It had not occurred to me that she still dreamed of going back to what she was before she found herself on the farm, where she had slowly been taken apart by life. The farm had been called 'Kermanshah' after her best time. The nasty little bungalow, so hated by her and my father, which was what they could afford from the sale of the farm, was called 'Kermanshah' too. But these shadow 'Kermanshahs' were only changeling houses, temporary homes, and then real life would begin again. I, her so painful daughter, had married this cold Prussian, but her son would marry a nice English girl and then . . . perhaps there was a girl he had met while he was in hospital, just like herself in the Royal Free, and then she, Emily Maude McVeagh, would find herself . . . (She had been christened *Emily* Maude, but dropped the Emily.) And yet she knew, must have known, that her son dreamed of one thing, one thing only, how to get back to his right and true place, which was wandering about the bush, the veld, in old khaki shorts and shirt, for he didn't care about success any more than her husband had.

That was a cruel scene, and I did not really see how cruel. And I did enable Harry to marry Monica without scenes and reproaches. Not that he would not have married her in any case, for his way of dealing with opposition was still – was always – simply not to notice there was any. There was a wedding, and everyone but my mother was pleased. She said to me, full of a sorrow that fed truth into the old jingle, 'A son is your son till he gets him a wife.'

'But,' I said, exasperated, 'what did you expect to happen?'

And now to the improbable juxtapositions of my life were added visits to the Allans, on Sunday afternoons, with the baby. We might drive out with a full car, but never with anyone who might upset Mamie Allan, who was not one to put up – for instance – with Kurt. I found her hard to get along with. She could scarcely approve of me, and as for Gottfried . . . The war put together many people who never would otherwise meet, but perhaps those afternoons on the Allans' farm represented a quintessence of improbability. I look back and see Gottfried in his always perfect clothes, his smooth lacquered cap of black hair, his face that seemed to demand a monocle, sitting smoking from an amber holder, answering Mamie Allan's brisk disapproving questions as she handed him cups of tea or a scone neatly set out on a little doily, while she sat upright in her little 'tailored' dress, her hair in a coiffure like a newly brushed terrier's.

'And you were brought up in Germany?'

'But Mrs Allan, you see, I am a German.'

'Then why weren't you fighting for your country?'

'For one thing, not all of us liked Hitler.'

'So you were a conscientious objector? We used to put them in prison, and quite right too.'

'I think there are different opinions on that subject.'

'No, I don't agree, not for right-minded people.'

With David Allan I had the same kind of relationship as with the other 'old men'. But I was growing up and no longer thought of them like that. He enjoyed talking to me about religion. I was still a militant atheist, and had all the arguments, which I now think jejune, lined up in my brain like toy soldiers. He was Low Church, and undeviating from the path of godliness. It was he who had an extremely valuable bull, just flown in by air, shot because it killed a careless black minder. On hearing of this death sentence people drove out to beg for the animal's life: they had arrived by the carload at weekends simply to see the beast, in the animal kingdom the equivalent of the Taj Mahal. But it was no good. 'He has done wrong, he must be punished.' 'But he didn't know he was doing wrong,' said Gottfried. 'The law says that to be punished for murder you have to know you are committing murder.' 'An eye for an eye, a tooth for a tooth,' said David Allan.

Time dragged on and on . . . My life was all orts and fragments, bits and pieces, no coherence except for one thought: Soon (but when?) I'll be out of here. In the evenings, there were now fewer people dropping in. Our Jewish friends had gone to Israel. Simon was dead. The RAF at last began to leave. Even more we seemed to be on some vast stage, like a desert, without edges, where people drifted across, and past . . . Sometimes, ill with restlessness, I would leave Gottfried reading histories of Byzantium, studying Russian, or talking to Hans Sen, and I went walking up and down the streets and avenues. There were few street lights. There were few cars. The shallow little town shrank unto the earth under the pressure of the stars and a moon that was always in rapid movement from somewhere to somewhere. I walked along under the jacarandas and the cedrillatoonas, passing houses that spilled out light and music and voices from the radios. You could walk for an hour, or two, up and down, back and forth, and hear the same tune coming from every house.

There's a small hotel, with a wishing well,
I wish that we were there, together . . .

Love walked right in and drove the shadows away
Love walked right in and brought my sunniest day
One magic moment and –

From every house, at the same moment, into all those receptive brains, and into mine too, feeding my longing for love and for escape. Then, at about ten, the houses went dark, they were dark streets I walked along, only the street lamps making pools of light I stepped through from dark to dark. Silence. A silent town. I stood under great trees watching moonlight sift through the leaves. Dogs barked. Soon, soon, I would be away from here . . . *Night and day I think of you* . . . I never thought of difficulties. I would launch myself into space, into London, but on my own wings. I would have no money? Really, how petty, not to say petit bourgeois, can you get? You will be in London, with a small child, without money? Well, it will all work out. It did, but now it seems to me remarkable that I took it for granted it would. It's going to be very difficult – surely someone must have said, but if so I didn't listen. I was writing love-and-friendship letters to two of the RAF, now in England, and in their letters to me were descriptions of post-war Britain, but these failed to dismay me, because any suggestion of hardship fed the exhilaration of my confidence. I would quickly make friends, I would find a lover, and besides Gottfried would be in London. I expected that we would get on very well when we were not living together.

When I got back home Gottfried would still be reading. He looked up with the glitter of his lenses that so intimidated people, and drawled, 'Where have you been?' 'I was walking, that's all.'

Or he might have gone to bed, and there lay watching me undress, flinging off my clothes. Seeing his expression I hastily gathered them up and put them outside of sight somewhere.

Once, when he had gone out to dinner, I needed to move, so I put the baby in the pram and went out walking in the already dark streets. When I got back Gottfried uncoiled from where he had been sitting white-faced on the edge of his bed. 'Where have you been?' 'Walking, that's *all*.' 'But you can't take a baby out at this time of night.' 'Why not, it's very warm. And he's asleep.' 'You can't do that kind of thing,' says Gottfried. An echo of my mother's 'But it's simply not *done*.'

You can't do that and *Why ever not?* punctuated our arguments and left us staring at each other, frustrated, blocked in our deepest selves. I used to wonder why I hurt him so, when his wayward and impulsive mother would certainly have taken a baby to a dinner party if she had taken a fancy to the idea. That is, if the nanny let her.

There is really nothing much we can do about what we are born with.

'But why not?' I would cry, and then, he: 'If you can't see why not I'm afraid I can't help you.' And he would turn away, cold, angry. 'Well then let's talk about it. Let's try and make some sense out of it. For God's sake, Gottfried, we might be stuck together for years yet. Hundreds of years . . .' 'No, I do not suppose it will be as long as that.' 'All *right* – but what's the use of just glaring at each other. Even for six months.' 'I was not aware that I glared.'

When things were at their worst between us we mended them by bridging our differences just where they were worst, that is, his dislike of literature, my love of it. We found a book we could agree on, *The Story of the Holy Grail*. He read a bit. I read a bit. We sat in our stuffy room, the baby probably not asleep, but amusing himself on the floor, and we read to each other. When people dropped in they laughed at us, because of the improbability, and because of Gottfried's German accent, and my Rhodesian accent, adventuring in high romantic places. On Sundays we went out to picnic at a vlei near Salisbury, now built over, then with trees over the water course where hornbills balanced and squabbled around their nests, and we read to each other and to anyone else around, Hans Sen, Gottfried's girl, the RAF, visitors from South Africa. Quite amusing really, everyone thought, but we were keeping ourselves sane.

The streets I went walking through, night after night, for weeks, for months, with never a thought of danger: now it would be impossible for a young woman, black or white, to go strolling there so casually. They are dangerous streets at night, these days. Now every house is locked and double-locked and guarded by dogs, all the windows are barred, the verandahs made into cages. Inside these little fortresses black and white families watch television, the same programmes in every house. The cars standing along the street edges are locked and chained. Nothing was locked in the old days, not houses, not cars. A young white woman might go wandering around until long after midnight. And in London, when I at last got there, I used to walk for miles by myself at night, and it never crossed my mind to be afraid. I do not believe that what has happened in our cities – and in the countryside too – has much to do with the political or racial complexion of governments. Something else is at work.

What?

What is it?

Is it possible – and I know this mad hypothesis is asking for ridicule – that we are poisoning ourselves with music? Our lot, my contemporaries, from our adolescence on, we listened to dance music, day and night, and it was all of it romantic or sentimental. It yearned, it wanted, it longed, it needed – and expected, too, for somewhere, some time, a promise had been made. *Some day I'll find you* . . . We were immersed in dreams. But since then music has changed. Its rhythms no longer swoon or sway or linger, they beat and pound and drive and the sound is so loud you have to hear it with your nerves. I was once leaving a party in New York because the music was so loud it was literally making me sick, and a black woman coming in said, 'What's wrong with you, honey?' and I told her. She said, 'But you don't listen to this kind of music with your ears, you hear it with your whole body, you listen with your nerves.' Which nerves? So my question is, when some person goes out to kill or torture or maim, can one reason be that he or she has been set for the crime by music that has driven them mad? Shamans have used music for thousands of years to create special moods, young men are prepared for killing by stirring marches, churches use inspirational music to hold their flocks together, and it is known that real spiritual teachers use music, but this is so delicate a thing that it is used carefully, by specialists, in special circumstances. But we deluge ourselves with music, of every kind, soak ourselves in it, often feed it direct into the brain with machines designed for this purpose – and we never even ask what effect it may be having. Well, I, for one – and I know there are others – think it is time we do ask.

19

I SUPPOSE WHEN WE MOVED AGAIN it was because this time it was a whole house, a small one, on the usual pattern of two rooms and verandahs back and front, but there was a garden for the baby, with a big jacaranda in it. At the back of the house, in the boys' kia, was Book, for we again had a servant. Not to have one was more trouble than it was worth. The word went around They don't have a boy – and they came to beg, plead, please, baas, please missus . . .

The white population of Salisbury was then 10,000, the black population was believed to be 100,000, and it seemed that the ambition of every one of the 100,000 was to work in a white house. At the least, it meant you had a legal place in the city, with its delights, you were fed, you had a place to sleep, you had a little money. Book was a clever young man, who had the house clean by nine in the morning and at once learned to cook. We decided this was a waste of natural intelligence and offered to pay for him to go to night school. We imagined him, in two or three years' time, as a book-keeper or clerk, earning several times what we paid him. But he resisted all attempts to make him want to better himself. He was twenty-two. Getting this job with us seemed to him the ultimate in good fortune. Why should he spend his nights at school? He enjoyed himself, for he was a champion dancer. He had a smart girlfriend. And here were these earnest whites with their ideas like missionaries. What he wanted was for us to pay him more money. We already paid him several times more than the usual amount, making him promise not to tell the other servants. Of course he did. Again we were attacked by angry householders complaining that if we had money to throw away on these cheeky munts, they didn't. They said to our faces and behind our backs that this kind of thing was only to be expected from the people who had organized the Meeting in the Location. We were all amazed, we were dismayed, at the way this Meeting in the Location went on and on as a symbol of everything dangerous, insurrectionary, revolutionary. No use saying to these frightened people, 'But I was at the Meeting in the Location and not a word of Revolution was spoken. Nothing was said there you couldn't hear any day from any black person, if you actually talked to them.' They never did talk to black people, only as servants.

Easy to say now that this level of white stupidity made their overthrow inevitable, but to us, then, who were in such a minority, 'white supremacy' seemed invincible. Once, and not long before, that elation, exuberance – vitality – which is fed by improbability, had kept us laughing, for instance at Charles Olly, or when we passed about among us – laughing – a cutting from the *Herald* that began, 'It is not generally understood that the Native stomach is not adapted to a European diet. Green vegetables can only cause stomach upsets and . . .' The fact was, our morale was low. Just to get out of this stupidity, get out – *get out* . . .

Perhaps because we were so low, and because of the hunger of everyone in the world for good news, after so much horror, there was such excitement over the news about penicillin, the new medical discoveries. Of all my memories of that time this one is the strongest for it seemed that every other programme on the radio was about the imminent end of malaria, of insect-carried diseases, of syphilis. We even organized a public lecture, given by triumphant doctors, to audiences that reminded us of when we took it for granted any meeting we organized would be a full house.

This is how I personally experienced this new dawn. It was cold, dry, dusty, mid-winter in June or July of 1947. One morning the baby's face and arms and then, soon, legs and body, were mottled red and almost at once erupted into weeping sores. He was eight months old. He had not been sick before. The doctor came, a locum, just graduated. He said, 'And now we are going to see a miracle.' In an ordinary enough bottle was a lotion made of penicillin, and he swabbed it all over the baby's body. 'I'm coming back in a couple of hours to have a look. I tell you, I've never seen anything like it.' The baby grizzled and complained, for the sores must have tormented him horribly. Then . . . it was indeed a miracle, for the sores stopped running. Then they were dry. The doctor came back, bounding up the steps, his eyes already on the baby, and when he saw him, he laughed in triumph, picked him up and jigged all around the verandah with him. 'Do you know what this means? Do you realize what we are seeing? It's the end of disease. Malaria – that's finished. Yellow fever . . . bilharzia . . . cholera . . . VD . . . TB . . . we've got them on the run. Can you believe it! Sometimes I can't. But it's true.' He swabbed the baby all over for the second time. 'Didn't I say it was a miracle? You're going to have to put up with me coming back again.' By nightfall there were dry scabs on still reddened skin. In rushed the doctor, hardly able to contain his exultation. 'I'm using it up at the African hospital. They say it is magic. So do I.' Next morning the scabs had dropped off, the skin was a clear pink. Less than twenty-four hours had done it.

Gottfried, still doing two jobs, was overworked and – more to the point – discouraged by the long wait and, like me, was sleeping too much, and always tired. We went with friends to visit Zimbabwe in Fort Victoria, the ruins that would forty years later give a name to the whole country.

When I walked into the lounge, I knew this was my ideal home. Or one of them: another is the palace in Granada: from one extreme to the other. You don't need to calculate rent, or degrees of common sense and suitability, if it is a question of dream palaces. The main room was wide and long, like a hall, and the thatch made a high vaulted ceiling. The floor was of stone flags, with a great fireplace and windows opening all down one side to show kopjes and bush. It was the house I was brought up in, but grand and solid, not likely to succumb to white ants. The bedroom block was separate, like the hotel in Macheke, rooms built in two strips back to back with verandahs all round. Here I watched a horse being brought up the hill for Gottfried, watched Peter, a brave baby, being handed up to Gottfried by a black groom, and there he leaned back into his father, and held on tight to the front of the saddle, his face stretched in ecstasy and terror, letting out cries of delight, of alarm, and then off they trotted, man, baby and horse, into the trees.

It was very hot. One afternoon Gottfried went off riding and I laid the naked child on my shoulder, lightly covered with fly gauze, and sat with him on the verandah. He was asleep and I dozed. When I opened my eyes I was looking into an open bedroom door. On the bed a half-naked young woman crouched over a naked baby boy of about four months. Her yellow hair was loose on her shoulders, her green eyes from time to time met mine, as she slowly licked the baby, like a cat with a kitten, a female leopard with a cub, his little arms, his legs, his stomach, then she turned him over and licked his back and buttocks. She finished, flung back her mane of hair and laughed at me, showing her white teeth.

In the dining room she sat opposite her husband, the baby in a pram beside her. She wore white slacks and a checked shirt. Her yellow hair was in a chignon. There was something of the official about her smart young husband. Perhaps he had just been demobilized. As she left the dining room, pushing the pram, she gave me the swiftest of comradely smiles.

I did not tell Gottfried: he would have been sarcastic, huffy, threatened. Among our companions was a young RAF who was still waiting to go home. He was working-class, from a large cockney family, very handy and helpful with our baby. I told him I had seen that young woman over there lick her baby all over like a cat but he said, Wouldn't it have been easier just to wash it?

Zimbabwe itself – the ruin – in those days was supposed to have been built by the Arabs. I climbed the hill that overlooks the ruin, and sat on a rock. I saw Gottfried riding away, far below me. The car was under the shade of some trees. In the car was the baby, asleep in his basket, and the RAF, asleep too. The silence of mid-afternoon on the veld, doves cooing, the cicadas, the crickets. There was another sound that has haunted me ever since. Somewhere down there, in a hut I could not see, or from under a tree where someone sat came two notes on a drum, a high note, then a lower, then an interval, then the two notes again. These notes are not to be found on a piano, and the interval between them has its justification in a region unfamiliar to a European ear. Like two raindrops falling, tap – tap, then the silence; tap – tap, and silence. On and on. On and on. Soon everything – the ruins, the landscape, the rocks, the hazy plains of the hot sky populated with afternoon clouds – all seemed absorbed into these two notes that repeated, and repeated, and went on, and were going on still when I climbed down the hill a couple of hours later to find the sprawled young man asleep, flushed with heat, and the baby asleep in his basket, the net tucked in tight, the flies scattered on it.

Those two notes . . . I wrote a play, which they introduced, hoping that what I would hear forever in my mind would somehow get heard by whoever wrote the music. The play was pure and simple and unashamed agitprop. At that time there was a syndicated cartoon in the newspapers, featuring a caricature black man, rather a coon, a wog, or nigger, or kaffir, embodying every nasty racial stereotype. It was in the newspapers every week, for years. I set the play in the deepest levels of the Rand gold mines, and the hero, a miner, on whose face was clamped the mask of this cartoon kaffir, and his mates, with similar masks, organized a strike. (Strikes on the Rand have always been suppressed and punished without mercy.) The plot was simple. When they stood up to the mine owners, represented by a brutal overseer, the masks loosened, and when they capitulated, they tightened again, and the miners stood groaning, trying to prise them off. I used the Zulu war dancing, tamed and sanitized, as shown to tourists, to start, but then this dancing, uninhibited, became part of the finale, which was the fight between the miners and soldiers in the depths of the mine. The masks loosen, fall, are gone. The trouble was, I needed a musician, I didn't know any, certainly not African musicians. I took the draft of this play to England. But the place for it was South Africa, not even Southern Rhodesia, still sunk in colonial attitudes. I sent it to Brecht, and got a letter back saying he liked it but he was being criticized by the Party for expressionism and formalism and similar vices and could not afford to do a play with masks.

I showed the play to Dorothy and Nathan Zelter. They always disapproved of what I wrote on ideological grounds. No, they did like a little over-whimsical sketch I wrote when I was eighteen. A short story would be handed back to me with the remark, 'I am afraid I have to tell you we are both very disappointed in you.' Clipped neatly to the manuscript would be a note: 'Paragraph three suggests the Africans are superstitious. This kind of thing is ammunition for our enemies.' About my agitprop play: 'We both feel masks are *by definition* reactionary.'

When I first met Nathan I was nineteen, and he was an almost too beautiful youth, a fervent lover of humanity and of women, and I knew him when he was very old, looked like Methuselah, and still dreamed of creating an ideal community in freed Zimbabwe. The people he proposed as fellow souls had long since decided we had nothing in common. He is difficult to write about because he was – well, in a word – preposterous, but in choosing this word I feel guilty, just as I did always, with Nathan, because he was so kind and so generous. 'But damn it, Gottfried,' I might wail, 'No, I *won't* go to supper, he's impossible.' 'Yes, yes,' says Gottfried, 'that is true, but for all that, we shall both go to supper.'

He came from Romania before the war, a refugee, and worked for a relative in an import-export firm. I hope by now I do not have to say he was a Communist. The Left Book Club was only temporarily his spiritual home, for it was too tame. There he met Dorothy, a New Woman, truly an exotic in 1930s Salisbury, a slim dark girl in long skirts, often of green linen, embroidered blouses, a chignon. She had small blue honest eyes, and on either side of an intelligent and homely face swung enormous exotic earrings. She was a very good teacher, and often confronted by parents who wanted to know why she was teaching their children these advanced ideas. She smoked cheap cigarettes in an expensive amber and silver cigarette holder. Their love affair scandalized everyone. This was the epoch of Free Love which was to end the falseness of conventional morality. But Free Love had not taken root in Salisbury. The beautiful young man, like a leveret, with his golden eyes unblinkingly fixed on the Truth, and this bluestocking girl, loved well and publicly and refused to get married, she to annoy her mother, he out of principle. He had many and would have gone to the stake for any one of them. It was he who, seeing I had tacked up – I was still married to Frank then – an Augustus John reproduction, tore it off the wall and ripped it in two, not on aesthetic grounds, but because a woman doing the washing should not be made a subject of art. It was he who, seeing I was reading the amiable essays of one Lin Yutang, took it out of my hand, because the Long March had cancelled China's long history, which henceforward would be the story of the peasants. It was he who, when he saw on my wall a genuine and valuable Japanese print given me by my Cape Town artist lover, said I should be ashamed to own the representation of a courtesan – i.e. an exploited woman. Later when I asked him to look after it for a while he destroyed it.

I imagine him as a minor character in *The Possessed*. Kirilov is sitting alone, it is late at night. A knock. 'Come in, Nathan.' – But he is so deep in his thoughts he hardly sees his guest, who is in a state of deep emotion. 'What is the matter, Nathan? Have you eaten? There's some bread in the cupboard.' 'No, no, no, I can't eat,' says Nathan, absently opening the cupboard, staring at the bread, shutting the cupboard door. 'Kirilov, I've been walking . . . I've been walking all night.' 'The moon is shining,' says Kirilov.

'I saw it earlier. Do you think the moon is inhabited, Nathan? If so, is it full of worms and criminals like us? What do you think?' 'No, no,' whispers Nathan, his tears streaming. 'I've just understood. That's what I've come to tell you. Quite soon life will be beautiful! The end of crime . . . the end of cruelty . . . no more poverty and hungry children.' 'Do you think so? I used to think so too,' says Kirilov dreamily. 'Kirilov,' says Nathan, pouring kvass into a dirty glass, 'I'm so happy, so happy . . .'

Nathan might have left the Communist group on principle, thereby becoming wishy-washy, temporizing social democrat and lackey of the ruling class and so forth, but he always came to our meetings, and would accost us in the street with 'Here is the traitor Nathan. Will you come to supper on the Thursday?' He admired Gottfried, that is, his mind. This poor boy from some Bucharest slum could never forgive the rich Berlin boy, and when he had become a very old man, would still say, 'But he used to put on a hairnet when he was dressing. I saw him.' And I would say, 'But Nathan, that was a decadent time in Berlin, have you forgotten?' 'A *hairnet*, Tigs, a *hairnet*.' Until he died Nathan insisted on calling me Tigger, a name I refused to answer to from the moment I left Rhodesia. 'Nathan, why do you, when I hate it?' '*I* don't hate it,' says he, calm and in the right.

He never laughed. No, that is not true. He laughed. But always sarcastically, angrily, contemptuously, or with a trembling sadness; recognizing inevitability. *'Well, and what else?'*

Nathan worked hard for the Southern Rhodesian Labour Party, and for Race Relations, and when Charles Olly put filthy anonymous letters under the doors of his enemies, attacking Mrs Maasdorp, he used to go personally to argue the rights and wrongs of the case with citizens uneasy because confronted with an impassioned and exotic foreigner, who might say, 'Well, if you won't subscribe to socialism, though one day you will see it is the only *possible* solution, how do you reconcile this behaviour with Christianity – which is the foundation of this culture. *So-called* culture.'

He kept minutes and accurate notes about everything that ever happened, meetings, events, scandals, conversations. We used to say that if Nathan recorded something, every syllable and fact would be there, but his interpretations of it, his reporting, would be nonsense: for he never understood anything that was happening.

He also started a magazine, for which I contributed cooking notes, short stories, and got advertisements from amazed businessmen. 'Why should I advertise in a magazine that advocates my abolishment?' 'Oh come on, why not, you can get it off income tax.'

He worked, we thought, probably eighteen hours a day. This was because he had married (they married when she got pregnant) not only Dorothy but also her mother. Dorothy was married to her mother: this is what we said, interpreting this relationship, and many others, with the aid of D. H. Lawrence. At nine o'clock every morning, Harty (the nickname suited her) arrived in her daughter's house and stayed there the rest of the day. Rigid with resentment, meek with the psychological inhibition I understood so well, Dorothy sat opposite her mother, smiling bitterly, both women smoking elegantly, posing their long amber holders. Harty was large, sporty, loud, and drank a lot of gin and tonic, which Dorothy did too, from the strain of it all. Harty was always rude to Nathan. Thus: 'Dorothy only married a Jew to annoy me.' We told him he should throw her out, tell her she must not spend every waking moment in his house. He said that it was for Dorothy to throw her out, not him. There were three women among us who had never as it was put 'stood up to their mothers', a ritual act necessary for their psychological salvation. We knew what was wrong with these mothers who could only live through their daughters: they needed jobs and a life of their own. We were right. But they belonged to a different generation. Either way the daughters suffered. If they did 'stand up to' their mothers, like me, they knew they were unkind. If they did not, they were like rabbits hypnotized by a car's headlights.

Dorothy had no vitality. She suffered from headaches, migraines, interminable periods, sometimes for weeks in the month, and was always ill. And often unhappy because of Nathan, who travelled for his firm over Southern Rhodesia and Northern Rhodesia and Nyasaland. He had announced when first in love with Dorothy that no one human being could stay in love with another for all of a life, and he had every intention of having affairs, but only when he fell in love. He would reaffirm this statement of principle before leaving on a trip and on returning report to Dorothy.

We actually discussed this in the group and collectively informed Nathan that his behaviour was unsocialist because it was unkind.

'One does not do this kind of thing,' drawled Gottfried, speaking not from a socialist, but a very different tradition.

'Then I'm sorry,' says Nathan. 'As a human being I have my rights in this matter. I believe in honesty in all matters, particularly between a man and woman.'

Dorothy asserted herself in ways that we all understood and explained to Nathan, who sometimes did and sometimes did not see the obvious. 'Thank you, I understand why this woman has to have a period lasting sometimes three and a half weeks.' Dorothy would not allow any food in the house that could be called foreign, particularly not garlic, or herbs. Nathan bought herrings, showed me how to prepare them, and would drop in to eat herrings on rye bread and, wafting garlic everywhere, arrive home to say, 'No, thank you, Dorothy, I have already eaten. You and Harty may eat the cold beef.'

He was extremely generous. He gave money to anyone who asked. He educated the children of his servants. When Northern Rhodesia became Zambia, he went there, and set up a cooperative with Indians and Africans, giving them equal shares though his was the money and experience. He was, much to his surprise, his sorrow, cheated by all of them. The enterprise failed. When I came to England and was having a poor time of it, I asked the richest man I knew to lend me £100, but got a gracefully apologetic note back. He did not wish to lose my friendship, which he valued: he did not lend money, on principle. Nathan, who at that time had very little, sent me £100 by return of post. It was to the Zelters I wrote, from London, for a year or so, breathless chattering schoolgirl letters, rather like Sylvia Plath's to her mother, letters designed to hide behind, while offering a progress report.

Nathan was in love with me – thus. 'I find I am physically attracted to you. You must not think I am in love with you. That is a very different matter.' 'No, I promise you, I won't.' 'But I have to make myself clear. If we were to find ourselves in the same town at the same time, without either Gottfried or Dorothy, then I could not answer for the consequences.' 'But Nathan, luckily I am sure I could.' 'What is that? You are saying – of course, I understand! You are so much in love with Gottfried you could not be unfaithful to him. Anyone could see that at a glance.'

I was in fact being unfaithful to Gottfried. I was having that classical love affair every woman should have just once. He was the man who ran the radio station, on a level somewhat higher than most of the citizens deserved. He would come to the house in the afternoons, walking up the path under the big tree looking neither to the right nor to the left, knowing that curtains were twitching all down the street. We talked literature, laughed a lot, and made love while one ear listened for someone coming to the house: after all, people always dropped in. Once it was my mother. Once it was Nathan. My lover hid in the wardrobe, which was shaking with his laughter, while I told lies. This affair couldn't have been more satisfactory for either of us, except that I am ashamed to say I fell in love. This appalled me. All plans of leaving the Colony as soon as I could, for a new life in London, had faded into the need of marrying this man who would leave his wife and children. Did I believe this? I did not, but watched this pattern running in one part of my mind while another commented on it, in derision, just as he – my partner in guilt – did. 'You are crazy,' said he. 'Stop it at once.' One may say – women may say – lightly or otherwise, 'We are in the hands of our biology', but it takes an experience like this to know how merciless an imperative it is. It was time I had another baby. Nature was saying so.

Round about then dragging back pains took me to the doctor, who said I had a retroverted uterus, which should be stitched up, and while he was about it, he might as well take out my appendix, which was of no use to me and he recommended tying up my fallopian tubes 'while he had me open'. He gave me forty-eight hours to think about it. Gottfried said that I might later marry again and want children with my new husband. But that, as he did not understand at all, was the trouble. I knew that by the time I reached the menopause I would have been in love several times and each time have had a baby. My deepest nature was against me, *my* nature in league with Nature. I was being given the opportunity to outwit the almost-certainty that I would be the old woman who lived in a shoe. I was intrigued that Dr Rosen could see this. He could hardly approve, this good Jewish family man, of a woman who had left one husband and two children and then almost at once started all over again. I did not care about his motives. This was probably the most sensible thing I have ever done in my life. A deeply buried instinct for self-preservation was working for my good. I was not thinking of a convenient sex life: for my sex life was, when it was given a chance, satisfactory.

While I was being wheeled into the operating theatre I saw this was what I would feel like on my deathbed, not because of the operation, which I did not fear, but because I was shaved, trussed, bound up like a corpse, hair bundled out of sight by an eighteen-year-old nurse, as fresh and plump as a baby of two months. For the first time I understood why in nineteenth-century novels twenty-eight-year-old women could be described as old.

When I was resting after the operation a young woman who had also been operated on by Dr Rosen was still half-conscious, and for half an hour or so she raged hysterically against Jews, and particularly Dr Rosen, who was in the ward. Never in my life had I heard anything like the filth emitted by this conventional Rhodesian matron. Twenty or so women lay in their beds and listened to her ravings in silence. Later a nurse told her she had been saying 'nasty things' about Dr Rosen. She was embarrassed, and when he came on his round, said, 'I believe I was rude to you. I'm sorry.' He said gravely that patients recovering from chloroform often said things they did not mean. And went on his way. But if she had these attitudes, why choose a Jewish doctor? She did not know that inside her was this well of disgusting substance, like shit full of cholera.

20

EVERY GROUP, OF WHATEVER KIND, however it starts, will end as a religious or mystical group. So the psychologists claim.

Ours began as a mix of left-wing people from every part of Europe, became for a short time strictly Communist, at least in theory, but as it lost Communist purity became more a branch of social welfare and charity.

For the last couple of years before we left it could be said there was no group at all. Gottfried would certainly agree with that. But is a group to be called one only if its composition remains the same? Ours was more like that standing wave which retains its shape while the water rushes through it. Of the original members Gottfried and I and Nathan remained. But Nathan was involved with Labour Party politics. Various refugees, who began as Communists, had done well and acquired liberal views. Charles Mzingele and friends dropped in, when they could, but really they came for books and information. There were others I have not mentioned: they might not want their revolutionary past remembered.

There appeared among us a young man who when we walked in the park, if we went to a restaurant, or were in the street with him, caused angry and disgusted glances to be directed at us. He was slight, burned brown, wore frail white very short pants, golden sandals, earrings, and golden hair to his shoulders. Now no one would turn their head, but he was the first portent of a long and motley summer. He was intelligent, well-read, understood music, could not live a day without our company, and was more than a little mad. He lived alone in a bedsittingroom where the electric light fittings sent him messages from the KGB, who were controlling his thoughts. I had never met anyone like this, was fascinated, and always ready to hear the latest communiqué from Moscow. It has been my fate to be involved with quite a few people eccentric in various ways, and by only ten years later, if someone said casually that they were being spied on by the KGB or CIA through the light bulb, I would have learned to say, 'Oh dear, are you sure? Well, never mind.'

This man appeared each evening, ate up whatever I was cooking, sat tapping his gold-sandalled foot, waiting for that moment when the conversation turned his way, which it did almost at once, for it seemed everyone wanted to offer anecdotes about haunted houses, table-turning, and witch doctors. Kurt could not keep away. The girl who was wistfully in love with Gottfried, and her half-section, were always with us. People we had not seen for months or years came again. Poems were read from that heady period in Russia when revolution, personified as mad horsemen, mystic monks and sibyls, raised temperatures in Moscow and Leningrad thirty and more years before. Anyone who has lived in Africa knows something about witchcraft and shamanism. These evidences of Higher Thought had been kept at bay by the demands of the class struggle, but now it was evident that everyone believed in the occult even more than they did in socialism. There has never been more than a short step from socialism and Communism to some level of mysticism, perhaps the most visible exemplar being Annie Besant, who began organizing the match-girls and ended by inventing Krishnamurti.

Our period of popular mysticism did not last long. The stage that follows chat about seances and ghosts must be practice. Someone knew of a medium, and so people drifted off and away to find more exciting evenings.

It cannot be said that much 'analysis of the situation' went on. Gottfried was already living, in his mind, in London. So was I. Now I think the only real use we were to any of the black people – by 'we' I mean here any of the 'progressives' – was that our books were lent and given to anyone who asked for them. Charles brought friends who brought friends, and before they had even sat down their eyes were straying to our bookshelves. They were not easy, these visits to our house, because Book our servant must not know of them. He would have at once informed the whole neighbourhood that his employers were entertaining blacks, and that would have meant Charles would have even more trouble with the police. They never arrived all at once, but in ones and twos over an hour. They might carry a piece of wood and a saw, or pretend to be sellers of something or other. I made them tea and then sat on the verandah to make sure no white neighbour would interrupt, and so they could take their time. It was always painful seeing how these men – always men then – touched the precious books, and how, when they saw some book they had never heard of, but which spoke to them, handled it with a delicate reverence for its possibilities, for they were in search of the education which most of them had been denied. I would glance through the window from the verandah and see them bent respectfully over the books which we took so much for granted. We also ordered them books, saying we knew of a fund that paid for books for Africans. Not long after I got to London I got a letter, 'Do you remember me? I am So-and-so, you gave me books and I passed my examination.'

Who now fills this function in a country doomed always to be book starved? Often it is the British Council, which in Harare maintains rooms filled with textbooks and videos and every kind of teaching device. They are always besieged by black people, mostly young, just as hungry for education as the men we knew in our time. The black government has done nothing to provide books, fund libraries. In the long run this oversight – though sometimes it is hard not to believe it is an actual policy – will turn out to have been the most stupid of its mistakes.

It must not be imagined that I sat on that verandah full of adult and benevolent thoughts. I watched those decent brave men leaving, one by one, trying to seem like house servants or messengers, looking nervously about for some lumpen white housewife who might start screeching at them – and I was in a simmering rage. (It is a pity that word 'lumpen' from the Communist vocabulary has gone out of use: it so perfectly expresses its meaning.) This kind of anger is damaging. An ebullient, exuberant and generous anger, is fed by a belief you can do something, change something. But there is an anger like acid spurting into the stomach, locking you in cynicism and impotence. Stupidity, stupidity, stupidity, I was muttering, several times a day, as if I had been ordered to stare, never for one moment blinking or averting my eyes, at deliberate cruelty, like a small boy torturing a bird. I certainly could not share this affliction with Gottfried, who would have reacted with 'Come the Revolution . . .'

Revolution in Europe: he had given up hopes for Africa. 'Perhaps in a hundred years . . .' he might pronounce. The 'correct' formula was still that only a black proletariat could free Africa. But there was no black proletariat in the sense meant by the Fathers of Theory. There were black miners and some workers in secondary industry, but Black Nationalism was, quite simply, a reactionary deviation – I omit all the other adjectives. Moscow's 'analysis of the situation' had come up with this formula. I remember Gottfried drawling that it would do the Comrades there some good to actually experience Africa.

Yet we were always following up rumours, leads, items in the *Herald*, or in gossip, about possible black leaders. Charles Mzingele did not know of anyone. Now it could be said that he was rehearsing his role at Uncle Tom, for while he approved of any black leader, and had done everything to help anyone with aptitude, he found the violent language of nationalism distasteful. How could it not be so? For years, decades, he had kept his position, but only just, and so precariously, under attack from the whites as an incendiary agitator, Communist etc., by tact, good humour, patience. He was by temperament gentle, thoughtful, considerate. And, too, he was getting old, and sad. Although the Roman Catholic Church had threatened him with excommunication he depended on his religion. 'Only God can help us,' he would insist in a room full of atheists-on-principle. 'Only God and his angels.'

When there was the Strike in the Location, as famous as the Meeting in the Location, we were taken by surprise and so was Charles. The leader of the strike was in Bulawayo. They were striking for a minimum wage of £1 a month. This was the wage then earned by a cook, but supplemented by rations and gifts. No one could live on £1 a month. The demand for a minimum wage was rejected by Parliament. That the whites should be in their usual stage of outrage, of alarm ('They will rise and drive us all into the sea') struck the familiar note of farce. But behind that demand was such suffering, such misery, such brutalities – as we knew well, but it seemed the whites did not. The munts could not possibly be striking out of their own initiative: it must be the work of Communist agitators. Then the almost inevitable, clinching note of farce. The whites did something very stupid. Stupid from their own point of view. In those days, few black people were politically motivated or informed about their own situation, let alone other countries. They hardly knew about trade unions. Also, for the sixty years of the white occupation they had been told day and night that they were monkeys, ignorant, backward, inferior. When the authorities were sure the entire labour force of the city was inside the Location they locked the gates, and put guards all around the perimeter. The people inside could not get out. No food or supplies were allowed in. The whites were starving their rebellious subjects into submission, and made no bones about it. But inside the fence were also some 'agitators' who made good use of their time with these captive, frightened audiences. For five days, increasingly angry crowds listened to inflammatory descriptions of their situation, not Communist, pure nationalism, but with comparisons made between their conditions and workers in other countries. It was a short course in political theory and practice. We could not believe that Authority could be so stupid. But Authority, when scared, is usually stupid. The whites were being faced with their nightmare, always latent, that the blacks would rise and cut their throats.

I think it is likely that this strike, and the locking up of the black people to starve and listen to agitprop, was a real turning point. The famous Meeting in the Location only gave the whites some enjoyable short-term *frissons*; most black people had never even heard of it. But the Strike demonstrated to everyone how cruel the whites really were, and above all, how ignorant about their sufferings.

Since then I've seen the same phenomenon many times and in many contexts: people in power, in authority, never seem to know how the people they govern are living and feeling. It is as if there is some mechanism in the brain that separates them – by the mere fact of being put into power, or a position of responsibility – from the ruled, from an imaginative understanding. Otherwise how can one explain it? Obviously it is in the interests of the people in power to know the situation of their citizens. Several times in London I have told friends in high positions that such and such a thing was going on in the lower reaches of their departments or fiefs, only to hear, 'Oh no, that's impossible, my employees could not behave like that, you are exaggerating.'

Only once have I read about rulers who understood this thing. It was in the Middle East and the Middle Ages. The rulers made sure that there were always inspectors, pretending to be ordinary people, applying as petitioners or working as employees, to find out how officials behaved. If incompetent or cruel, they were removed. But the point is, since every official or person in power knew that the person standing in front of him might be a government inspector in disguise, they tended to behave better than if they had unsupervised power.

Gottfried: 'But when we have a Communist society, injustice will no longer exist.'

It will be asked, 'But surely you did not still believe in perfections of Communism? Surely you know about –' the forced collectivization, the trials, and so on. Round about then there circulated a book called *I Chose Freedom* by Kravchenko. When I say circulated, people like Gottfried and Nathan said they were not interested in anti-Soviet propaganda. Some of us read it, and discussed it. The trouble was, the picture drawn by Kravchenko was so different, in fact, was the opposite, of everything we had read or heard about. Of course there were difficulties, problems, troubles, in the Soviet Union . . . but that it was a *total* tyranny? I watched this exact situation repeated in the mid-1980s. A Russian girl had married an Englishman, and had come to live in London. She often visited Russia. She kept saying that when she returned to London, or to Russia, it was like becoming part of the negative of a photograph: everything said in the West about the Soviet Union was the opposite of what the Soviet Union said about itself. On arriving there, on arriving back here, she felt as if her brains were being turned inside out. We felt, reading Kravchenko, as if our brains were being turned inside out. I remember exactly, no false memory about it, how I felt: if what I was reading was true, then *nothing at all* of what I believed to be true, could be true.

Yet it could be said that something like 'an alternative reality' was building itself slowly, in my mind at least. Koestler's coins of true belief were falling fast from our pockets. If there is no such thing as a sudden 'conversion', that is, a sudden reversal of opinion, without a lot of little impressions accumulating in the brain first, then there is no sudden reverse conversion.

The word 'paranoid' always appears in discussions on this theme.

Dr Jerrold Post, an American expert on political paranoia, quoted in Tom Mangold's book, *Cold Warrior*, on James Jesus Angleton who ran the CIA for years (although he was quite mad) defines paranoia as 'a fixed conclusion searching for confirmatory evidence and rejecting disconfirming evidence. Paranoia is an adoptive mechanism. It is socially induced and learned in a family environment from early childhood. It develops as a defence against insignificance and being ignored. Paranoids feel it is better to have people against them than to be ignored. They also feel it is better to have an organized view of the world than to have chaos. A clear, organized, conspiratorial view of the world is easier for them to have since it gives them a sense of psychological security.

'Paranoia is not fixed in time, it is dynamic and changes over a lifetime. A paranoid's mind-set is that he is maintaining a lonely vigil and pursuing a lonely task. The weight is on the paranoid's shoulders.

'Paranoids are always the last persons to know that they are troubled. And if they have problems, they believe it is always someone else's fault. Perhaps the most important audience for a paranoid's thinking is in his own head . . .'

With this definition it is easy to see half the human race as paranoid. (What, only half?)

The point is, these processes have nothing to do with the rational mind. We are dealing in religious attitudes thousands of years old, burned deep into us, sometimes literally, by the fires of the Inquisition.

It took me four or five years from my first falling in love with Communism, or rather, ideal Communism, in 1942, to become critical enough to discuss my 'doubts' with people still inside the Communist fold. In another two or three years I discussed with other Communists facts and ideas for which in a Communist country we would have been tortured or killed. By 1954 I was no longer a Communist, but it was not until the early 1960s I ceased to feel residual tugs of loyalty, was really free. That is, it took a good twenty years for me to no longer to feel guilty, to shake it all off. I remember with shame how difficult it was to say what I thought with people who were still the Faithful.

I was able to be freer than most because I am a writer, with the psychological make-up of a writer, that sets you at a distance from what you are writing about. The whole process of writing is a setting at a distance. That is the value of it – to the writer, and to the people who read the results of this process, which takes the raw, the individual, the uncriticized, the unexamined, into the realm of the general.

Perhaps the largest of the coins that dropped from our pockets was the earliest, for the Soviet Cultural Attaché and his wife came from South Africa to add authority to Medical Aid for Russia. We dreamed for days about meeting these representatives of the beautiful new world coming into being over there in the Soviet Union. The word disappointment is simply not adequate. They were the essence of commonplace. The woman, who we women were looking forward to meeting, as the latest of New Women, was like any Salisbury young matron, but dressed 'Johannesburg'. ('You can never have on too much gold.') He was all hair-grease and easy good looks. They liked popular novels and bad films. They knew nothing about the politics of Southern Rhodesia and not much about South Africa. They said the black people gave them the creeps – she with a delicate little grimace of distaste. They were the essence of what most of us had reacted against.

1947 . . . 1948 – this was certainly the worst time in my life. Bad times that seem to be endless make of the heart a kind of black hole, absorbing all life, all energy. How it did all go on and on. There is something in my Fate, or my destiny or perhaps my character, that takes me into backwaters where I lie becalmed. I wait. I am very good at waiting. This, like everything else, has two sides to it, the twist in its essence. You wait on events, because you know they will swing your way, or rather, an inner logic of inevitability is slowly working its way out. Then, when things change, you take hold and move. But this waiting can also become a lethargy, a missing of opportunity. Well, I had no alternative but to wait, then, at that time, that slow, muddled, grinding post-war time.

I was not sleeping well. Neither was Gottfried. We would lie awake in our separate beds, knowing the other was awake and unhappy. Or we smoked, while our cigarette tips brightened and darkened like fireflies. We felt as friendly to each other during those long bad nights as ever we did in our marriage.

He would drawl, 'Yes, well, this time is not of the most pleasant.'

And I said, 'I suppose it *will* end?'

'Well, yes, I think we may safely suppose that.'

I used to listen to the trains shunting and panting a half a mile away, and the long trailing shriek of parting as one set off somewhere. I used to hear the milk cart come along the street, starting, stopping, the horse shaking its harness, the voice of the milkman, soft, persuasive, in the early light, because of the sleepers. He talked Shona to the horse.

Towards the hooves and clanking of the milk
The hours began to chime together
And there, dawn-lit, the church stood high against a sky of silk.

I dreamed those lines. I was always waking with lines of verse on my tongue. What a pity I am not a real poet. If I were, that filter or sieve through which sounds must fall from the sea of sound to become words would be set finer and subtler. I used to think, If I am going to dream sequences of words, then why not much better ones? Now this really is looking a gift horse in the mouth.

We were going in for social activity of the kind that earlier we would have described as 'bourgeois' and therefore contemptible. We danced at the Highlands Park Hotel, a few miles out of town, on Saturday nights. This was a house near where I had been a nursemaid with the Edmonds opposite Rumbavu Park. A big room for dancing, a bar, and a court under the musasa trees where you sat between dances. Gottfried was his Russian self, and smashed glasses and told the black band they must learn to play gypsy music. I sat under the trees and, pretty drunk, worked in my mind on how the atmosphere of this place might be conveyed when I wrote about it: *Landlocked*. Athen Gouliamis and his friend, just off to Greece, came once, and said that they understood now why the bourgeoisie were so reluctant to part with their privileges. The moon, a sorrowful, flying apparition, was in attendance on these revels.

Once, among us dancers, a provincial lot, arrived a woman in a wonderful black lace dress, in the arms of a man twice her age, still in uniform – our army, not Britain's. She was quite lovely, but her face was sliding out of shape, like a heated wax mask: she knew I understood her situation which was the familiar one, an Englishwoman marrying a Colonial to secure her future. We stood together on the verandah in the moonlight and she said, 'I was a model in London.' Then, 'I can't wear this dress here, can I?' 'Why not, it gives us a bit of a lift.' Then, suddenly, an appeal and a statement together, 'Am I going to be happy here?' And then she went on dancing round and round in those proud elderly arms.

I read, I read, I read. I was reading to save my life. How hard it is to convey the suck and drag of bad times that seem as if they will never end, fit only to be contemplated by that ancient unblinking lizard's eye. I hinted at the dreadfulness of it in *Going Home*. But enough. I was reading poetry, chanting – silently as it were, under my breath – lines of Eliot, of Yeats, like mantra. I read Proust, who sustained me because his world was so utterly unlike anything around me. Among all the other delights, he supplies a non-literary one, more like history. Proust describes, in an eleven-volumes-long irony, how the aristocratic Guermantes at last absorbed people they had despised so much they would not even meet them. The daughter of the courtesan, Odette, married an aristocrat, the vulgar Madame Verdurin became the Princesse de Guermantes. We are invited to observe that ever-repeated process – one of the long, slow, rhythms in society – how the rejected and despised rise, and how they in their turn despise those who will supplant them. I could have used this exemplary story to cheer myself up about the apparently indestructible structures of 'white supremacy' but I was not so intelligent.

I dreamed every night about the sea, washing in and out of my sleep in sad slow tides of nostalgia, of longing. The title *Landlocked* came from then.

I began taking Afrikaans lessons. What could have been more ridiculous? 'Well, if I'm never going to get out of here I might as well learn . . .' My teacher was a young Afrikaans woman and we sat in the café in the park and laughed a lot. Her husband, a teacher, had written a novel which he brought me to criticize. It was in that category, almost very good, but it needed rewriting. It was full of the lyricism and love of nature and of women that is the Afrikaans writer's gift. But I said to him, 'Look, you simply cannot write a book that is a copy of *Farewell to Arms*.' Even the end, rain falling on the mourning lover, was the same. He had never read *Farewell to Arms*, nor heard of it. That was not the last time I have been sent an apprentice novel by an innocent plagiarist.

Outside Salisbury was, and still is, Mermaid's Pool, where rocks slope into water, surrounded by bush. It was and is a picnic place, though now you have to pay and they sell hot dogs and Coca-Cola. On Sundays an improbable association of people might meet there, as many as twenty or thirty. Harry and Monica. Dora and her two children. If she was in town, Mary with her two children. Frank Wisdom and Dolly, who was now married to Frank. Jean and John. Gottfried and myself and our child. Hans Sen. There were usually one or two of the girls in love with Gottfried, and some of the RAF, either our friends or my mother's. Also, my mother. When she heard of these picnics she was appalled, but then began to orchestrate them. I remained appalled. At first, the picnics were Gottfried and I, the baby, Hans Sen, our friends, and Dora. She told Frank of the picnics and he at once said it would be nice for the children to see their mother in this way. Meanwhile my mother had been arranging occasions when I might 'see' the children in carefully controlled situations. Controlled by her. I found them almost unendurable, and bitterly – but silently – accused her of enjoying the situation. If she was thinking *Serves her right*, then ninety-nine out of a hundred people would have felt the same. Not Dora, dear Dora, who never, ever, criticized me, and whom I met often for bulletins about what went on with the children. We sat for an hour or so over ice-creams in Pockets tea room where the civil servants' wives whose morning tea parties I had shared sent me measured smiles, and then discussed me in low voices. Dora put off her old wife's persona, and shed twenty years. Her bright hazel eyes were kind and shrewd, her pink lips were twisted in satirical enjoyment of her malice, she was a handsome funny woman few people ever saw, certainly not her husband. (But the joke was, he would have liked her very much.) 'My dear! How utterly utterly brilliant, how too too clever of them, of course poor John and Jean should see their mother with a new husband and a new baby. The poor babes don't understand a *thing*, but what does that matter? Those Wisdom men, they are so *clever*, they are quite ga-ga, but never mind. My dear, shall we go absolutely *mad* and have another ice?'

Certainly it would be hard to think of a more puzzling situation for the two little children: there, inexplicably and suddenly, was their own real mother, but in her arms was a new baby and she had a new husband not their father. Later, asked about these occasions, they remembered mostly that they found Gottfried frightening. Why did I go along with it? For one thing, when one is so deeply in the wrong, it is not easy to assert oneself. For another, I felt dragged along by powerful currents, contradictory ones and that was the point. I felt – and still wonder if this wasn't right – that it would be better to make a clean break with the two children until – the formula used for these situations – they were old enough to understand. I pictured myself in a home of my own, but this formula did not mean merely a flat or a house, rather a feeling of myself solidly based somewhere, it didn't matter where, nor was this base money, or respectability, but that I should have earned an identity, that would justify my having left them. Meanwhile, these picnics were a nightmare, and I have had doubts about the joys of the extended family ever since. As for the two children, the worst they ever said to me was when John remarked – but he was an already middle-aged man then – 'I understand why you had to leave my father, but that doesn't mean I don't resent it.'

My mother had no idea how bitterly, if silently, I was accusing her, or of my cold and frustrated anger. Why should she? She was only doing her best, doing her duty as she saw it. She was always rushing in to wherever I was, on the impetus of an imaginary dialogue with me, where the rightness of her arguments floored me, brought me to agree with whatever plan for me she was hatching. But the moment she arrived, her poor old face fell into disappointment because the young woman who confronted her in no way resembled the one she had been flattening in argument. 'Do sit down, Mother. Have a cigarette? A cup of tea? . . . No, Mother, no, no, I'm sorry, *no*.' She knew by now that Gottfried and I were divorcing, and secretly I am sure she was pleased, because now I could marry a nice Englishman. She said she was pleased my father was dead, because this new blow would have killed him. She had decided it would be a good thing to live with me and run my life, because I was always so feckless and irresponsible. Really, though, she wanted to live with my brother and run his life and Monica's. She really did not see why she should not do this.

Harry and I were cruel in the way grown-up children often are. We told her she should get married again. There were a couple of not inconsiderable men who wanted to marry her. Harry and I approved of them. She said, 'But how could I marry someone else after being married to your father?' She said, 'Once you've been really married to someone you can't marry someone else.' A view of marriage or of her marriage not likely to be understood by many people now. She was still in her early sixties, a handsome, well-dressed, dryly humorous woman, efficient, practical, and full of energy. She was living as she would have done in England: went in for charitable work, and visited nice people. She was never invited to Government House, which was her dream. Frank gave her financial advice, and so did Gottfried. 'A funny thing,' she would say, with her dry little sniff at the world's ironies, 'but both Frank and Gottfried are so good with money, and you're so hopeless.'

What this really meant was, 'I don't understand why you don't care about your future – about security – about appearances. And why you are content to live in this graceless way?' Yet Frank, with whom I had lived in several places, gracelessly, was now the essence of conformity. She was confused. And so was I. When I first met Frank he claimed to despise anything that might be called 'bourgeois', from mortgages to fidelity.

People in their twenties do not find it easy to believe that their friends, layabouts or adventurers, these often directionless, inept or revolutionary mates or playmates, are going to turn into the Fathers and Mothers of the city, and generally run the world. Frank, whom I had met less than ten years before as a very junior civil servant, was now getting on for forty and rapidly on his way up. Later he became Master of the High Court. Then he was Secretary of Justice. When he was Secretary of Native Agriculture, his policy was too progressive for the time. He did all he could to increase the number of black Master Farmers. When Zimbabwe got independence, the number of these trained and skilful men (and women too) meant that Zimbabwe's black agriculture was successful when neighbouring countries were doing badly. The Minister of African Agriculture was Graham, Duke of Montrose, and 'thick as two planks' – this potted biography of his father was dictated to me by John Wisdom in 1992: he was proud of his father, if he found it hard to get on with him. The Minister did not approve of Frank's progressive ideas, nor of the other men in the Department who shared them. Frank was 'kicked sideways' back to Secretary of Justice, and then kicked upstairs to the Public Services Board. Several of the men in PSB objected to policies which prevented black civil servants from ever getting higher than a certain level: the famous 'ceiling' here was evident as law, not something merely understood. Frank found it intolerable and resigned, aged fifty-three. Had he stayed, he would have had another twelve years in the Civil Service. He joined an old Sports Club mate, one Chippy Pringlewood, in a firm called the Guardian Trust, and they worked together as independents for the High Court.

Frank, then, in his later years was a frustrated man, in disagreement with Native Policy. If his views and those of men like him had been accepted, Zimbabwe's history would have been different. For one thing, there would have been a stratum of black people trained in administration.

21

AT THE END OF 1948, suddenly everything began to move. Gottfried became a British citizen, and I did too. There is no way of exaggerating what I felt about losing my British nationality on marriage, and having to apply for it again. This goes much deeper than words, tears or – well, what? These processes go on well out of sight and of understanding. I was left with a feeling about my British passport that the most simple-minded patriot would applaud. The law has been changed since. The proceedings for the divorce were started. Since Gottfried was so well known in the legal community, it was a formality, and went fast, although the courts were clogged with wartime divorces. I deserted him or he me, I have forgotten, but we went on living together throughout. Nothing could have been more amiable. I would have custody of the child until he was fifteen, thereafter Gottfried would; there would be access for both parents, and Gottfried would pay a small monthly sum for maintenance – it was he who insisted on this, for some legal reason. Both of us assumed that, since we would be living in the same city, London, and both earning, money would not be an issue. I still like the terms of this divorce. In those days it was taken for granted in progressive circles that goodwill should rule divorces, which in any case were only a formality required by the law, which – it goes without saying – is an ass. When I see the avaricious or vindictive divorce terms so often demanded by feminists now, all in the name of progress, I think our generation was more likeable.

Meanwhile, neither of us had much more money than our fares to England. Gottfried had created a fine and prosperous legal firm literally out of nothing, but had never been given as much as a kind word in return. I had been earning small sums for short stories and pay as a parliamentary typist.

The Grass is Singing had been bought by a Johannesburg publisher. When Juliet O'Hea of Curtis Brown in London saw the contract she was furious, said the publisher should be exposed as a criminal, and sent him a telegram to this effect. For one thing, the publisher would get fifty per cent of the royalties. His rationalization was, I heard, that this was a risky book and he should be rewarded. In the event he made no attempt to publish it, and relinquished it on receipt of Juliet's telegram. There had never been any suggestion of an advance. Juliet sold the book to Michael Joseph almost at once.

I would go to England first, with the child, and when Gottfried arrived, he would get a good job and help me with the support.

It was still not easy to get a berth on a ship, certainly not from Rhodesia.

A friend of Gottfried's from Johannesburg came up to visit, several times. He was rich. He advised us to give up this little house, saving the rent, and I should stay at his house, until we were able to negotiate a berth on a ship. Gottfried should stay with friends. This was done. At last, I left Salisbury, goodbye, goodbye. I was getting *out* – and then I was in Johannesburg in a large house in the same rich suburb I had been in in 1937, guard dogs, barred windows, night watchmen, wealth. But these were Communists, last time it was the Chamber of Mines. No difference in the style of living.

The Nationalists were in power, and some former Communists were in a panic, burying books and being careful how they met each other. The atmosphere was to say the least different from the exuberant confidence of only two years before.

One day I might write a little book called, *Rich People I Have Known*. The family I was now with will have star roles. He, the husband, made scenes about a few pence too much on a pound of tomatoes, while the wife tried to laugh. He insisted the car should be driven miles to a market where vegetables were a little cheaper. She was a cockney, from London's Unity Theatre, famous for its political reviews and its left-wing plays when the Cold War was at its worst. Many famous actors and actresses were schooled at Unity Theatre, later to lose glamour, because in the late 1950s and 1960s, socialism was fashionable again and Unity Theatre was not alone. One cannot now describe someone as a cockney. That pretty pert clever little cockney girl, where is she? Once she was in books and plays (*Pygmalion* for one) recognizable from the first word. If she exists, she is no longer considered representative. This cockney girl, in this house, now a rich woman, was going mad with boredom just like the woman in the other house. She was also quite mad with jealousy, of me. I heard her – I was meant to – making telephone calls to her friends when the words were repeated, shouted, 'And she is here, in this house.' Slowly it dawned on me that she meant me. It had not occurred to me to be in love with her husband. For one thing, I was too full of anxiety. Was he in love with me? If so, it could not have been described as more than the mildest *tendresse*. Doors slammed, telephones shrilled, husband and wife screamed at each other. I said I would leave at once, they said nonsense, I must stay, but went on screaming. There was no news yet of the boat arriving in Cape Town. I arranged berths on the train south, but before I left that most unlikeable city, two events. One was when a medical student took me on a Saturday night to a clinic that provided free medical treatment in a black suburb. Every Friday and Saturday night this big, bare, poorly equipped room filled with the victims of knife fights. I sat on a stool in a corner and watched for hours as drunk black men staggered in, or were brought on stretchers, each one slashed with knives and pangas, and streaming blood. These were tribal fights. Some wounds were frightful. A couple of men died. Four decades later I met a young white doctor from Johannesburg who said he spent weekends at a clinic for black people who came in or were brought in with knife wounds, drunk or, more often, maddened with drugs. Blood flowed, said he, in rivers. I described what I had seen in 1949. Nothing had changed, except that now it was drugs as well, then only alcohol.

The other thing I remember is a lunch given for me by – but I can't remember, only that at that table were the literary and political 'names' on the left at that time. Uys Krige, the poet, was there, and the editors of magazines who had printed my short stories. There was also Solly Sachs, the trade unionist, and a couple of other trade unionists. For a short time in South Africa – the Nationalists ended it – there were trade unionists who united poor white women workers and Indian workers and Coloured workers, to improve conditions. Impossible – so you'd think, but it was the personality of these men who did it. What did we discuss at lunch? I don't have to remember: we all discussed, all the time, the Nationalists coming to power and what this would mean for South Africa, and the Communists coming to power in China.

Later, in England, I got a letter from one of the trade unionists that went like this: 'Comrade! My life has been spent in the service of suffering humanity, uplifting the lives of the wretched of the earth. At all times my eyes are on the glorious horizons to which all mankind is marching. I can say that I never spare myself, everything I ever do or think is for the good of all and . . .' This went on for some sixteen pages, and it was only at the end I understood it was a proposition for a shared life or at least a bed. I was surprised because I had not even sat next to him at that lunch. Later, though, I accumulated quite a collection of similar letters: it was the spirit of the times. A style, though, appropriate only for certain nations: one can hardly imagine Anglo-Saxons going in for this kind of thing. Two Poles, three Yugoslavs, two Afrikaners and a revolutionary from Chile; but the letters could hardly be told apart.

In case the cycle brings this around again, I have some useful advice for women. The last thing you should do is anything as crude as Oh, you fancy me do you? All right, let's meet and we can see how things go. No, you should write at least the equivalent number of pages of similarly elevated sentiments, ending with, 'We shall always be together in the struggle.'

This advice will be useful in a parallel situation. You get a letter many pages long (the writers always have plenty of time) on these lines: 'The cosmic perspectives of eternity beckon me towards you and I feel we must meet and share our thoughts about . . .' The reply should be, 'You and I will always be together on a higher plane, what need for a meeting in the flesh?'

I found the cheapest boarding house in Cape Town. I described it in *In Pursuit of the English* – biography written in the comic mode, and why not? – but in fact it was a dispiriting time, which went on and on. The boat was dawdling around the coast. The agent said Yes and then No. He was waiting for a bribe but this never entered my head. Six weeks was a high price to pay for honesty. The boarding house, built of wood, seemed to cover an acre, and was crammed, not only with country people, the relatives of the Dutch woman who ran it, using the place as an extension of their own homes, but English war brides. There were two couples bound for the Great Peanut Disaster in East Africa. One of the young husbands would die there a year later of malaria. (This was the scheme that cost many millions of pounds and failed at once, leaving samples of the most advanced farming technology rusting on the edges of fields rapidly filling with weeds and young trees. It seems no one bothered to ask the local people for their advice.) Meanwhile, all four were full of idealism. The English war brides were brave and anxious. Some had waited years to get a place on a boat, and now had to meet husbands or fiancés last met in feverish wartime Britain. Some had small children. Twice in those six weeks English women arrived off boats, while others left on the interminable rail journeys to Northern Rhodesia, Southern Rhodesia and Nyasaland. It was only then I really understood how fortunate I was to have been brought up in Africa and not in the Home Counties. These women seemed to me ignorant, innocent, insular. I was feeling protective, as if they were children. But the main thing was this: what they knew, what they had done, was determined by their class. Even here, in this place you might think was an excuse for fraternizing, officers' ladies and other ranks' women were apart, just like in Kipling. The middle-class women, and an Honourable or two, sat in a tight defensive huddle in one part of the verandah, lowering their high bossy voices a decibel when they commented unkindly on the children of the lower classes. They were being observed by people wondering – not for the first or last time faced by this phenomenon – just how they saw themselves, that they were able to be so arrogant. Not one of them could have held a conversation for five minutes with the working-class or lower middle-class (sorry, but this is Britain) men who were products of an, alas, now dead culture – killed by television – the working men's colleges, the socialist, liberal, Communist and pacifist study classes, summer schools, night schools, literary groups, who had come to lectures and study classes in Salisbury and Bulawayo. Yet they would have patronized these men. I watched and swore, 'I will not, I will not' – meaning I would not let myself become a player in the class game. Within a couple of years I was gridlocked into it. Within a couple of years I would write a book review for *John O'London's Weekly*, where I would innocently remark that England was as caste-ridden as India – only to be showered by letters saying there was no class system in England, all written by middle-class people. Within a couple of years I would sit in a London courtroom at the time of the big Ford strike, watching a lout of a judge sneering and jeering at one of the strike leaders for his accent and his grammar, deliberately humiliating him. As I write this, snide jokes about our current Prime Minister, who is working-class, and his wife, and their demotic tastes, enliven our newspapers. Yes, yes, we are not the only country with snobbish people, but is there another which could produce that pitiful little game, Non-U and U, and find it funny? Find it funny for decades? Poor England, but it is no use, apparently there is nothing to be done.

The weeks I spent in that boarding house taught me how circumscribed I was going to be, with so little money and a small child. Here I was in glamorous, beautiful and – as I had learned – 'boheem' Cape Town, but only with difficulty did I find enough time to get out of the boarding house to queue at the shipping office. Since the child was born, my husband had been a good father, I had friends who competed for the pleasure of baby-sitting, and my mother felt she was not often enough made use of. There was, too, a servant. Now I was on my own. Not then or later did I feel I had made a mistake. This is a question, I suppose, of temperament. I have always felt it is silly to say, Oh why did I do that? But I never did lead the adventurous life I had dreamed of, exploring wild Africa or the Gobi desert, never did I loose-foot it around the Mediterranean or enjoy café life in Paris. Everything was determined by the fact I had a small child. I intended to earn my living by writing and I did, but it was a poor one for a time: not till ten years after I got to London did I earn as much as the average worker's wage. It never occurred to me to regret that, for everyone I knew was poor. Now young writers talk first about advances and security, but our lot thought differently, perhaps because of the war. We wanted to write, to succeed on our terms, to keep our independence and our privacy. No writer can do that now, for our personalities, our history, our lives, belong to the publicity machines.

It took me a long time, a very long time, to see something obvious. The child – the hair shirt, the 'burden', which was how other people saw my situation – was what saved me. (This was why I so much liked Margaret Drabble's first novel, *The Millstone*.) If I had arrived in London alone, this is what would have happened. Soho was then an attractive, not to say seductive, place. I would very soon have found my way to it. For the third time in my life I would have been the new girl in town. This particular subculture was not kind to women. Girls were gobbled up like chocolate drops – or like gins and tonics. In Daniel Farson's book, *Soho in the Fifties*, women hardly appear. The photographs of Nina Hammett, once beautiful and a serious artist, old, drunk, incontinent, begging for drinks or a handout, says it all. In the early 1960s Elizabeth Smart, described by Daniel Farson as a citizen of Soho, came to have lunch with me. She drank and wept and wept and drank from midday to seven at night and was savagely witty about her life and the lives of women. I would not describe her as an advertisement for the *joie de vivre* of Soho. Of Soho's habitués surely only Francis Bacon thrived. Daniel Farson took himself out of it. There was a constellation of talents, but mostly they drank and talked their gifts away: the world may be well lost for gossip. (It has been well lost for much less.) The current equivalent, the Groucho Club, a sanitized bohemia, is as much an eater of talent, but the conversation in the old clubs was better. That was the attraction of the Mandrake, the Gargoyle, the French Pub, the Colony. I was taken to each of these places – once – by John Somerfield the writer, who said I ought to know how the other half lived. I thought they were sleazy, but recognized the atmosphere and its attraction. These were all eccentrics and misfits, oddballs and originals, and they had created a world where they called the tune. Boheem, in fact. Just like us in Salisbury, Southern Rhodesia. How could I not understand that people could get lost there? I would have done too, I am pretty sure. I can too easily see myself, again drinking too much, as I had done hardly at all since 1942 and the end of my first marriage. And then I would be in love with one of these poets and painters. Not because they were glamorous, but because they were lost souls. Irresistible. But these were not men to be in love with. Not unless one had a talent for suffering. Believe it or not, I had not yet understood I had been conditioned for tears. No, Soho might very well have done me in, and I was saved because I had a responsibility, the child. It was a heavy one. This was a child of great good nature and sociability, but he was not a sleeper. He woke at five and went to sleep at nine or ten at night, and never slept in the day. This until he was nine or ten. This meant I woke at five too. In those days mothers took it for granted they woke and got up when their children did, but now women may leave their children to care for themselves until they want to get up. Sometimes for hours. 'It is my right.' Autre temps, very much autre mères.

Peter enjoyed – for eighteen hours a day – that great boarding house, like an assembly of Brobdingnagian packing cases, and the great garden full of fruit trees and all the other children. As for me I watched and waited. I have spent a good part of my life waiting. Women do, more than men. There is this famous question of women's passivity, but it is often a protective mechanism. And perhaps it is protective, too, when you look ahead and plan, but your plans are based on illusions. I did not expect an easy time in London, but believed Gottfried would be somewhere around, a father to the little boy, and my good friend. I did not expect to bring the child up by myself, which is what happened. Had I known, I would have been, at the very least, apprehensive. But I was not at all afraid, sitting on that verandah, watching all the little children playing among ancient fruit trees. They were war children, but talk of the Trenches would not shadow their childhoods. I sat day after day, watching the children, and listening to the war brides camped on either end of the verandah talk of their futures in Africa, and I was matching their expectations with what I knew they would find. And I wondered how Gottfried did with his plans for living in London. Had he heard yet from the prospective employers he had written to, mentioning the earlier distinguished Lessings who had already lived and worked in London?

This is what in fact happened to Gottfried Lessing. Shortly after I arrived in London, he came. Dorothy Schwartz had decided to try her luck in London. She had a flat, and Gottfried used a room in it. He was confident that the worst time in his life was over and he would at once get a good job in London. There was no reply to his applications. He kept himself working for the Society for Cultural Relations with the USSR and waited. Difficult to imagine a time when it was harder for a German and a Communist to find work in a respectable firm. The joke is, ten years later nothing could have been more chic than to employ a German, and even a Red, for Communism became fashionable again, and once again, as always when the heat was turned low, people called themselves Communists – for the thrill of the thing, for the fun of thumbing a nose at Mummy and Daddy. Such people never take out party cards. Many 'Communists' had no idea what Communism was. I remember lunch with a prominent film-maker who was proclaiming the virtues of Communism and the Soviet Union, called himself a Communist, and asked me if it was true that a Communist had to be an atheist. When I said there was something called dialectical materialism he said he thought people should not be concerned for their material welfare. This sort of ignorance was typical of the fashionable Communists.

Gottfried was discouraged, depressed. He got jaundice, he could not work. All this time our relations were excellent. He saw a lot of his child, particularly as I broke my shoulder soon after I arrived. He got a visa to visit his sister and her husband – 'the eternal student' – now working in the Kulturbund in Communist Berlin. When he came back he was elated, all his optimism had returned. He said he was going to live there, and he wanted me to join him. This frightened me: never, not once, had we talked of staying married or living together. He said, 'They live very well over there. They have good flats and cars and chauffeurs.' He also ridiculed them for their exaggerated attention to security. 'They are crazy,' said he, 'they think there are spies under the beds, and they wouldn't let me say anything in the car because the chauffeur might hear.' When he laughed at them they said he had spent too much time in the West to understand.

Now he formally applied to the East German government to be allowed to return as a citizen. Nothing happened. He applied again. Silence. He could not understand it. 'Of course they have so many problems, one has to take these factors into consideration.'

At this point, enter Moidi Jokl, who had a quite astonishing effect on my life, in several different ways, but I will confine myself here to her influence on Gottfried's. She had been well known in Vienna before the war. A very young woman – a girl – she had created a radio programme unique for then. She talked, sang, joked, clowned, projected herself in a way that gained her a very large audience: there was a match between this new thing radio and her temperament. She was, of course, a Communist. She was a friend of the German Communists, then living hand-to-mouth, hiding, on the run, or in the Soviet Union, certainly 'dead men on leave', who became the government in East Germany. She went to East Germany to live. Then came Stalin's purge against the Jews throughout the Soviet Union and the satellite Communist countries, referred to by them as 'The Black Years'. She was thrown out of East Germany together with dozens of other Jews. The young policeman who took her to the frontier was in tears and said, 'If they are expelling people like you then there's something very wrong.' Perhaps he was one of those who danced when the Wall came down thirty years later. She was now one of the first refugees from Communism, in London, but there were refugees from everywhere. They lived as they could, God knows how they kept alive; sometimes there were ten or more in a room, or they slept on sofas in friendly flats, moving when they had worn out a welcome, and earned a living translating, dressmaking, hairdressing, anything they could find. All the cities of Europe were full of people like Moidi. When I told her about Gottfried, waiting for a formal reply to his formal application, she simply laughed, said that he did not understand the first thing about Communism. He should make his way to Berlin on a temporary visitor's visa and then pull strings. He had relatives in important positions? I relayed this information to Gottfried. He was coldly and contemptuously angry: being poor, being without real work, being afraid and uncertain, had made him even more of a Communist, even more narrow, suspicious, paranoid. He said he wasn't interested in listening to anti-Soviet propaganda – that was the phrase then used for even the mildest criticism of Communism. He went on waiting. Moidi said, 'If he'll meet me I'll tell him what Communism's really like.' First, he refused, but time went on passing while he waited for every visit of the postman. I arranged a supper. This was a meeting even more abrasive than the supper for the Freudian and the members of the Soviet Trade Delegation.

Moidi was a large, colourful, exuberant, extrovert, scornful, funny woman, who looked like a gypsy – a style always useful when hard-up. She was the kind of woman least likely to appeal to Gottfried. He sat there, elegant as always, looking like a diplomat, not a hair out of place. He ate nothing, while Moidi ate with relish and told him what Communism was like, and he drawled out the Communist lexicon of abuse, anti-Soviet propaganda, hirelings, running dogs, jackals, and so on and so forth. When Moidi told him he understood nothing at all about Communism, he said he understood her and her type only too well. Moidi went off laughing. He went off saying I was contaminated by class – imperialist – capitalist – ideology. He was very very angry. I had never seen him like this, and he really frightened me. But, judged by its results, the meeting was a success, for only a few days later he told me he had applied for a visa to visit his sister, and there he would 'see what could be done'. He never again mentioned Moidi. He said when he was in Berlin Peter could visit him for holidays. I said that it would be very bad if he began something that he could not keep up: Moidi had said he was mad to think he could keep up contact with the West, the penalty was death and worse. People who had Western backgrounds or contacts were always under suspicion. I relayed her messages to Gottfried, he sent insults back.

When I, and the child, and Dorothy Schwartz, saw him off at the station it was a cold, bleak, grey day and he already looked foreign, estranged, in his karakul lamb cap.

He got a job in the Kulturbund. The 'joke' was that, since Hitler had killed off the Communists, any that remained were bound to get good jobs. Not only Communists: people who knew the Communist hierarchy in Berlin as it was later, might be surprised to hear that their early days were open and flexible. A German businessman working in London, who went over on a business trip, was invited to meet a selection of high-ups who asked him to come and work re-building Germany. He said he was not a Communist, was not interested in politics: they said it did not matter, they wanted people of ability. But that was before Stalin's final flares of lunacy, before East Germany hardened and became a death-box for Westerners.

Soon Gottfried wrote and asked Peter to go over for the summer. This was the most frightening thing I ever did, but I saw no reason not to trust him. The child was four. He went for two months, spent them with his cousins, had a very good time, and returned having forgotten English, chattering in German. He had also been taught to eat with his left hand resting beside the plate on the table and to click his heels and bow when spoken to.

Within a couple of weeks the little German boy disappeared, and the English boy came back. Gottfried had sent a letter with him saying that Peter should spend summers there. And then – nothing, silence. The child had had a good warm close father, whom he had visited and found with him another family, and now this father had gone. I wrote to Germany, I sent messages by people going over. Nothing. I wrote and said the child was desolated, asked after him, wept himself to sleep. Could he not even write letters? But nothing. Then I went to Berlin, and tried to contact Gottfried. But he would not answer telephone calls or reply to messages. This was before the Wall was built. I already had a publisher in East Germany. I asked him to put on pressure for me. He did. I don't know what kind of pressure. I was too angry to care if this was dangerous. I went to one of the new and hideous blocks of flats and there was Gottfried, there was his sister Irene, in a smart, new, but not large flat, full of the clean new furniture of the type then called Swedish. Both of them looked as if the war had never happened, as if they were still in the life Hitler interrupted. They were elegant, worldly, with that half-cynical joking style the rich and successful often use. They certainly were not rich. Both made a point of saying that weekends they went off 'to the people' to work on building sites, or something of the kind. I told Gottfried he had made promises to the child he had not kept. Gottfried was airy and arrogant, as if nothing important had happened. He must have been very frightened, I had no idea then how frightened. He gave me a small amount of money, enough to buy a toy. I said I was not interested in the money, I wanted him to keep contact with his son. This was one of the worst experiences of my life. I could see that my visiting him was not going to change anything.

And that was it. Soon Gottfried was the equivalent of the President of the Board of Trade, a more political position than it would be here. People returning from East Germany told me Gottfried had to be approached through rooms full of underlings. He sent me airy goodwill messages. Some people who 'knew the score' said that of course he could not keep contact with the West, the price was death, particularly for those who had spent the war not in the Soviet Union, but as refugees. Others who 'knew the score' said that the Party, always concerned with humanist values, would understand his need to keep contact with his son. The proof of the pudding is in the . . . I did not care if I never saw him again – and I didn't, but it mattered very much about his son. By then I had switched off, an inner door had slammed shut, I 'didn't want to know' – a most accurate description of my state of mind.

Meanwhile, Gottfried had married Ilse Dadoo. Dadoo, an Indian, was one of the people who, in that remarkable and short-lived climate just before the Nationalists took power, organized – like Solly Sachs – poorly paid workers, Indians, Coloureds and whites, all together. There was a little girl from this marriage – Ilse's with Dadoo. She said she wasn't going to bring up a half-Indian child in fascist South Africa and went home. There she met Gottfried. They married, I feel, probably because of their shared African experience. Both must have felt very exotic fish in those grey waters. Gottfried was a loving step-father to Ilse's child, as he had been a loving father. And now something pretty hard to explain, that is if one is using ordinary ways of looking at things, discounting Communist paranoia. Gottfried was 'demoted' and forced to spend a year at a Communist re-education school. Why should he need it? How could he possibly be a more loyal Communist than he was? But he was contaminated by Western thoughts, and needed some intensive brainwashing. He did not go back to Trade, at least not directly. He was sent to Indonesia, as a diplomat in fact, though East Germany was not recognized as an independent state, and there he represented trade. He had much influence in local politics. He got ill there: the climate and the food did not suit his liver. He returned to East Germany. Then he was sent to Tanzania. He was a well-known figure in East Africa, and had more than local influence. These two, Gottfried and his wife, were well placed. Their knowledge about Africa was of the kind few people can have had then, in Communist Germany. Precisely what made their position precarious was also what made them valuable. Ignorance about Southern Africa was general in Europe. When I came to Britain and for some years afterwards I and others who attempted to say that South Africa and Southern Rhodesia were not happy lands full of smiling and satisfied blacks were told we exaggerated, or were wrong-headed. In the 1950s when I lobbied prominent members of the British Labour Party about Southern Rhodesia, then generally held to be a just and fair place, simply because it was British – by the few people who had ever heard of it – these politicians literally did not know where it was, believed it to be part of either South Africa or Northern Rhodesia.

There is a vivid glimpse of Gottfried and Ilse in Dar es Salaam. A friend of mine, a friend of Gottfried's – a prominent African politician – dropped in late one night, in the informal African way. There was a long wait, frightened voices, then Gottfried came to the door, obviously afraid, while his wife stood guard behind him, making gestures for him to be careful. When my friend, still standing on the doorstep, teased Gottfried for being so cautious, Gottfried too made gestures, indicating that the place was bugged, while joking loudly back, and his wife scolded the visitor loudly – for the benefit of the microphones – about being so careless, so inconsiderate. 'But this is Africa,' my friend protested, 'this is Africa.'

Meanwhile East Germany was advising various African countries how best to install prisons, secret services, torture, informers, on the Communist model. Somalia was one. Uganda another.

Later Gottfried became Ambassador in Kampala. He went there with his third wife, Margot, Ilse having died, disillusioned with Communism. Probably this third wife was the first wife he really loved. Pictures of her remind me of his Viennese merry widow. She looks a warm, kind, pleasant woman. She was not an intellectual and certainly not political. The Party did not want him to marry her, told him to get a wife who now would be called politically correct. He had to fight the Party – and this could not have been easy for him.

The Soviet 'line' was to support the butcher Amin. They supported him until just before he ran away. That meant East Germany had to support him. When Amin fled, the Tanzanians came in to restore order. All the embassies had left some days before, in convoy, on the open road to Kenya. All except for the Iraqi ambassador and the East Germans – Gottfried, his wife, and two staff. On the night before the Tanzanian troops marched in, Gottfried telephoned the Iraqi ambassador, a personal friend, saying, 'Shall we go together tomorrow on the open road to Kenya?' He said, 'Are you mad, are you crazy, we have been told to get inside our houses, bolt the doors, and keep our heads down.' Next morning Gottfried put his wife into the car, together with the two members of his staff, and drove it straight past the square where the Tanzanian troops were, drunk and trigger-happy. They were shooting at anything that moved. They turned flame-throwers on the car. This information came from Tony Aberfan, of the *Guardian*, who was in Kampala, and from East African friends who made enquiries. The East German Communists put up a plaque in Berlin with four names on it over a 'grave' that could have nothing in it.

Gottfried, whatever else, was not stupid. The plots evolved to explain this stupid behaviour outclass James Bond. People who knew East Germany well said obviously it was the long arm of the KGB: several diplomats of the Soviet bloc were murdered in mysterious ways around that time. Gottfried was a member of the KGB. So it was rumoured. For years I did not 'take this in', and always said I thought it improbable. But really, I didn't want to know. Then it was confirmed to me. By whom? By my son John Wisdom. He had close friends in the Southern Rhodesian secret police. (These people continued to work for the black government. That such a thing could be possible is just another little symptom of the general lunacy of our times.) John wanted to find out about

Gottfried Lessing, his mother's second husband, and because of his Zimbabwe secret police contacts was able to make contact with someone in the South African secret police. This man said Gottfried was a member of the KGB and that his influence in East Africa and much further afield was extensive. True or false? Who knows, in such a muddy, dirty, murky area. Associates who knew him well in East Germany say it is impossible, because he was a good man, not the kind of person likely to be employed by the KGB. They relate – with emotion – that people used to say of him, 'Gottfried Lessing is not a Communist, he's a real human being, he is kind and generous and good to people in trouble.'

He was certainly a Communist for a long time, for many years. The Party was for him, as it was for certain kinds of people, a kind of absolute, a God. The psychologists say that there is a proportion of people who, once they have acquired a set of beliefs, are incapable of changing them. They never do. They never will. Their minds are set in cement, for once and for all.

But wait . . . a man who knows the Party is always right (even when it isn't – merely a minor and temporary lapse), a man who obeys the Party, like his own conscience – such a man does not fight the Party to marry a woman considered unsuitable, considered a threat. So there is a truly terrible possibility: suppose Gottfried was no longer a one hundred per cent Communist, suppose the cement had in fact cracked, suppose he was in a position where he had to carry out orders he hated? This was not exactly unknown in the Communist world . . .

Gottfried was murdered in 1979. In 1949 he was planning his future as an adopted Englishman. In 1949 I was sitting on this verandah in Cape Town, and looking forward to meeting Gottfried soon in London. For six weeks I sat there, watching, listening, planning . . . six weeks is a long time, when you are young, not yet thirty.

I was looking forward, with never a glance behind me. I was waiting for my future, my real life, to begin. Behind me a door had slammed shut. Doors had been shutting behind me all my life. The worst – that is, of what I can remember, is when I was sent to boarding school before I was seven. I knew all about the mechanisms of the shut door, recognized it not by some loud external bang but by what went on inside me. If it is a person who is being left behind, then that door closes of itself. Ah, I think, the door has shut itself, has it? Thereafter I expect nothing, though I may be behaving as I always did, I hope more or less well. But how old is that person who approaches another with such confidence, such trust, such optimism that here is a friend, a real friend, and for life? One much too young to have learned the folly of expecting too much. 'Ah, the door has shut? Interesting . . .' There is nothing more implacable than this process over which one has no control.

In this book I have been presented – I have presented myself – as a product of all those McVeaghs, Flowers, Taylers, Batleys, Millers, Snewins and Cornishes, sound and satisfactory English, Scottish, Irish compost, nurtured by Kent, Essex, Suffolk, Norfolk, Devon and Somerset. I am slotted into place, a little item on a tree of descent. But that is not how I experienced myself then. That's all over, I was thinking, that's done with, meaning the tentacles of family. I was born out of my own self – so I felt. *I didn't want to know*. I was not going home to my family, I was fleeing from it. The door had shut and that was that.